Birds of the Antarctic and Sub-Antarctic

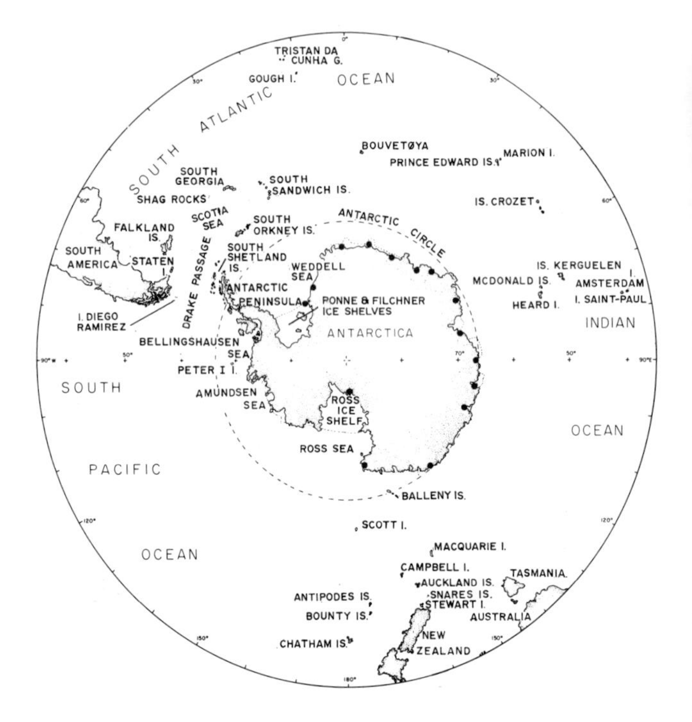

Frontispiece. South polar projection to 35°S showing major landmasses, ice shelves, and seas in the Antarctic and sub-Antarctic. The dots around East Antarctica are the sites of active coastal research stations.

Birds of the Antarctic and Sub-Antarctic

George E. Watson

in collaboration with
J. Phillip Angle
Peter C. Harper

illustrated by
Bob Hines

ANTARCTIC
RESEARCH
SERIES

BIRDS OF THE ANTARCTIC AND SUB-ANTARCTIC

George E. Watson

ISBN 0-87590-124-7
Library of Congress 75-34547

Published by the
AMERICAN GEOPHYSICAL UNION
1909 K Street, N. W.
Washington, D. C. 20006

Printed by
THE WILLIAM BYRD PRESS, INC.
Richmond, Virginia

THE ANTARCTIC RESEARCH SERIES

The Antarctic Research Series is a medium for authoritative reports on the extensive scientific research being done in Antarctica. The series has elicited contributions from leading scientists; it seeks to maintain high scientific and publication standards. The scientific editor for each volume is chosen from among recognized authorities in the discipline or theme that it represents, as are the reviewers on whom the editor relies for advice.

Research results appearing in this series are original contributions too long or otherwise inappropriate for publication in standard journals. The material is directed to specialists actively engaged in the work, to graduate students, to scientists in closely related fields, and to laymen versed in the environmental sciences. Some volumes comprise a single monograph. Others are collections of papers with a common theme.

In a sense, the series continues a tradition dating from the earliest scientific expeditions: expeditionary volumes that set forth everything that was seen and studied. This tradition is not necessarily outmoded, but these days, one expedition blends into the next, and it is no longer meaningful to separate them. Nevertheless, antarctic research has coherence, and it deserves a modern counterpart of the earlier expeditionary volumes. The Antarctic Research Series is such a counterpart.

Aided by a grant from the National Science Foundation in 1962, the American Geophysical Union began the Antarctic Research Series and appointed the Board of Associate Editors. Supplemental grants received in subsequent years and income from sales have enabled the series to continue. The response of the scientific community and the favorable comments of reviewers cause the board to look forward with optimism to continuing success.

PREFACE

Antarctica is the fifth largest of the continents, and it has the highest average elevation of all the continents. It is enveloped in a remnant of the Pleistocene ice sheet, which permanently hides 98% of its land area. A phantom secondary coast of sea ice in winter extends the continental area as far north as 64° south latitude. The land hub of the southern hemisphere, Antarctica is surrounded by a world ocean that is the foraging area for myriads of flying and flightless seabirds. The area contrasts sharply with the Arctic, an ocean almost wholly enclosed by landmasses that extend north to about 83° north latitude. Penguins, petrels, and skuas are nearly ubiquitous in the Antarctic; other birds, notably loons, waterfowl, birds of prey, shorebirds, jaegers, gulls, auks, and several songbirds, are characteristic of the 'Arctic.' Only one species, the Arctic tern, is widespread in both places. There are also a number of interesting ecological equivalents, i.e., snow petrel-ivory gull, diving petrels-auks, and skuas-jaegers.

For more than half the year, and over most of its area, continental Antarctica is nearly lifeless. The exception is the emperor penguin, which comes to the edge of the continent in early austral winter to breed on the sea ice. With the return of the sun in the austral spring, even before leads have split the pack ice into floes, Adélie penguins begin their trek over the frozen sea toward snow-covered breeding grounds. About the same time, skuas and giant petrels, predators and scavengers, appear over the colonies of emperor penguins and their young, which have survived blizzards, low temperature, and the darkness of the polar winter. As the days grow longer, a stream of small and large seabirds converges on Antarctica from all points, and the silent continent once more awakens to the sounds of life. What it lacks in species, the antarctic avifauna makes up in numbers, and it constitutes an important part of the antarctic biomass.

Marine birds come to Antarctica in the austral spring and summer to incubate their eggs and to care for their young, but even during this period they spend much of their time at sea. Away from the continent, seabirds are widely scattered, their distribution being controlled in part by atmospheric and oceanic circulation and by the availability of food. Shipboard observers are well aware of the concentration of seabirds along the convergences and in regions of upwelling waters. But it is a rare day when strong-flying seabirds cannot be seen following ships anywhere. Japanese observers on the *Umitake Maru* en route from Fremantle to Cape Town via South Georgia, in December 1971 and January 1972, counted 19,541 seabirds, most of them along the Antarctic convergence.

My first awareness of this phenomenon was in 1959 when Carl Eklund and I accompanied U.S.N. *Arneb* from Wilkes Station to the United States. The voyage across the South Pacific, Indian, and Atlantic oceans was a memorable experience because of the never ceasing presence of birds and the constantly changing avifauna as we steamed from one region to another. Later, as Chief Scientist on U.S.N. *Eltanin*, I took the opportunity, when the ship was conducting scientific stations, to collect birds at sea for George Watson's study. Collecting was carried out from a dory several miles from *Eltanin*, and as the dory rose up and down the ocean swells, I had the opportunity to see many of the smaller seabirds at close hand. Both these experiences pointed up the need for a reliable pocket guide.

There are abundant descriptions and a wealth of published information on the relatively small list of antarctic avifauna. But these data are scattered and are in several languages. Two landmarks in southern hemisphere ornithology are Robert Cushman Murphy's (1936) *Oceanic Birds of South America* and Sir Robert A. Falla's (1937) 'Birds,' which appeared in the reports of the Australian-New Zealand Antarctic Research Expedition of 1929-1931. Interestingly, a field guide for voyagers preceded both these works. This work, *Birds of the Ocean*, by W. B. Alexander (1954) has served as a companion to mariners and biologists alike since its publication in 1928 and in subsequent editions. One can only sympathize with George Watson's predecessor, Herbert Friedmann, in his predicament when he was asked by the U.S. Antarctic Service Expedition in 1939 for information on birds of Antarctica: 'I made a quick mimeographed list of birds known to breed in Antarctica and illustrated it with crude pen and ink diagrams to enable potential users to identify the birds; to write and illustrate took me all of one afternoon.'

Two field guides that describe some antarctic species have appeared recently: *A Field Guide to the Birds of New Zealand and Outlying Islands* by R. A. Falla, R. B. Sibson, and E. G. Turbott (1966) and *Guide des Oiseaux et Mammifères des Terres Australes et Antarctiques Françaises* by Jean Prévost and Jean-Louis Mougin (1970). These well-illustrated and inexpensive pocket guides attest to the wide popular interest in subantarctic and antarctic birds.

Fortunately, George E. Watson, curator of birds at the Smithsonian Institution, and others brought many new data and background facts together in a 1971 study, 'Birds of the Antarctic and Subantarctic,' published by the American Geographical Society (New York) as Antarctic Map Folio Series 14. The folio, edited by Vivian Bushnell, comprises large maps with locations of at-sea sightings and breeding distributions and is indispensable for students of antarctic birds and seabirds generally.

The present publication is an outcome of discussions that I had with Herbert Friedmann, Philip S. Humphrey, and Waldo L. Schmitt, staff members of the Smithsonian Institution, in 1962 and 1963. At the time, George E. Watson was preparing his 'Preliminary Field Guide to the Birds of the Indian Ocean' for use during the International Indian Ocean Expedition. That work, which deals with both land birds and seabirds, suggested a style and basis suitable for an antarctic bird handbook.

George E. Watson's *Birds of the Antarctic and Sub-Antarctic* is an up-to-date, scholarly contribution that both professional biologists and the increasing number of laymen visiting Antarctica will appreciate. It is well documented by the author's research and has been enhanced by the contributions of colleagues from many countries. It is beautifully illustrated by the art of Bob Hines of the U.S. Fish and Wildlife Service, whose well-executed color plates and drawings prove again that one picture saves a thousand words. The strength of this field guide comes from the firsthand experience and observations of Dr. Watson on expeditions and cruises to the Magellanic region, Drake Passage, islands of the Scotia ridge, and the Antarctic Peninsula.

This work was sponsored under the biological program of the Office of Polar Programs, the National Science Foundation. Publication by the American Geophysical Union was supported under a National Science Foundation contract.

Much remains to be done: the life histories of many antarctic birds are little known. The physiological adaptations to cold are still poorly understood, and the behavioral relationships of long-lived colonial breeders present many challenging and tantalizing problems. We know almost nothing of the distribution patterns of birds over pack ice or the dispersal of young. There is a critical lack of winter observations. This handbook will be a useful companion and guide to professionals and amateur biologists who may have the great good fortune to visit south of the Antarctic convergence.

George A. Llano

Chief Scientist
Office of Polar Programs

ACKNOWLEDGMENTS

My interest in antarctic and subantarctic birds was first stimulated during a 4-month visit to southern Chile, the South Shetland Islands, and the Antarctic Peninsula as U.S. observer with the 18th Chilean Antarctic Commission in 1963-1964. The late Henry M. Dater, staff historian to the U.S. Navy Antarctic Projects Officer, arranged my participation and shared with me his enthusiasm for birdlife in the far south. After crossing the notorious Drake Passage 4 times, entering Port Foster, Deception Island, through awesome Neptunes Bellows, and living for several days in solitude with penguins, petrels, and skuas at Harmony Point, Nelson Island, and Coppermine Peninsula, Robert Island, I was hooked.

The inspiration to write this handbook came from George A. Llano, Program Manager, Polar Biology and Medicine, in the Office of Polar Programs (formerly U.S. Antarctic Research Program) at the National Science Foundation after the discussions, mentioned in his preface, that he had with Herbert Friedmann, Philip S. Humphrey, and Waldo L. Schmitt. Dr. Llano and Dr. Schmitt have continued their active interest in and encouragement of the project.

Funds to produce the manuscript and color plates and to print the plates were provided from National Science Foundation grant GA 169 to the Smithsonian Institution.

J. Phillip Angle joined me at the Smithsonian Institution as research assistant in 1964 to gather initial data for an identification guide. Together we visited the Antarctic twice to obtain additional firsthand information on its birdlife. We made observations and collected specimens at sea aboard the 'baby' aircraft carrier *Croatan* as it cruised rapidly down the west coast of South America to 60°S in March and April 1965, pausing every 10° to launch space-probing rockets for the National Aeronautics and Space Administration. Dr. Llano organized a biological collecting expedition aboard the U.S. Coast Guard icebreaker *Eastwind* in January and February 1966 that permitted us to study birds in the Drake Passage (6 crossings), Weddell Sea, South Orkney and South Shetland islands, Palmer Archipelago, and Antarctic Peninsula as far south as Marguerite Bay. I am grateful to the captains, officers, crews, and scientists aboard those two ships and the earlier Chilean ships for aiding and facilitating the fieldwork and providing camaraderie.

At the Smithsonian, Angle and I, and later Peter C. Harper, a New Zealander who had just spent nearly 2½ years aboard the U.S.N.S. *Eltanin* observing birds in the far south, prepared a first draft of the species accounts for the handbook. Simultaneously, along with Margaret A. Bridge, Roberto P. Schlatter, W. L. N. Tickell, and Maria M. and John C. Boyd, we compiled distributional information on antarctic birds that we used to draw preliminary small-scale distribution maps for the handbook and subsequently published in 'Birds of the Antarctic and Subantarctic,' Antarctic Map Folio Series 14 of the American Geographical Society.

First drafts of the species accounts and the maps went to 30 colleagues in 10 countries. The following sent back detailed helpful comments based on their experience with antarctic birds: J. R. Beck, W. R. P. Bourne, Louis J. Halle, Holger Holgersen, Robert Hudson, Christain Jouanin, Karl W. Kenyon, Fred C. Kinsky, Richard Liversidge, C. C. Olrog, Kaijiro Ozawa, Jean Prévost, R. W. Rand, M. K. Rowan, Ian F. Spellerberg, Bernard Stonehouse, Eduard M. van Zinderen Bakker, Richard W. Vaughn, Karel H. Voous, John Warham, and Robert C. Wood. Copies of initial and later drafts also went to the Antarctic with several ship's crewmen, scientists, and travelers, some of whom reported their experiences and reactions to us.

On the basis of these comments plus considerable newly published information I revised the species accounts and the distribution maps and wrote the introduction and geographic accounts. Some of the original correspondents and all of the following colleagues answered a host of detailed queries and provided unpublished information for the revision: Braulio Araya, Roger S. Bailey, Jon C. Barlow, David Bridge, G. M. Budd, Robert Carrick, P. A. Clancy, Theresa Clay, Rodolfo Escalante, Sir Robert A. Falla, J. R. Furse, Susan E. Ingham, Allan R. Keith, Jean-Louis Mougin, G. W. McKinnon, M. D. Murray, Paul Peterson, Roger T. Peterson, Olin S. Pettingill, C. J. R. Robertson, Finn Salomonsen, Kurt V. Sandved, Waldo L. Schmitt, Walter R. Seelig, M. Segonzac, William J. L. Sladen, E. A. Smith, Ian J. Strange, W. L. N. Tickell, P. S. Tilbrook, J.-F. Voisin, Milton W. Weller, and Nixon Wilson. Without the willing help of these colleagues this book could not have been nearly so comprehensive.

I used reference specimens from several museums to supplement the extensive antarctic collections at the National Museum of Natural History, Smithsonian Institution. I am grateful to Charles O'Brien (American Museum of Natural History); Melvin A. Traylor (Field Museum of Natural History); Raymond A. Paynter, Jr. (Museum of Comparative Zoology); Ian Galbraith and David W. Snow (British Museum (Natural History)); Christain Jouanin, Jean Prévost, and Jean Dorst (Muséum National d'Histoire Naturelle); and Fred C. Kinsky and Sir Robert A. Falla (National Museum of New Zealand) for loans or for reporting specimen information.

The penultimate drafts were reviewed in detail by George M. Sutton, Joel Hedgpeth, Stephen V. Shabica, George A. Llano, and Philip S. Humphrey and were tested in Antarctica by David F. Parmelee and Stewart D. MacDonald. They caught inconsistencies and provided me with new critical insights. Fred G. Alberts checked all locality names and made helpful comments on the geographic accounts. On the basis of their valuable help I prepared the final draft.

The color plates, the black and white vignettes of vagrants, and Figure 3 are the work of Bob Hines of the U.S. Fish and Wildlife Service, who responded willingly and patiently to my requests for revisions and the addition of new vagrant species. Jack R. Schroeder and Paul Mazer drew the frontispiece map and the rest of the figures. Vivian C. Bushnell of the American Geographic Society provided distribution base maps.

The numerous drafts and revisions of the manuscript were typed by Martha P. Lanum, Barbara B. Anderson, Barbara Googe, Joanne F. Williams, Hazel W. Fermino, and Kathryn Martin. L. Elaine Kennell has been a most patient and educational editor in tying all the loose ends together. And lastly my wife Terry, who helped on the revisions, endured with me a prolonged scientific accouchement.

TABLE OF CONTENTS

Frontispiece ii
The Antarctic Research Series v
Preface vii
Acknowledgments xi
Introduction 1
Avifauna 2
Antarctic and Subantarctic Environment 4
  Sea Environment 5
Land Environment and Vegetation 16
  Climate 21
  Adaptations of Birds to Extreme Climatic Conditions 24
Species Information 26
  Identification 26
  Flight and Habits 32
  Voice and Display 32
  Food 33
  Reproduction 34
  Molt 36
  Predation and Mortality 37
  Ectoparasites 38
  Habitat 39
  Distribution 39
Antarctic and Sub-Antarctic Life Zones 44
Banding 52
Recording Observations at Sea 54
  Species Log 55
  Environmental Log 55

Preserving and Shipping Specimens 56
Collection and Importation Permits 58
Conservation 58
Penguins: Spheniscidae 63
Albatrosses: Diomedeidae 85
Fulmars, Prions, Gadfly Petrels, and Shearwaters: Procellariidae 99
Fulmars 100
Prions: *Pachyptila* spp. 114
Gadfly Petrels 125
Shearwaters 139
Storm Petrels: Oceanitidae 152
Diving Petrels: Pelecanoididae 161
Cormorants: Phalacrocoracidae 166
Herons: Ardeidae 173
Waterfowl: Anatidae 176
Birds of Prey: Falconiformes 187
Rails and Coots: Rallidae 188
Plovers: Charadriidae 193
Sandpipers: Scolopacidae 194
Phalaropes: Phalaropodidae 200
Sheathbills: Chionididae 201
Skuas and Jaegers: Stercorariidae 206
Gulls and Terns: Laridae 214
Gulls: Larinae 214
Terns: Sterninae 220
Land Birds 232

Shag Rocks 249
South Georgia 260
South Sandwich Islands 262
South Orkney Islands 263
South Shetland Islands 266
Antarctic Peninsula 271
Antarctic Continent 275
Scott Island 284
Balleny Islands 285
Peter I Island 285
Tristan da Cunha Group and Gough Island 286
Bouvetøya 289
Prince Edward Islands 290
Iles Crozet 293
Ile Amsterdam and Ile Saint-Paul 296
Iles Kerguelen 298
Heard Island 303
McDonald Islands 304
Macquarie Island 305
Analysis of References 308
References 315
Variant Names 339
Index 346

INTRODUCTION

This handbook is intended to help scientists, ship's crewmen, and travelers identify most of the birds observed at sea and on land in the antarctic and subantarctic regions and to indicate what is known of their distribution and biology. The book was originally planned merely as a guide with coverage restricted to species that occur in Antarctica. Because few species occur exclusively in Antarctica, however, and because people who visit, or are stationed there, pass through remote and little-studied areas not adequately covered by other identification manuals, the geographic limits have been extended northward into the sub-Antarctic. Furthermore, because of the potential contribution to knowledge that can be made by even casual bird observers it was decided to present synopses of the biology of the major species and to suggest what further information is needed.

The antarctic and subantarctic regions covered in this handbook comprise all waters, lands, and islands south of 55°S latitude, roughly corresponding to the line of the Antarctic convergence, plus the cold sub-Antarctic islands near the convergence and the temperate sub-Antarctic islands in the central South Atlantic and Indian oceans. The specific lands and islands included are the Antarctic continent and Antarctic Peninsula; the maritime Antarctic islands of the Scotia ridge (South Shetland, South Orkney, and South Sandwich islands, plus South Georgia and Shag Rocks); Bouvetøya (Bouvet Island) and Balleny, Scott, and Peter I islands; the cold sub-Antarctic islands of the Indian Ocean (Prince Edward Islands, Marion Island, Iles Crozet, Iles Kerguelen, and Heard Island); Macquarie Island; and the temperate sub-Antarctic oceanic islands of the southern Atlantic and Indian oceans (Tristan da Cunha group, Gough Island, Ile Amsterdam, and Ile Saint-Paul). The New Zealand sub-Antarctic Islands, the Cape Horn and Falkland islands (Islas Malvinas), and Tierra del Fuego are omitted because of their relatively extensive land bird faunas and because the birds are already well covered in several recent guides and handbooks. Otherwise all species that have been recorded within the Antarctic and sub-Antarctic are included in this handbook. Those that are regular residents and visitors are treated in greater detail than occasional vistors, but the known occurrences of vagrants are fully documented in order to facilitate confirmation of new additions to the avifauna.

This handbook is divided into three sections. The first introduces the antarctic and subantarctic environments and provides general information on how to identify and study birds. The main body of the book consists of species accounts, illustrations, and distribution maps. The accounts present identification information and what is known of the birds' biology and distribution. Each species is illustrated, either in the colored plates in the center of the book or in a black and white text drawing. Maps show the distribution of most of the regular species. In the third section the environment of each of the major land areas is discussed briefly, and tabular checklists show the bird observer what species to expect on the continent and various islands or in different sectors of ocean.

English names in this handbook come from a variety of sources. In cases in which regional names differ, one has been selected on the basis of the most widespread usage, and some of the others have been listed in the index as alternatives along with Spanish and French names. Scientific names are based on the latest revisionary work and field studies. They reflect a preference for 'lumping' in the same species similar forms that replace one another geographically and for considering as separate species those whose breeding ranges overlap. Geographically differentiated subspecies are cited and described in the text only when they are recognizable in the field on the basis of morphological characters.

AVIFAUNA

The majority of the birds in this area are seabirds that derive all their food either directly from the sea or, in the case of scavengers, predators, and pirates, at most one step removed from it. Even the few land birds that occur on islands near the Antarctic convergence move to the shore or coastal waters for feeding in winter. Some species, most notably albatrosses and giant fulmars, are wholly pelagic during the first few years of their lives while they circumnavigate the globe several times without ever touching dry land. After they attain adulthood, most seabirds still depend wholly on the sea for food, but because they must come to shore periodically for breeding, they may not range as widely. Seabirds are therefore full-fledged members of the marine ecosystem, and in order to understand their distribution and ecology one must be aware of relevant oceanic environmental conditions as well as those ashore.

The extraordinary richness and great extent of antarctic and subantarctic waters support vast numbers of seabirds. The nesting grounds available to them, however, are limited. In the far south, relatively few localities around the shores of Antarctica provide snow- and ice-free nest sites, even in summer, whereas near the convergence the widely scattered subantarctic islands present little total land surface. As a result, particularly favorable breeding localities teem with birds concentrated from vast reaches of open ocean.

In contrast to the abundance of individuals, the variety of antarctic birds is restricted: only 10 species are hardy enough to breed regularly on the continent, and only 17 breed on the peninsula and adjacent islands. The avifauna of the subantarctic islands, even though it is about triple that of Antarctica, is nevertheless poorer in number of species than that of temperate and tropical areas.

Although the penguin is the unofficial 'national bird' of Antarctica, tube-nosed petrels of the order Procellariiformes constitute the majority of species that regularly occur throughout the area. In number of species the penguin order Sphenisciformes runs third to the shorebird-skua-gull-tern order Charadriiformes. Two species of cormorant, four of ducks, three of rails, and a small assemblage of subantarctic songbirds complete the regular avifauna.

In size the seabirds vary from the large emperor penguin, which stands 4 ft (1.22 m) tall and weighs up to 41 kg, and the wandering albatross with a wingspread of over 11 ft (3.35 m) to Wilson's storm petrel, which weighs less than 60 g and has a wingspread of about 16 in. (41 cm). Some of the land birds are even smaller.

Remoteness has made Antarctica and the subantarctic islands inaccessible to terrestrial predators except to those unthinkingly introduced by man, himself the most effective land predator of all. The absence of native land predators fosters a remarkable tameness in the birds that permits close observation and study. Communal nesting habits, lack of wariness, and, in the case of penguins and petrels, extreme adaptation to pelagic life have left the birds highly vulnerable to introduced predators on land and in the past provided an open invitation to exploitation by man.

The climate, especially in the far south, is severe and inhospitable to man, but the birds that frequent the shores of Antarctica and the surrounding islands are physiologically adapted to the harsh environment. Indeed, many of them cannot live in less extreme climates. Their low stature and behavioral adaptations permit them to frequent microclimates for breeding or molting that are far milder than those that man experiences in the same localities. Nevertheless, even though nests are sheltered, eggs are closely brooded, and most chicks are covered with thick down at hatching, egg loss and chick mortality may be high during storms and unseasonable freezes.

One of the most remarkable features of the environment in high latitudes is the continuous daylight of the summer months. In winter, by contrast, days are short, and south of the Antarctic circle the sun never even rises above the horizon in June and July. Therefore in midsummer, when the majority of species breed, adults can gather food for their chicks almost continuously. Chicks grow fat, mature relatively rapidly, and leave the nests as inexperienced neophytes while food is still abundant. In winter, when temperatures fall, thick ice covers the sea, and daylight is restricted, food becomes scarce in the far south, and most species depart for the north either on long-distance migration or at least to the pack ice or open sea.

Several of the larger and more conspicuous species of seabirds have been studied intensively on land. In general, we know their displays, breeding habits, food given to the young, and, to a certain extent, population fluctuations. What is almost totally lacking, however, is biological information on the birds during that portion of the year when they are continuously at sea or even while adults are foraging during the summer in waters near the breeding grounds. We also know little about the abundance and distribution of various food species, how the different bird species capture their food at sea, how far individuals range in seeking food for their young, and what actually controls mortality, either during the critical fledging and first at-sea periods in young birds or during the time that adults are at sea between breeding seasons. Shipboard observation and specimen collection will help, but these studies must be coordinated with long-term marking programs that will permit identification of birds from different breeding localities at a distance without the necessity of capturing them.

ANTARCTIC AND SUBANTARCTIC ENVIRONMENT

The core of the Antarctic is the high ice-covered continent. Its eccentric roughly circular outline ranges between 65° and 75°S with deep embayments to 78°S in the Weddell and Ross seas and with the northward projection to 63°S of the mountainous Antarctic Peninsula. Only a few gravel beaches, offshore islands, coastal cliffs, and mountaintops are free of permanent ice and therefore accessible to birds for nesting. The mountains of the peninsula are linked through the ornithologically important Scotia islands and Tierra del Fuego with the Andes of South America. Antarctica is girdled by a ring of ice that is far more extensive in winter than in summer, when more moderate temperatures melt the ice and allow offshore winds to open up stretches of coast to penguins and other seabirds for landing and inshore feeding. To the north lies an expanse of open ocean with a sparse scattering of small islands between 55° and 35°S.

The sea and its resources provide food for seabirds all year, and the land provides breeding sites during a critical part of the annual cycle. Climate influences the distribution of the birds and their utilization of both sea and land, but because some species have been able to adapt to the harsh antarctic environment, they can take advantage of its food abundance.

Sea Environment

The southernmost reaches of the Atlantic, Indian, and Pacific oceans form a continuous circumpolar system around Antarctica, varying in width from 1050 km between South America and the Antarctic Peninsula to 2600 km south of Tasmania and New Zealand and 3900 km south of Africa.

Most of the sea is 4000-6000 m deep; its deepest trench, which is nearly 9000 m, lies just east of the South Sandwich Islands. Complex underwater ridges and plateaus, which run both north to south and east to west, break the sea up into several wide basins. The highest peaks of the ridge and plateau system emerge as islands.

The predominant surface currents and winds are from west to east. Water temperatures are generally cold with a regular latitudinal gradient of increasing temperature from south to north. Between 50° and 60°S an abrupt 2°-3°C change in temperature marks the meeting of two circumpolar rings of water masses having different characteristics. Antarctic surface water and sub-Antarctic surface water mix and sink there at a well-defined oceanic frontal system called the Antarctic convergence (Figure 1). The northern limit of sub-Antarctic surface water is also marked by an abrupt temperature change of 4°C at another oceanic frontal system near 40°S, the Subtropical convergence. The surface boundary area is marked by a visible line of current disturbance and very rapid temperature and salinity changes.

Water characteristics. The physical and chemical characteristics of these zonal water masses and their movements affect the distribution of seabirds, either directly or indirectly through the availability of their food. Although this fact has been recognized for some time [*Murphy*, 1936, pp. 59-110], good data for correlations have been slow to accumulate. Ornithologists making bird observations aboard oceanographic vessels are in a particularly favorable position to draw conclusions about ecological requirements of species, for they can compare bird distribution information with water condition and food abundance data collected at the same time.

Although the surface conditions may appear relatively simple, they are determined by a complex of physical and chemical conditions in the subsurface layers. The most easily measured characteristics of the zonal water masses are temperature and salinity. The Antarctic surface water immediately surrounding the continent is cooled by low air temperature and the presence of ice. This cooling creates strong temperature stratification during the warmer part of the year. Temperature soundings show a thin surface layer of warmed water below which temperature drops to a minimum near freezing at 100-300 m. From June to October or November, however, the entire upper 100 m is uniformly cold. Surface temperature ranges from below freezing at the continent to 1°-2°C at the Antarctic convergence in winter and 4°-5°C in the upper 50 m in summer. Antarctic surface water, which is low in salinity, varies from 30.5 to 34.5 ppt (parts per thousand) and averages about 33.9 ppt. It is much less saline in summer, when land and sea ice melts and precipitation is high, than in winter, when freezing leaves surplus salt in the water.

Sub-Antarctic surface water is appreciably warmer, from 5° to 10°C in winter and from 8° to 14°C in summer. Salinity averages about 34.3 ppt. There is an additional small frontal system south of Australia and New Zealand, marking two subzones within the subantarctic waters at about 52°S between the longitudes of 100°

Fig. 1. South polar projection to 35°S showing zonal water currents (arrows), the Subtropical and Antarctic convergences, the Antarctic divergence, and average seasonal limits of pack ice.

and 167°E. North of the front, salinity is greater than 34.5 ppt. There are thus two types of sub-Antarctic surface water: an extensive subzone of circumpolar sub-Antarctic water and a limited subzone of Australasian sub-Antarctic water of higher salinity.

North of the Subtropical convergence, Subtropical surface water is even more saline, averaging 34.9 ppt; temperatures are above 14°C in winter and above 18°C in summer.

Water circulation. There are three components of water movement in the oceans surrounding Antarctica: zonal, meridional, and vertical.

Circumpolar zonal movements: The simplest movements are the wind-driven circumpolar zonal currents from east to west near the Antarctic continent and from west to east farther north. These southern circumpolar currents facilitate exchange of waters between the northern oceans. Most of the surface water, both north and south of the Antarctic convergence, flows eastward, driven by the westerly winds of the West Wind drift. Bottom contours, even in very deep water, affect the direction of flow in this Antarctic circumpolar current. Over a rising bottom, such as that approaching a submarine ridge, the current is deflected north by the Coriolis effect of the earth's rotation, and as the floor deepens in the lee of a ridge, it is deflected south. Landmasses in the New Zealand region and near South America cause other irregularities in the eastward current (Figure 1). Part of the water veers north into the Pacific Ocean through the Tasman Sea, and another portion turns north into the Scotia Sea to enter the Atlantic west of South Georgia as the Falkland current. The flow also turns sharply north as it encounters the South Sandwich ridge, the very cold water of the Weddell Sea gyre thus being able to penetrate north as far as 53°S near the South Sandwich Islands. The rate of flow of the current is relatively high near the convergence and throughout the Drake Passage, averaging about 1 km/h, but the rate of flow falls off north and south of the Antarctic convergence.

Five to ten degrees south of the convergence, but generally north of the permanent pack ice, where the prevailing wind is from the east, there is a westward-flowing countercurrent, the East Wind drift, which is especially marked at the surface. It lies somewhat south of 63°S in the Atlantic sector and at about 68°S in the Pacific sector, but its exact position is variable depending on meteorological conditions (Figure 1). Along the coast of East Antarctica, where there are no pronounced coastal features, the East Wind drift is only a narrow coastal current. On the western sides of the deep embayments of the Weddell, Ross, and Bellingshausen seas the East

Wind drift is deflected northward and results in large clockwise cyclonic gyres. This northward deflection of very cold water is especially marked in the Weddell Sea gyre, where it prevails all the way to the bottom and reaches north to 53°S and east to 25° or 30°E. Because of heavy ice conditions, little is known of the East Wind drift along the continental shelf in the Bellingshausen Sea in summer or anywhere at all in winter.

The gyres cause local deviations from regular zonal distribution of temperature gradients. The Pacific sector is warmer than the other two sectors, and there the convergence averages 5° farther south. The coldest water is present in the Weddell Sea, and after it escapes from the Weddell Sea gyre near the South Sandwich Islands, it flows east with the West Wind drift through the Indian and Pacific sectors, where it is gradually warmed. A weaker less-pronounced cold-water gyre also accounts for the northward flexure in the surface isotherms in the Ross Sea region.

To the east of New Zealand a counterclockwise South Pacific subtropical gyre penetrates southward and extends eastward across the South Pacific between 40° and 50°S. The warm water in this gyre mixes with colder water from the Antarctic circumpolar current before it returns north along the west coast of South America as the Peru current (Humboldt current).

Meridional and vertical movements: Meridional and vertical movements transport a greater total mass of water than zonal movements do. They are thus of great importance in the economy of north to south water exchange between the oceans of the world, and they determine the frontal characteristics of convergences and divergences. Both meridional and vertical movements are ultimately related to differential densities of water masses or layers determined by thermal and salinity gradients.

South of the Antarctic convergence three distinct vertical layers of water are present: north-flowing cold surface and bottom layers with a warmer south-flowing layer between them (Figure 2). The Coriolis effect imparts a northward component to the wind-driven eastward current in Antarctic surface water. The water, which is cooled by low air temperatures, moves northward to the Antarctic convergence. There it meets and mixes with the south-flowing sub-Antarctic surface water and sinks to form sub-Antarctic intermediate water. The Antarctic surface water layer is relatively shallow, up to 250 m deep at the convergence and only 50 m deep farther south at the boundary of the East Wind drift.

At the beginning of winter the westward- and southward-flowing water near the Antarctic continent becomes more saline as drainage from the land ceases and freshwater separates out as ice. The more

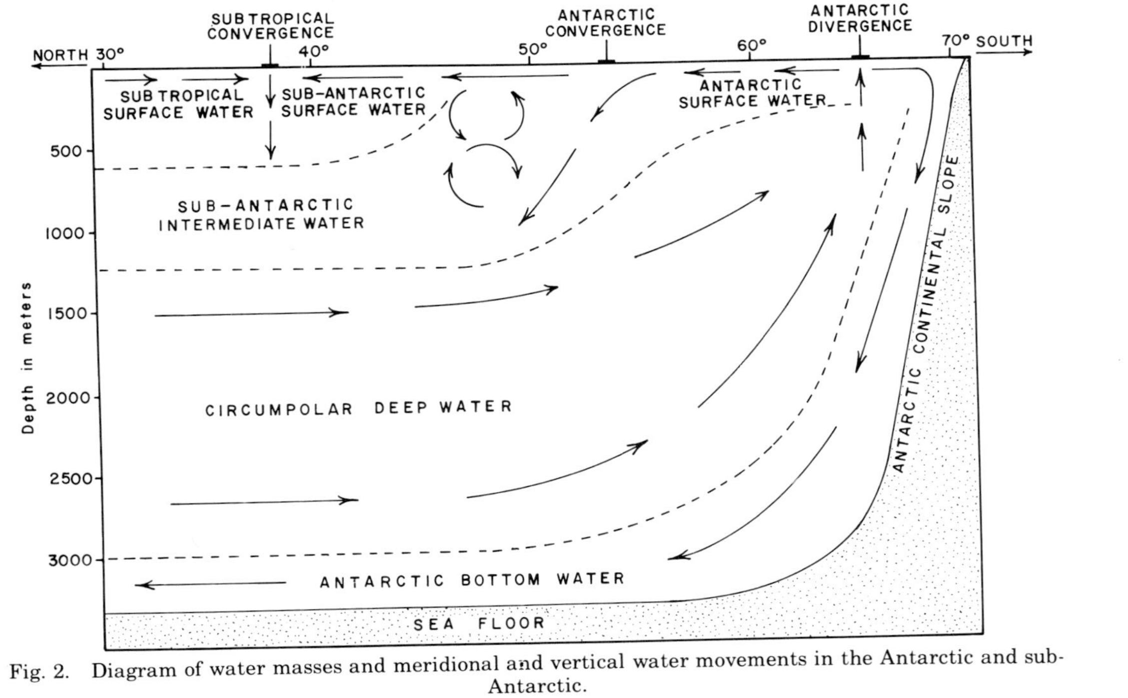

Fig. 2. Diagram of water masses and meridional and vertical water movements in the Antarctic and sub-Antarctic.

saline (34.6-34.74 ppt), colder, and therefore denser water that remains tends to sink down the continental slope to the bottom and forms Antarctic bottom water, which spreads northward into the northern hemisphere. Most of the very cold highly saline Antarctic bottom water forms in the Weddell Sea and flows westward around the continent, but some may form locally elsewhere, especially in the Ross Sea.

Between the two northward-flowing cold layers of Antarctic surface and bottom water there is a thick layer of dense, warm, and saline Circumpolar deep water, which originates in the northern Atlantic and to a lesser extent in the Indian Ocean. This layer flows southward and spreads eastward, mainly between depths of 2000 and 3000 m, gradually mixing with the layers above and below it. South of 35°S, Circumpolar deep water begins rising from 2000 m to only about 200 m under the shallow north-flowing layer of colder, less saline, and less dense Antarctic surface water. The boundary between east- and west-flowing currents near the continent marks a zone of upwelling and divergence of warmer saline (1°-2°C, 34.72 ppt) nutrient-rich water. At this Antarctic divergence (Figure 1), upwelled deep water replaces or mixes with surface water that has either a marked northward flow toward the convergence or a southward flow toward the continent. Most of this upwelled water probably mixes with the north-flowing Antarctic surface water and is additionally cooled by radiation and atmospheric exchange, its salinity being diluted by rain, snow, and meltwater from sea ice. If, however, it is cooled without being diluted, its density increases, so that it sinks near the continent to form Antarctic bottom water.

Immediately north of the convergence the water column is well mixed with only slight vertical changes in temperature and salinity. This mixing is due, in part, to the sinking of Antarctic and sub-Antarctic surface water to form Antarctic intermediate water, which extends northward at 500-1500 m in all three oceans, and, in part, to the high winds and violent eddies at the northern margin of the convergence. Still farther north in the sub-Antarctic zone there is evidence of the southward flow of warm more saline subsurface water, which helps warm the surface layer through reduced vertical mixing.

Antarctic convergence. Meridional and vertical movements of water layers with differing temperature and salinity characteristics contribute to the oceanic frontal system of the Antarctic convergence. The whole system is analogous to the discontinuity of an atmospheric polar front. North-flowing Antarctic surface water sinks at about the same latitude as that at which warm south-flowing Circumpolar deep water rises over the very cold north-flowing Antarctic bottom water. This area of discontinuity in meridional temperature, salinity, and density has both surface and subsurface expressions (Figure 2).

At the surface the convergence shows as an area in which temperature changes 2°C within a half degree of latitude. Midpoint temperature in the convergence gradient varies seasonally, being coldest in June and warmest from November to March. The subsurface expression is best demonstrated with a bathythermograph as a sudden change in the depth of the temperature minimum layer from 200-300 m south of the convergence to 500-600 m immediately north of it. Farther north still, mixing of Antarctic and sub-Antarctic surface water obliterates the temperature minimum layer. Usually, the two expressions of the convergence lie close together latitudinally, but local wind or current variation may cause a warm south-flowing tongue of sub-Antarctic surface water to ride 2°-3° south of the subsurface expression in the temperature minimum layer of the Antarctic surface water.

The convergence area is not completely stable. Local meanders and eddies at the surface may grow into small gyres, both cyclonic and anticyclonic, 65-80 km across. The gyres account for observed local irregularities and even reversals in temperature gradients during north to south crossings of the convergence. Nevertheless, its mean position is relatively constant, generally within a 100-km span, and only occasionally does it deviate as much as 2°-4° in latitude. The convergence reaches about 50°S in the Atlantic sector but lies farther south to 62.5°S in the eastern Pacific sector.

The convergence zone branches in the central Pacific sector near 140°W, its secondary southern branch extending east to about 108°W. The water between appears to be an area of divergence with upwelling of warmer deep water. This is in the area of a fracture zone in a submarine ridge. Another area of divergence may occur between 76° and 83°W and may be continuous with the more westerly one (Figure 1).

Abundance of food. The basis for the abundance of higher life in the sea is the plankton, which floats or drifts passively in the open ocean. It is composed of minute plants (phytoplankton), mainly diatoms in antarctic and subantarctic waters, and the somewhat larger animals (zooplankton) that feed on them. Phytoplankton abundance is in turn related to the amounts of various nutrient salts, such as nitrates, phosphates, and silicates, dissolved in seawater. These are taken up from the water by the phytoplankton and converted to new plant cells by the sun's energy in photosynthesis near the surface.

Nutrients from deeper waters are brought to the surface in areas of upwelling, such as the Antarctic divergence, and are consumed by the phytoplankton as the water moves away from the upwelling zone. The southward and upward movement of deep water provides abundant nutrients that permit vast blooms of phytoplankton during the long days of the austral summer. Furthermore, the great turbulence of the 'roaring forties,' Drake Passage, and some other ice-free waters also helps to recycle nutrients, so that locally, the nutrient and plankton levels are higher than they are in temperate and tropical zones. Contrary to earlier belief, however, not all antarctic seas are uniformly rich in nutrients and phytoplankton, nor are they uniformly high in productivity.

In general, the areas of the Antarctic divergence and coastal inshore waters are extraordinarily rich, whereas the open ocean, especially the central Pacific sector, is relatively poor. Antarctic waters south of the convergence are richer than the subantarctic waters to the north, and both are on the average richer than temperate waters. Areas of particular abundance are the Drake Passage, the eastern and southern portions of the Scotia Sea, the west coast of the Antarctic Peninsula, the northern and southwestern Weddell Sea, and the coasts of the Ross Sea, Tasmania, and New Zealand. There is undoubtedly a 'landmass effect' of upwelling on the upcurrent side of islands and continents that also results in increased local richness, but this enrichment has not yet been specifically demonstrated.

The areas of greatest phytoplankton abundance support large populations of zooplankton and of seabirds that feed on them. In addition to the areas already cited, the Antarctic convergence appears to be a circumpolar zone of abundance for some bird species. In this area of convergent sinking waters there is a surface concentration of buoyant planktonic organisms at which prions and storm petrels may gather to feed.

Photosynthesis, which makes phytoplankton growth possible, decreases or ceases altogether in winter owing to low light levels and the filtering effect of surface ice. In the ice-free waters near the convergence, phytoplankton production increases rapidly from September to December and then gradually decreases until April. It remains low in the sea and pack ice, farther south, until the breakup in December and January. Then because of high nutrient concentrations near the divergence, plankton becomes exceedingly abundant until ice re-forms in the fall.

The most common and widespread bird food organisms in antarctic and subantarctic plankton are euphausiid shrimp or krill. These 25- to 50-mm-long crustaceans form dense swarms near the surface in summer that may turn the water red. There is a short time lag between phytoplankton bloom and maximum abundance of zooplankton. The krill become abundant in January near the convergence and from January to March farther south. They live under the pack ice in winter and become available to birds only when the ice melts. The main species near the convergence are the large krill *Euphausia superba* and *E. frigida*, but other small species are dominant to the north, *E. vallentini*, and in shallower shelf waters near the continent, *E. crystallorophias*. The krill are best known as the food of the baleen whales, but they also serve as the main staple for many species of birds, nototheniid fish, squid, and the crabeater seal.

Krill is probably so abundant that it is in no danger of serious depletion by the birds and other animals that prey on it. There is always a potential danger of overexploitation by man, but although it has been discussed, commercial krill fishing in antarctic waters is not yet a reality. Great reduction in populations of large baleen whales in the Antarctic has cut down on the major consumers of krill. This presumably has increased the food available to birds and may possibly have allowed some species to increase in numbers. There is, for instance, evidence that the krill-feeding chinstrap penguin may have become more abundant in recent years and extended its range virtually around the Antarctic [*Sladen*, 1964].

In addition to krill, other crustaceans, particularly amphipods, copepods, and isopods, as well as pteropods, squid, and fish are important in the diets of seabirds. Little is known of the biology of antarctic and subantarctic Crustacea, other than krill, but they are abundant and available near the surface in summer. Squid of a variety of sizes are found throughout the antarctic area and probably come to the surface at night, whereas fish, largely nototheniids, are relatively common inshore but are scarce in the open ocean. No marked seasonal increase such as regularly occurs in phytoplankton and crustaceans has been reported in Antarctica in either squid or fish populations. On the other hand, much of the zooplankton, other than krill, and the squid and fish that feed on it are found in warmer water and at greater depths in winter than in summer. Consequently, they are more available to the deeper-diving penguins and cormorants, which can therefore remain in the Antarctic all year, than they are to surface-feeding petrels and terns, which are forced to migrate northward in search of food in winter.

The relationships between local phytoplankton productivity, zooplankton concentrations, and seabird abundance are in need of further research. There are obvious general correlations between the seasonal abundance of food and the distribution, migration, and breeding activities of seabirds. Much more needs to be learned, however, about the species composition, seasonal distribution, and vertical migration of plankton, squid, and fish before more precise correlations can be made. Nevertheless, estimates of abundance of birds at sea and observations of their feeding behavior still constitute valuable information for comparison with productivity and food organism data collected by oceanographers.

Ice. The climate and biological environment of the Antarctic are profoundly influenced by the presence of ice. Permanent flat ice shelves occupy one third of the antarctic coastline, predominantly the Ross, Ronne, and Filchner ice shelves in the Ross and Weddell seas (see frontispiece). They are the floating seaward extensions of vast continental glaciers but are also augmented by accumulation of snow on the surface and freezing on the underside. Their average thickness is over 0.8 km at the land junction but only about 180 m at the seaward end, where melting and calving of icebergs take place from terminal ice cliffs rising from 2 to 50 m above sea level. Although low ice shelves serve as habitat for a few emperor penguin rookeries, they are otherwise largely devoid of birds. Icebergs, both large and small, are attractive floating perches for many seabirds, notably Antarctic and snow petrels, and their margins provide a sheltered habitat for planktonic food organisms.

In contrast to shelf ice, which is of land origin, sea and pack ice are composed of seawater, which freezes when its temperature falls below −1.5° to −1.9°C. Sea ice that is attached to the land or ice shelves is termed 'fast sea ice'; most of the emperor penguin rookeries in Antarctica are on fast sea ice. Broken-up sea ice forms the drifting pack ice, which constitutes important feeding and molting habitats for a number of antarctic birds.

In the fall, new ice is relatively soft sludge, but it becomes harder and thicker with a further drop in temperature. Snow falling on its surface and, to a lesser extent, water frozen on its underside add to the ice, which attains an average thickness of about 1.5 m in a year. The colder portions of the southern Weddell and Ross seas and open leads in the pack ice of the previous year freeze earliest, from late February to April. Ice formation spreads northward as a nearly continuous fringe around the continent, its northern limits varying from year to year depending on the severity of the winter.

From September onward the ice begins melting at its outer edges, and it disintegrates as leads open up. Melting continues into March, when its surface is reduced to one half or, in very mild years, to one third of its winter maximum. Because of the large melt each year most of the antarctic pack ice is less than 2 years old and consequently rather soft and greenish. An exception is the thick pale blue multiyear ice of the Bellingshausen Sea, which makes the coast between Alexander Island and the Ross Sea virtually useless to birds as well as inaccessible to ornithologists who might study them.

The edges and undersides of ice cakes and small floes in the pack provide shelter for huge numbers of planktonic and nektonic organisms. The continual movement and grinding action of the pack expose and break up the animals, so that they are available for predatory birds. The fulmarine petrels, especially Antarctic and snow petrels, are particularly adapted to take advantage of this feeding niche. The same source of food partly accounts for the attraction of birds to ships moving through ice: propeller action overturns ice cakes and churns animals to the surface.

Ice is also responsible for the virtual lack of intertidal and shallow sublittoral organisms on antarctic coasts. The scouring action of grounded ice would crush any sessile invertebrates, and anchor ice forming on the bottom would trap other benthic animals and eventually lift them to the surface. Thus the beach scavenger niche is essentially absent from most of the coast of Antarctica, and the lack of gulls is thereby partly accounted for except on the peninsula, where the large limpet *Patinigera polaris* is common on rocky shores.

In severe winters, fast sea ice is present around most of the continent, in the Weddell Sea, and south of about 65°S in the western peninsula, and the passages of the northern peninsula and water around the islands of the Scotia ridge are cluttered with broken pack ice. In mild winters, on the other hand, all passages to the north of the Antarctic circle may be largely ice free.

The more open pack ice, which is broken up by sea swell, moves westward along the coast under the influence of strong easterly winds near the continent. Movements and stresses produce thick pressure ridges and open up ice-free leads, especially around headlands. Westward drift in the western Weddell Sea is blocked by the Antarctic Peninsula. This landmass induces a clockwise movement in a tongue of Weddell Sea ice that extends north and east in the Scotia ridge to about 55°S.

The pack ice influences local conditions at sea. There is a noticeable drop of 4°C or more in temperature as one approaches the pack, and it reduces surface waves and sea swell. Long before the ice itself is visible its white reflection or 'ice blink' shows on low overhanging clouds, whereas extensive leads in the pack show up as dark areas of 'water-sky' on low clouds.

The climate ashore on the Antarctic Peninsula and maritime Antarctic islands is profoundly affected by the pack ice. After the pack ice has melted, the peninsula and islands bask in 4 or 5 months of relatively moderate summer temperatures, well above freezing on clear summer days, but from May to December they suffer harsh winter conditions due partly to the insulating and cooling properties of the surrounding pack and fast sea ice. Warm winds from the north are cooled in passing over the ice before they reach the islands, and south winds are not subject to warming over open seawater as they are in the subantarctic islands that year-round are ice free and enjoy a more moderate climate.

Land Environment and Vegetation

The land surface actually available to birds for nesting in Antarctica and the subantarctic islands is severely limited. Although the total area of continental Antarctica is about 14.3 million km^2, less than 8000 km^2 is free of a thick permanent cover of snow and glacial ice. Only the relatively few offshore islands, beaches, coastal outcrops, and inland dry valleys and mountaintops are biologically important, and some of them are too remote from the sea or too inhospitable to serve as breeding sites for seabirds.

In the far south the scanty vegetation is of no significance to birds. They nest on shingle beaches, in rock debris, in moraines and screes, on cliff ledges, and even on fast sea ice. They cannot burrow in permanently frozen glacial sand and gravel, and the only nesting material available to them is gravel and pebbles, which they use to raise eggs and chicks above meltwater.

On the peninsula and the islands of the Scotia ridge much the same conditions prevail, although more ground is exposed. In addition, thick moss banks thaw sufficiently in summer to provide nest materials for skuas and gulls and to permit Wilson's storm petrels to burrow.

Farther north on the small and widely scattered subantarctic islands near the convergence, permafrost is limited to higher ground; at lower elevations, deep peaty soils are suitable for burrowing, and thick vegetation provides nest sites, material, and shelter for a greater variety of breeding birds. Few species can find sufficient insect and plant food on these islands to be independent of the sea, even for part of the year. At the northernmost fringe of the sub-Antarctic, woods on the islands in the Tristan da Cunha group furnish breeding habitat for subtropical noddy terns and are frequented by resident passerines.

The stunted trees and woody shrubs of the Tristan da Cunha group, Gough Island, Ile Amsterdam, and Ile Saint-Paul mark these islands as being botanically temperate in comparison with the other islands and continental land lying south of the limit of tree or woody shrub growth. The more southern area may be divided into two botanic zones: the sub-Antarctic botanic zone, in which phanerogams (flowering plants) are dominant from sea level to about 300 m, and the Antarctic botanic zone, in which cryptogams (mosses and lichens) are dominant at all altitudes. Closed phanerogamic communities are present only on the low-latitude islands that have a relatively mild climate. These botanically temperate and subantarctic islands present a distinctly green vegetated appearance at lower elevations in contrast to the more barren ice-capped antarctic islands and continent farther south.

Temperate oceanic islands and sub-Antarctic botanic zone. The vegetation in the temperate oceanic islands and in all sectors of the sub-Antarctic botanic zone is remarkably uniform, both in appearance and in floristic composition, some species and many genera being common to South Georgia, the Tristan da Cunha group, Gough Island, the Indian Ocean islands, and Macquarie Island. On the basis of the growth forms of the constituent plants five types of vegetation or patterns of plant communities may be distinguished: fern bush, tussock grassland, tundra meadows and herb fields, wetlands and heath, and feldmark. Fern bush and heath are confined to the temperate oceanic islands, whereas only feldmark communities of low-growing cushion plants are present in the Antarctic botanic zone.

Fern bush: In protected glens and ravines on Tristan, Nightingale, Inaccessible, and Gough islands and Ile Amsterdam, evergreen thickets of *Phylica arborea* grow to about 8 m high. The main trunks are generally prostrate on or under the peat surface on which *Empetrum rubrum* shrubs and various ferns, including *Blechnum*, form an understory. Lichens and other epiphytes hang from the branches to the ground. Trees are restricted to steep inaccessible ravines on Tristan Island and Ile Amsterdam, where they are cut for firewood. They are widespread up to about 450 m on Gough Island and are most fully developed in the lowlands away from salt spray. The only other tree is the arborescent legume *Sophora macnabiana* found sparingly on Gough Island. Noddy terns, thrushes, and buntings nest in the *Phylica* woods of the Tristan da Cunha group and on Gough Island, but there are no resident land birds that exploit the Ile Amsterdam woods.

Tussock grassland: Flat coastal terraces and steep coastal and inland slopes that are protected from violent winds are covered with tussock grass communities dominated by one or more species of *Poa* or *Spartina arundinacea* on the temperate islands. The grass may form a continuous canopy up to 3 m high, even though the bases of individual tussocks are isolated on stools a few meters apart. If the canopy is not continuous, other plants such as the composites *Cotula plumosa* and *C. goughensis*, Macquarie cabbage (*Stilbocarpa polaris*), or small ferns may occupy the interstices. Growth of grass is inhibited by salt spray; by exposure to strong winds, which increases transpiration; and by a high water table and shallow soil. Bird species using tussock grass communities for breeding include gentoo and rockhopper penguins, wandering albatrosses, several species of mollymauks, burrowing gadfly petrels and shearwaters, and the gray-backed storm petrel.

Tundra meadows and herb fields: Slopes and flats that are subjected to only moderate winds and have a high water table form meadows covered by grass or sedge tundra or by a low thick carpet of large-leaved perennial herbs. The dominant plants are fescue grass (*Festuca erecta*), especially on South Georgia; the rosaceous plant *Acaena adscendens*; sedge (*Carex* sp.); the composite *Pleurophyllum hookeri* (on Macquarie Island, associated with Macquarie cabbage and taupeta, *Coprosma pumila*); and Kerguelen cabbage (*Pringlea antiscorbutica*). The composition of local communities depends on wind, water table, and degree of soil slippage caused by erosion. Birds breeding in tundra meadows and herb fields include the two sooty albatrosses, the northern giant fulmar, and several burrowing gadfly petrels and shearwaters.

Wetlands and heath: On those parts of the coastal terrace and valley floors where the water table is at or just above the ground surface there occur two types of wetland vegetation, fens or bogs, depending on water acidity. Where the water is either neutral or alkaline owing to contact with basic rocks and mineral soil there develops a low fen vegetation of plants less than 15 cm high, dominated by rushes, *Juncus*, associated with bulrush (*Scirpus* sp.), hair grass (*Deschampsia*), bent grass (*Agrostis magellanicus*), and various mosses. The composition of communities depends on depth of water, drainage, and wind. Peat bogs are found under similar water table conditions, but where the water is derived directly from rain, it is distinctly acid from contact with decaying vegetation and is low in dissolved nutrient salts. Rush (*Rostkovia* sp.) and *Sphagnum* moss bogs are common on South Georgia, whereas *Sphagnum*, other moss, and *Colobanthus* mat occur sparingly on some other islands.

In the temperate islands the thick *Sphagnum* peat bogs of the low flat grounds and plateaus have developed a cover of wet heath communities that are characterized by dwarf woody *Empetrum rubrum* shrubs up to 1 m high interspersed with grasses and sedges. Only ducks, rails, and the Kerguelen tern frequent wetlands, but some petrels and shearwaters may burrow in drier heath.

Feldmark: Rocky highlands of scree and moraine debris that are exposed to strong winds have feldmark communities of low-growing cushion plants. On most islands the cushion *Azorella selago* is the dominant plant, but the feldmark also includes a very few other phanerogams and many mosses and lichens. *Azorella* is absent, however, from the feldmark of South Georgia, and it is replaced by *Empetrum rubrum* and *Rhacomitrium* cushions on the Tristan da Cunha group and Gough Island. The dominance of *Azorella* or mosses and the luxuriousness of cushions depend on the local violence of the wind. Feldmark vegetation can occur as low as 180 m and is the exclusive vegetation type above approximately 300 m. Feldmark communities are used for breeding by southern giant fulmars, snow petrels (mountains of South Georgia only), prions, burrowing gadfly petrels and shearwaters, storm petrels, and diving petrels.

Antarctic botanic zone. Farther south on the other islands of the Scotia ridge and on the Antarctic Peninsula the vegetation in the Antarctic botanic zone is very similar to the feldmark vegetation in the highlands of the northern islands. In protected areas at low elevations, closed cryptogamic communities occur, whereas in exposed or elevated areas, open communities occur. The main ground cover is a tundra cushion of moss and lichens in which only two diminutive phanerogams occur: a grass, *Deschampsia antarctica*, and a pink, *Colobanthus crassifolius*. Under particularly favorable conditions the moss cushions may be more than 1 m deep. Bouvetøya may also belong to this low Antarctic botanic zone. The even more meager vegetation of the ice-free headlands and nunataks of the high Antarctic botanic zone, including the continent and Balleny, Scott, and Peter I islands, consists only of open communities of mosses, lichens, and algae.

Mammals and birds have a profound effect on the vegetation in the antarctic area through physical disturbance and grazing. Elephant seals flatten grass tussocks, break up the continuous canopy, and form wallow pools. Large penguin rookeries are generally devoid of any low vegetation owing to continual trampling and guano deposition, but gentoo penguins and rockhoppers nest in tussock grass on some islands. The burrowing activities of petrels and rabbits are not usually destructive to the vegetation and probably increase soil aeration and drainage. Herbivores, chiefly introduced rabbits on Iles Kerguelen and Macquarie Island and domestic sheep on South Georgia, have, however, drastically overgrazed the tussock areas and herb fields and caused severe erosion, making them less attractive to breeding birds.

Beaches and terrestrial food sources. The beaches in the far south differ from those farther north in two ornithologically important ways. The rich intertidal zone, which elsewhere provides such abundant forage for shorebirds and gulls, is essentially lacking in Antarctica. Grounded ice cakes and the lifting action of anchor ice scour the beaches and shallow sublittoral bottoms virtually clean of intertidal and benthic algae and sessile invertebrates out to depths of about 15 m. A few attached limpets in protected rock crevices are generally the only sign of life to be found on rocky antarctic beaches at low tide. Likewise, flotsam and jetsam are seldom found on antarctic beaches, an important forage ground for beachcombing scavengers farther north thus being absent. Nevertheless, sloping beaches and rocky coasts are important landing places for penguins that use higher raised beaches and coastal terraces for breeding and molting.

In contrast, the ice-free subantarctic beaches have rich intertidal and sublittoral faunas, especially associated with luxuriant shallow-water *Macrocystis* kelp beds, where rockhopper and other penguins and cormorants forage. Cast-up seaweed, intertidal organisms, and other beach debris above the surf line are worked over by wekas, sheathbills, gulls, and passerines.

The tiny antarctic land animals (the largest are wingless flies) and plants are of no significance as food for birds on the continent, peninsula, or most of the islands south of the convergence. On the subantarctic islands, including South Georgia, Iles Kerguelen, and Macquarie Island, but not Heard Island, land and freshwater plants and animals provide food that supports ducks, rails, the Kerguelen tern, and a few passerines. Also, scavenging sheathbills, skuas, and gulls may find some terrestrial food on the subantarctic islands in addition to their usual fare in bird colonies. Food on land may be less available in winter, and most species, except the passerines on the Tristan da Cunha group, Gough Island, and Macquarie Island, must turn to the shore or coastal waters for feeding during the coldest months of the year.

Permanent research stations operate at several locations around the periphery of Antarctica and on some of the subantarctic islands. They have affected the habits and distribution of birds in the vicinity, especially when installations are near breeding sites. Construction of the bases and daily activity have disturbed local bird populations to varying degrees, and in some instances, interference has been extreme (see the section on conservation). On the other hand, a few species have taken advantage of man's presence at permanent stations in Antarctica either to overwinter or to expand their summer ranges southward. Carrion-eating species, such as the southern giant fulmar, American sheathbill, skua, and southern black-backed gull, feed at refuse heaps in winter, and the gull may have invaded the far south only during the past 125 years to take advantage of offal near whaling and sealing stations.

Climate

The area covered in this handbook embraces a wide range of climatic zones: the temperate and cold sub-Antarctic and the maritime and continental Antarctic. The high ice-covered continental plateau that roughly fills the Antarctic circle and the broad belt of relatively warm water surrounding it impose circumpolar climatic zonation on the Antarctic and sub-Antarctic. Although temperature is the most obvious climatic factor influencing biological life in the area, precipitation, cloud cover, winds, storms, and photoperiod are also of significance to the birds.

The least rigorous of the climatic zones is the temperate sub-Antarctic, which includes the northernmost islands, the Tristan da Cunha group, Gough Island, Ile Amsterdam, and Ile Saint-Paul. Mean monthly temperature averages above 10°C; seldom, if at all, does the temperature fall below freezing in winter, whereas summer temperatures frequently range above 20°C. In the islands of the cold sub-Antarctic zone, lying immediately north and south of the convergence, no month has a mean temperature greater than 8°C, and less than 7 months are above 5.4°C. For at least half of the year, however, mean monthly air temperature remains above freezing. The Antarctic zone, in which no mean monthly temperature averages above 1.5°C and there is permafrost at sea level, comprises two subzones. In the maritime Antarctic islands of the Scotia ridge and on the Antarctic Peninsula, at least 1 month averages above freezing at sea level, and in winter, mean monthly temperature rarely drops below −10°C. In the cold dry continental Antarctic, no mean monthly air temperature is above freezing, and in winter it is well below −20°C.

The air isotherms over the ocean closely parallel those of sea surface temperature. They show a relatively orderly zonal spacing northward from the very cold central core largely because unlike the Arctic Ocean the antarctic seas are invaded by no great meridional current of warm surface water and because the prevailing westerly winds are also zonal. Deviations from regular spacing and from sea surface isotherms are most marked where an air isotherm crosses a land-sea interface and at the Antarctic convergence, especially where the latter shows strong meridional undulations.

The Atlantic and Indian ocean sectors are on the average colder than the Pacific at the same latitude, partly owing to the eccentric position of the Antarctic continent and its ice sheet with respect to the geographic pole and partly to the strong cold north-flowing current of the Weddell Sea gyre. This cold current makes the northern tip of the Antarctic Peninsula, South Orkney and South Sandwich islands, and South Georgia much colder than other land areas at similar latitudes.

Freezing of the seas adjacent to the continent in winter in effect doubles its size for climatic considerations. This, of course, provides increased cooling in winter throughout the Antarctic and sub-Antarctic, but its effect along the coast and, in particular, on the peninsula and the maritime Antarctic islands is most marked. The latter two areas enjoy a relatively moderate climate in summer but become far more continental in winter, when they are closed in by pack ice. The extensive ice shelves and pack ice in the Weddell Sea likewise cause the east coast of the Antarctic Peninsula to be far colder at all seasons than the west coast.

The annual range of air temperature is greater at inland continental stations than on the coast and islands, where open water has a moderating effect. At most coastal and island stations, temperature decreases sharply with elevation but not inland on the continental plateau, where there is a strong temperature inversion. Likewise, over land, local topography influences precipitation patterns, snow drifting, cloud cover, and winds, whereas over the sea, water currents and temperature are particularly important.

Although precipitation is nearly impossible to measure directly in the far south because blizzard driven drifting snow cannot be differentiated from true precipitation, average measurements suggest a regular zonal increase from south to north. Approximately 15-19 cm equivalent of water falls annually over the continent, the coastal areas of which receive far more (20-25 cm) than the interior (10 cm). The interior is a virtual desert due to the low water vapor capacity of extremely cold air and the prevailing anticyclonic polar highs. In the maritime Antarctic and cold and temperate sub-Antarctic, stronger cyclonic activity provokes frequent precipitation, but here again low temperature reduces the water vapor content of the air, and the intensity of precipitation is moderate. In the maritime Antarctic Peninsula and islands, 35-50 cm fall, precipitation occurring on more than two thirds of the days of the year, whereas in the subantarctic islands, 1-2 m fall, precipitation occurring on up to nine tenths of the days. Nearly all precipitation falls as snow on the Antarctic continent, whereas the Antarctic Peninsula and Scotia ridge islands show a south to north increase in the percentage of rain over snow from less than 10% at 68°S to 50% at

60°S and 75% at 55°S. Rain is also the predominant form of precipitation on the subantarctic islands.

Seasonal variation in amount of precipitation is minimal in the continental and maritime Antarctic, but if anything, there is less in winter. By contrast, in the subantarctic islands and at sea north of the convergence, winter tends to be wetter than summer. In this northern area there is a lag in cooling and heating, so that snow persists late in spring and has even been recorded in summer at most stations and rain is frequent in early winter.

The amount of yearly precipitation and that of summer melting combine to determine the buildup of snow and glacial ice on the land. In the far south, accumulation is low, but because melting is insignificant, glaciers are common at sea level. On the subantarctic islands, total precipitation is greater because temperature there is higher, but increased summer melting at low elevations limits glaciers to the higher mountains.

Coastal antarctic stations generally experience more cloud cover than inland stations, and the northern part of the peninsula, some of the islands, and the oceans, in general, experience sea fog or are frequently overcast, especially in winter. On an average day in this northern area, clouds cover seven eighths of the sky.

The Antarctic continent, especially its coasts and the slopes of East Antarctica, is one of the windiest regions of the world. The strong winds in exposed areas accentuate the cooling effect of low temperature and thus influence the life of birds, especially during breeding. Although the winds that pour down the slopes from the central plateau lose their force far out at sea, they drive the ice offshore locally and therefore make the windiest coasts those most available for landing by penguins and the scientists who study them. At 60°S, mean daily wind speeds of over 75 km/h have been recorded in both winter and summer. Likewise, the roaring forties of the open ocean are notorious for strong and persistent west winds. The winds are strongest in winter and only slightly more moderate in summer. The regions of greatest gale frequency are over the open oceans north of the Ross and Weddell seas, in the approaches to the Drake Passage, in the Indian Ocean sector between 20° and 60°E, on the Adélie Coast, and in McMurdo Sound.

South of the Antarctic circle there are 24 hours of daylight in midsummer, and a corresponding period of darkness exists in midwinter. Even at some of the more northern islands there may be more than 16 hours of daylight in December and January. Midwinter darkness is not total, however. As far south as 72°S there is 1 hour of twilight, and on the circle there are 2 hours of daylight and 2 hours of twilight. The warming effect of the sun at such a low angle of elevation is slight, but the light does allow some biological activity.

Adaptations of Birds to Extreme Climatic Conditions

In order to take advantage of the abundance of food in the far south, antarctic birds have adapted morphologically, physiologically, and behaviorally for living under extreme climatic conditions. The most obvious adaptations are those for keeping warm in a cold environment, but these same adaptations create problems when temperatures rise well above freezing on bright sunny days. Antarctic birds, particularly penguins and petrels, have a thick layer of subdermal fat, a coat of heavy down next to the skin, and a surface covering of closely overlapping contour feathers. The fat layer and the air trapped between skin and feathers are excellent insulation. Virtually no heat escapes from the body through the skin and plumage when a bird is at rest, the outer surface of its body feathering being nearly the same temperature as ambient air. The back of an incubating penguin, for example, may be so cold that snow falling on it does not melt but acts as an additional layer of insulation. As long as the outer feather surface remains intact, it is essentially impermeable to cold, wind, and water. During molt, however, surface continuity is broken, and heat can escape. Therefore most species that are resident in the Antarctic molt when the weather is still relatively warm. Penguins leave the water entirely during molt and seek sheltered locations on land or pack ice until the new plumage is fully grown.

In addition to their insulating properties, heavy subdermal and visceral fat reserves in penguins and petrels act as insurance against enforced periods of fasting in extreme cold during courtship, incubation, and molting in adults and during development and fledging in chicks. Male emperor penguins fast 3-4 months and lose almost one half of their body weight during courtship and incubation, which regularly take place under bitter winter conditions. The homologous early spring fast is 4-5 weeks in Adélies and shorter in other penguins. In order to reduce energy demands during incubation in cold weather further, emperor and Adélie penguins, and probably most petrels, can lower body temperature at least 4°C below normal resting temperature. All adult penguins endure an additional 3-5 weeks of starvation and weight loss each year during molt. In early fall, king penguins and wandering albatross chicks lay down heavy fat in preparation for long winter periods of starvation because of reduced parental feeding. Similar fat reserves allow nestlings of smaller species of petrels to survive periods of starvation during storms and when the adults normally cease to feed them just prior to fledging.

The bare skin of the upper foot and the lightly feathered underwing, which radiate heat readily and hence prevent overheating, would lose warmth too rapidly in cold weather were it not for a vascular heat exchange system between arteries and veins. Where the vessels run closely parallel in the extremities, heat from arterial blood warms the tissues as well as the cooled venous blood that is returning to the heart.

Ecological and behavioral adaptations also help to conserve body heat. Nest sites, except those of inland-nesting snow and Antarctic petrels and south polar skuas, are close to the coast and therefore benefit from the temperature-moderating effect of nearby seawater. The sites are generally protected from the full force of cold winds. Penguins, cormorants, albatrosses, and giant fulmars nest on open ground, where hazards of snowdrifts are reduced, and cliff-nesting petrels choose locations on northern faces exposed to the sun. They can thus take full advantage of the heat-absorbing properties of their dark dorsal plumage.

Nesting ordinarily takes place in the spring and early summer, when temperatures are moderate. This schedule also insures that chicks become independent when the food supply is still at or near maximum. Small petrels, nesting in rock piles and burrows that are subject to snow blockage, breed later in the season, when snow is lighter and less frequent. Parents closely incubate eggs and brood young chicks continuously until the chicks can thermoregulate effectively. Emperor penguins, which encounter bitter cold conditions while they are incubating, huddle together with their backs to the wind and thereby expose only one sixth of their body surface. Penguin chicks, during the crèche period, also huddle together for warmth.

On the other hand, antarctic birds are subject to overheating during strenuous activity, when they are excited or disturbed at the nest, and on warm sunny days. In order to cool off, penguins and cormorants sit upright, pant with open beaks, ruffle and open their feathers, and spread feet and flippers or wings to expose vascularized skin. Warm blood flushes to the surface, and excess heat radiates to the air. The underwings and feet are also exposed to cold water and therefore can shed excess body heat in swimming. The lightly feathered head, bare skin on the face and on the base of the bill, and the bill itself act as additional heat-dissipating organs.

Heat metabolism has been studied little in antarctic and subantarctic seabirds other than penguins [*Jarman*, 1973; *Stonehouse*, 1967; *Warham*, 1971*b*]. The field presents research challenges.

SPECIES INFORMATION

This discussion is intended as an introduction to the species accounts. The first three sections of each account (identification, flight and habits, and voice and display) present information for identifying species and age classes. The following five sections (food, reproduction, molt, predation and mortality, and ectoparasites) cover biology of the species, and the last two (habitat and distribution) tell where to find birds.

Identification

It is important to learn in advance what species to expect in an area and to know the identification characters of these species before making observations. Tabular checklists of birds that may be seen in Antarctica, on various subantarctic islands, and in different ocean sectors are provided in the geographic section. Knowing the species and their diagnostic characters in advance cuts down on time spent in leafing through pages of irrelevant species to identify an unknown bird.

Identifying birds at sea presents more difficulties than identifying birds in terrestrial habitats. On land, where there are few predators

---

Fig. 3. Topography of birds. Upperparts of gadfly petrel: (1) tail, (2) upper tail coverts, (3) rump, (4) back, (5) scapulars, (6) mantle, (7) hind neck (upper portion is nape), (8) head, (9) secondary coverts (ulnar or cubital portion of wing), (10) leading edge of wing, (11) alula, (12) primary coverts, and (13) flight feathers (tips form trailing edge of wing), composed of (*a*) primaries and (*b*) secondaries (speculum is in this area in ducks). Underparts of gadfly petrel: (14) undertail coverts, (15) foot, (16) leg (tarsus), (17) flank, (18) side, (19) belly, (20) breast, (21) side of neck, (22) throat, (23) chin, and (24) underwing, composed of (*a*) axillaries, (*b*) underwing coverts (underwing lining), and (*c*) bend of wing (wrist). Penguin: (25) bill, (26) forehead (nasal caruncle is in this area in cormorants), (27) eyebrow (superciliary), (28) crown, (29) crest, (30) cheek (moustachial streak is in this area in terns), (31) ear coverts (ear openings are hidden by feathers in this area), (32) shoulder, (33) flipper, (34) leg (tarsus), composed of (*a*) inner surface and (*b*) outer surface, and (35) foot, composed of (*a*) outer toe, (*b*) middle toe, (*c*) inner toe, and (*d*) web. Head of albatross: (36) eye (colored portion is iris), (37) eyelid, (38) maxilla ('upper mandible'), composed of (*a*) nasal tube (naricorn), (*b*) culminicorn (the culmen is the lateral profile of the dorsal ridge of the culminicorn and maxillary nail), (*c*) latericorn, (*d*) cutting edge, and (*e*) maxillary nail (unguicorn, which forms a hook), and (39) mandible (lower), composed of (*a*) ramicorn (colored longitudinal groove in this area in sooty albatross is the sulcus) and (*b*) mandibular nail (inferior unguicorn).

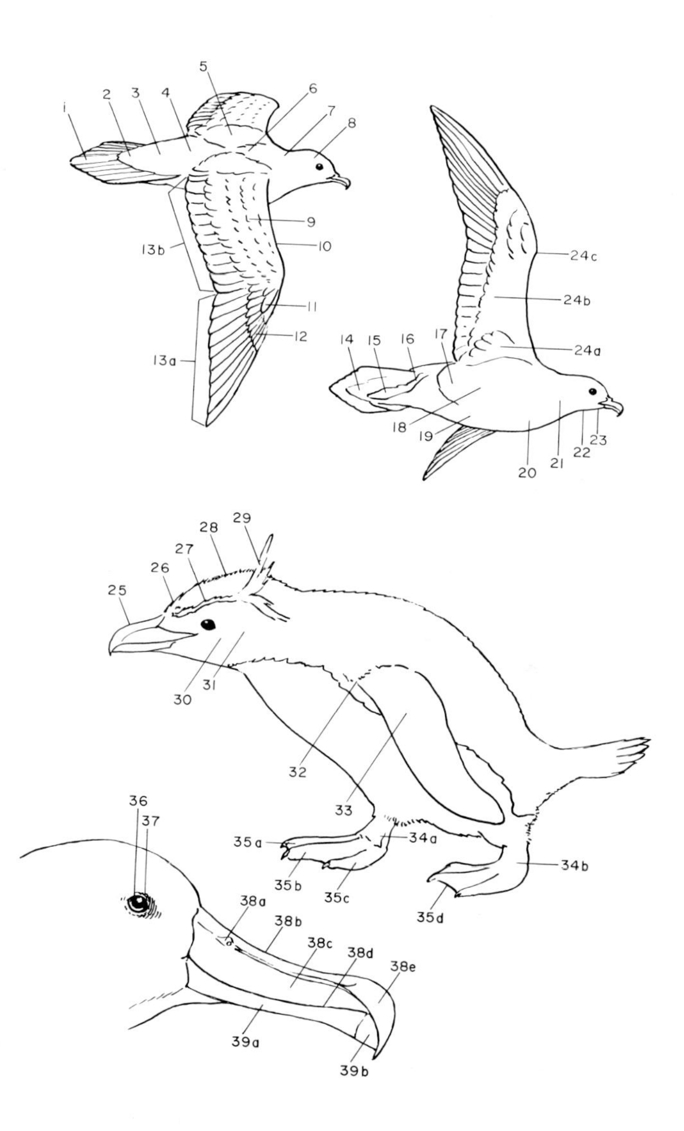
1
2
3
4
5
6
7
8
9
10
11
12
13a
13b
14
15
16
17
18
19
20
21
22
23
24a
24b
24c
25
26
27
28
29
30
31
32
33
34a
34b
35a
35b
35c
35d
36
37
38a
38b
38c
38d
38e
39a
39b

to disturb them, birds are generally tame and may be examined closely and at leisure. On the other hand, at-sea observations from a moving ship are frequently at long distance and provide only a fleeting glance before a bird disappears over the waves. If a bird cannot be identified immediately, it is useful to record as many characters as possible for later consultation of other reference books or with experts. The following hints on characters to look for are concerned primarily with at-sea observations and can serve as an outline for reference notes on unidentified birds.

Size. Although size is a useful seabird character, unless there is some standard, such as a well-known species or part of the ship nearby for comparison, it is difficult to assess. Distance greatly affects apparent bird size and is hard to judge at sea. Inexperienced observers tend to overestimate distance with the result that they usually underestimate bird size. Average measurements for each species are given in inches and to the nearest centimeter. The first two figures are overall length, and the figures following the slash are wingspread, except for penguins, for which the second figure is the length of one wing from shoulder to tip. Figure 3 shows the morphological features of birds.

Shape. Overall shape is useful for assigning a bird to a family and thus reducing the number of alternatives to consider for identification. Silhouettes of representatives of the major families and other groups are shown in Figure 4. The shape of a bird depends on the relative proportions and configurations of its head, neck, body, and appendages. The head may be large or small, the neck long or short, and the body stocky, slim, or elongated. Wing and tail shapes are particularly important for identifying flying birds. Wings may be long, short, narrow, broad, pointed, or rounded. The tail may be long, short, rounded, wedge shaped, pointed, square ended, or forked.

Color. Color is less important for distinguishing species of seabirds than it is for distinguishing land birds, since most seabirds are either uniform in color or a combination of white with black, gray, or brown. Furthermore, freshly molted individuals normally appear darker than worn individuals unless the new feathers are white tipped and thus impart a lighter or scaled appearance until the edges wear away. The worn tips of old feathers may also impart a scaled appearance to a bird; gray feathers frequently become brown when they are worn and faded.

Pattern. Color pattern, on the other hand, is of vital concern in species recognition, not only for bird watchers, but also for the birds themselves. Contrasting colors of back and underparts and particularly of underwings are visible at a great distance. Also look for the following key pattern characters: dark hood or cap, dark or light eyebrow or other contrasting pattern near the eye, white forehead or cheeks, light collar or rump, dark or light wing tip or band on upperwing or underwing surface, white flash at base of primaries, contrasting breastband, and dark or light tip on outer feathers of tail.

Bill. Shape of the bill, which may be long or short, heavy or slim, and hooked or pointed, is important in distinguishing families or genera, and its color in albatrosses, gulls, and terns may be useful for distinguishing species, age class, or reproductive state. Immature petrels, which generally do not differ from adults in plumage characters, have smaller weaker bills for a period after leaving the nest. Colors of feet, eyes, and bare facial skin, like bill color, are useful for species and age class recognition and in separating breeding and nonbreeding birds, but they are difficult to observe in flight. Closeup color photographs of head and feet are useful in documenting difficult-to-describe colors of captured birds.

Age and plumages. As a bird matures, it follows a regular developmental sequence of plumages, some of which may result in changes of appearance and are helpful in determining ages of individuals. Plumages are changed by molts (see page 36). At hatching all species in this handbook, except the cormorants, are covered by a natal or first down. In addition, the penguins and petrels have a second down coat that is longer and heavier but usually differs only slightly in appearance from the first. The down is replaced by a 'juvenal' plumage at the time that the young bird fledges and leaves the nest or colony. Penguins and petrels, in most of which the juvenal plumage differs little, if at all, from that of the adults, retain it for a full year or more. In other species this juvenal plumage, the first true feather coat, which is fluffier and less dense than that of the adult and which may differ markedly from that of the adult in appearance, is replaced soon after the bird leaves the nest with a first 'basic' plumage. Thereafter plumages are renewed once or twice a year. As adults all penguins, petrels, rails, sheathbills, and resident songbirds in this handbook retain the basic plumage all year. Cormorants, ducks, shorebirds, skuas, gulls, and terns have two body plumages a year, a basic or winter plumage and an alternate or nuptial plumage. The major wing and tail feathers, however, are generally molted only once, when the basic plumage is assumed. After the first year it is impossible to age most penguins and petrels on plumage characters, but skuas, gulls, and terns retain immature characters in their plumage for 2 or more years, whereas the wandering albatross and southern

Fig. 4. Silhouettes of representative birds, all drawn to the same scale: (1) emperor penguin, (2) Adélie penguin (two views), (3) wandering albatross (two views), (4) sooty albatross, (5) giant fulmar, (6) Antarctic petrel, (7) Cape pigeon, (8) Antarctic prion, (9) gadfly petrel, (10) greater shearwater (two views), (11) Wilson's storm

petrel, (12) diving petrel (two views), (13) shag (three views), (14) South Georgia pintail (three views), (15) Gough Island moorhen, (16) skua (four views), (17) southern black-backed gull (three views), (18) Antarctic tern (two views), and (19) American sheathbill (two views).

giant fulmar attain full adult plumage even more gradually in an extended sequence that takes more than 10 years.

Flight and Habits

Characteristics of flight, swimming, and walking and other habits are often of use in assigning a bird to a family and in separating two species of similar color and pattern at a distance. The methods of flight and swimming are usually correlated with means of searching for or securing food at sea. Notice whether the bird flaps intermittently and glides, soars, wheels, and banks like an albatross or petrel or whether it flaps almost continuously like a skua. Is flapping rapid and with stiff wings like that of a shearwater, or is it slow with flexible graceful wing movements like that of a gull? Is flight light, buoyant, and ternlike or heavy and ponderous like that of a giant fulmar? Is it rapid (shearwater), deliberate (skua), or leisurely (gull)? Is it erratic (prion) or direct (cormorant)? How high does the bird fly, just above the waves like a storm petrel, above the horizon like a gadfly petrel, or high over the masthead as a gull occasionally does? In swimming, does it 'porpoise' in and out of water like a penguin, or does it ride on the surface like an albatross? Does it dive or plunge deeply underwater like a penguin or cormorant or remain visible at all times like a gull or skua? Does it ride low in the water like a penguin or cormorant with only the head, neck, and upper back exposed, or does it bob buoyantly like a gull? How does it move on land, competently like a skua or awkwardly like a penguin or albatross? Is it aggressive, antagonistic, complacent, tame, or shy?

At sea some species, such as prions, are highly gregarious, whereas others, such as the gadfly petrels, are normally solitary. Birds that otherwise search for food alone tend to congregate in mixed flocks about ships in anticipation of finding discarded galley scraps or natural food churned to the surface by the propeller. Notice whether a bird follows the ship or pays it no attention.

Voice and Display

Voice and associated display postures serve to form and maintain bonds between mates, assert territorial dominance, and keep flocks of birds together. They are frequently quite different in related species and may be useful auxiliary identification characters. The elaborate displays of some species that perform in the open on land, such as the penguins, albatrosses, skuas, and gulls, have been studied, described, and analyzed well, whereas those of shearwaters and gadfly, storm, and diving petrels, which take place in burrows or at night, are virtually unknown and unstudied. Breeding grounds, both early in the season when mated pairs are displaying and later when chicks are being fed or prebreeding subadults are forming pair bonds, are noisy and active. The din in a penguin rookery may be deafening. Open ocean birds, which normally rely on vision to maintain contact, are silent at sea except when they are quarreling over food, whereas coastal species tend to be more vocal.

Food

Food and the method of capturing it not only are helpful in identifying birds but also play a central role in determining all other ecological requirements of a species. Most antarctic and subantarctic birds feed while they are swimming, either by actively pursuing their prey underwater as penguins, diving petrels, and cormorants do or by approaching their less active or already dead food on the surface as albatrosses, fulmarine petrels, skuas, and gulls do. Others, such as gadfly petrels, storm petrels, and noddy terns, pick food off the surface in flight, and black-capped terns after hovering over their intended prey plunge to capture it. Scavenging albatrosses, fulmarine petrels, skuas, gulls, and storm petrels may follow ships for galley scraps or other food stirred up in the wake or attend whales to feed on their excreta, whereas the giant fulmars, skuas, gulls, wekas, and sheathbills scavenge on beaches, hang about refuse heaps on land, or are predatory on other birds.

Food habits of antarctic and subantarctic birds are poorly known, and much more information is needed about methods of capture and the actual foods eaten. Closely related species may take different foods and thus not compete with one another. Likewise, immatures and adults of the same species may take somewhat different foods. Foods of a species may also differ at different times of the year. All records of feeding, both observation of live birds and analysis of stomach contents, are worthwhile.

Adult seabirds digest fish, squid, and crustaceans rapidly. Stomach contents of specimens collected at sea usually consist of only a few bones, otoliths, eye lenses, squid beaks, or parts of exoskeletons. On the other hand, adults that gather food for chicks at a distance from the nest carry it in an esophageal crop or in the proventriculus, where digestion is slow, while they return to the breeding ground, at times a matter of several hours. The physiology of inhibition of digestion in adult birds gathering food for young has not been studied much. Undigested food items intercepted just before they are given to the young [*Emison*, 1968] or retrieved from chicks immediately after feeding may be far easier to identify than partly digested mush in a feeding adult's stomach.

Reproduction

Some of the most interesting adaptations of antarctic and subantarctic birds are associated with reproduction, especially in choice of nest site and timing of breeding. In the far south, where blizzards may occur in any month, exposed headlands, cliffs, and islands that become snow free early in the season are preferred for breeding. The timing of laying and its synchrony are related to the strength of the bird, its nest site, and the probability of the entrance of the nest being blocked by snow. Early-breeding petrels, which are generally large robust species, lay early on exposed sites, and breeding is synchronous. Late breeders, the smaller weaker species that are vulnerable to predation, have hidden or burrow nests, and synchrony is low [*Beck*, 1970]. Albatrosses and petrels that have difficulty taking off from flat ground choose sites that have a clear runway into the prevailing wind or are on steep cliffs. Burrowing species on northern islands adopt sites with thick peaty soil, whereas in the frozen south the same or closely related species, if they can breed at all, nest in rock piles.

Nest materials vary locally. In the more northern colonies, penguins and petrels build substantial nests of plant materials, whereas in the more barren south they gather a few pebbles or lay on bare ground.

One egg is the rule in the larger penguins and in all albatrosses and petrels, whereas most other species lay two or three eggs. In those that lay more than one egg per clutch the last chick to hatch seldom survives in years of short or even normal food supply but may fledge in years of plenty. Laying its single large egg imposes a serious physiological demand on a female petrel. After selecting a nest both adults depart for 2 or more weeks at sea to feed heavily in preparation for laying and extended incubation. The male may engage in so-called 'courtship feeding,' thus helping the female get enough food to produce their egg. Penguins, ducks, skuas, gulls, and terns may lay replacement clutches if their eggs are lost early in the season, but petrels are usually incapable of producing more than one egg in a season.

With the possible exception of the ducks both parents participate in incubation and caring for the young, although the roles of the sexes differ from species to species. The respective roles are well known for the various penguins, a few of the albatrosses and petrels, and the skuas but need further study in most of the other species.

Incubation periods are long, generally 30-60 days in penguins and petrels and up to 80 days in the wandering albatross. The amount of time that the individual parent spends on the nest may involve long periods of fasting if the feeding grounds are remote. Even though the chicks of all species, except cormorants, are covered with thick down at hatching, they must be brooded to keep them warm as well as protect them from predators for several days or weeks before both parents can leave the nest simultaneously to forage for food. Most chicks stay in or near the nest until they are full grown and fully fledged, but young penguins gather into large nursery parties called 'crèches' out of which each pair of parents calls its chick for feeding. Young petrels and, to a lesser extent, penguins lay down a heavy layer of fat while they are in the nest. Prefledging petrels generally are so fat that they outweigh adults. Adult penguins and petrels leave the breeding grounds for molt and migration before the young fledge. The abandoned young live off reserve fat for the last week or two before their feathers are fully grown. They then depart to feed independently. Young skuas, gulls, terns, and some other species fledge early in relation to penguins and petrels but remain dependent on the adults for food for up to several months. They depart from the breeding grounds along with the adults.

Cormorants, ducks, rails, lesser sheathbills, some gulls, Kerguelen terns, and land birds are resident on or about the breeding grounds all year, and some petrels continue to visit their burrows during the off-season. In these cases, courtship activity and pairing rather than arrival at the breeding grounds are the best indications of renewed reproductive activity. Immatures of these sedentary species may disperse from the breeding grounds soon after fledging, and adults may frequent beaches or coastal waters as winter approaches, but there is no true postbreeding departure.

Breeding is highly synchronous in most colonies but may vary from colony to colony, with latitude, or between species. Closely related species of mollymauk albatrosses, giant fulmars, diving petrels, and terns may have quite different breeding schedules at the same locality. Most species breed during the austral summer and bring off their young when food is most plentiful. The emperor penguin, however, hatches its egg in the darkness of the antarctic winter, and the great-winged petrel and gray petrel are also winter breeders. Some species may take more than 12 months for reproduction. The wandering albatross may breed successfully only every other year, and at least on South Georgia the king penguin may rear only two chicks in 3 years [*Stonehouse*, 1960]. It is still unknown whether individual emperor penguins breed every year. A successful pair would have only 2 months for molting and feeding after its chick fledged before courtship for the new season began.

Mature adults return to the colonies early and reclaim the most desirable central nest sites. After the older birds have begun breeding, subadults visit the colony where they were born, form pair bonds, and select nest sites for future use. With increasing age, birds come to the colony earlier, stay longer, and occupy better nest sites. Younger breeders are generally less successful than long-established birds, and in many cases the first few nesting attempts may end in failure.

In species in which long-term banding programs have permitted such studies, there is a high degree of fidelity to natal site. Penguins, albatrosses, and skuas tend to return for breeding to the place at which they were hatched. Among breeding adults also there is a strong attachment for the colony and in some species even for the same nest site year after year.

Molt

All birds renew their feathers periodically by molting. The new feathers begin growth in follicles beneath the surface of the skin, pushing the old ones out of the follicles. In penguins, which are relatively inactive while they are molting on land or sea ice, the old feathers remain attached to the tips of the growing pinfeathers for several days and impart a puffy appearance to the bird before they are shed in patches. In flying species that are active during molt, however, the old feathers may be pulled out or knocked off before the new feathers emerge from the follicles.

Diving petrels and ducks become flightless during molt because virtually all flight feathers are lost at the same time. Most other antarctic and subantarctic birds molt the flight feathers more gradually, so that no more than three primaries are in growth on each wing at a time. Flight may be labored, but only in some shearwaters may flight be temporarily halted by a massive loss of outer primaries. Birds in wing or tail molt are usually identifiable as molting birds in flight, since ragged margins and missing feathers are apparent. In addition, some species, notably the southern black-backed gull, lose a majority of the wing coverts simultaneously and may show light 'windows' of whitish down feathers on the upperwings.

Reproductive activities usually postpone feather replacement. Molting birds about a colony during breeding are generally prebreeding subadults or individuals that have lost eggs and chicks. The larger petrels that breed early in the season are able to molt their primaries on the breeding grounds, whereas the smaller species that breed later and are still rearing chicks in March delay wing molt until they are on their winter quarters farther north.

The majority of species have one complete molt per year, usually after breeding, and retain the same plumage appearance all year. The only changes are due to wear of light-colored feather edges, making the bird darker, or to general fading, making it lighter. Cormorants, skuas, gulls, terns, ducks, and shorebirds have two molts a year, generally a complete one after breeding ('prebasic' or 'postnuptial') and a partial one, chiefly of body and head feathers, before breeding ('prealternate' or 'prenuptial'). The basic (nonbreeding or winter) plumages of these species may have a different appearance from the alternate (or breeding) plumages, especially about the head. Adult long-winged great albatrosses and giant fulmars molt the wing feathers continuously, a remarkably protracted procedure in which individual feathers may be renewed only every other year. The body molt in such cases, however, is usually on an annual cycle.

Predation and Mortality

There are few natural predators in the far south. Leopard seals, giant fulmars and skuas, gulls, and sheathbills are known to feed on adult birds, chicks, or eggs, and fur seals may also capture some birds at sea. Although the killer whale seems like a probable predator of penguins at sea and has been so mentioned by some authors, there are no confirmed reports of killer whales or any other cetaceans capturing birds in antarctic and subantarctic waters. Some prey species have evolved habits that help them avoid undue predation by other birds. Breeding in large rookeries or colonies exposes only those birds at the peripheries of colonies to predation. Adults brood eggs constantly and remain with small chicks until the young can defend themselves. Small petrels are generally nocturnal on the breeding grounds, and their late-season breeding may be partly an adaptation to insure that visits to the nest for feeding the young are made in darkness.

On the other hand, insular seabird colonies are highly vulnerable to introduced terrestrial predators such as rats, cats, dogs, pigs, and wekas. Some breeding populations have been eliminated entirely by introduced predators, whereas others, greatly reduced in size, are now confined to inaccessible steep cliffs or offshore islets. Man has been a serious predator of some species, nearly eliminating the king penguin from some subantarctic islands and perhaps entirely from the South Shetland Islands.

Most petrels, both adults and chicks in the nest, eject stomach oil either through the open bill or through the nostrils when they are molested on the nest. The oil, which is produced or stored in the glandular or proventricular portion of the stomach, is mixed with or derived from the bird's invertebrate or fish food [*Matthews*, 1949]. It is smelly and difficult to remove from clothing and presumably also from feathers. Spitting oil is effective in discouraging predation by skuas and other birds and makes it unpleasant to walk in open cliff colonies of fulmarine petrels. The ejected oil hardens in cold weather and, particularly at permanently frozen sites such as those in mountain ranges, may build up as thick waxy deposits called 'mumiyo.'

Mortality is low among adult seabirds. Some species of petrels may have an annual adult survival rate of over 90%, and individuals may live 20 or more years. In balance with low adult mortality, reproductive rates are low, and maturity is deferred. In the far south, clutches of one or two eggs are the rule, the smaller clutches being laid by young birds. Breeding in penguins, large albatrosses, petrels, and skuas may not begin until the bird is 5-10 or more years old. Young birds breeding for the first time are rarely successful, perhaps owing to their low status in competition for food near the breeding grounds. The consequences are delay in return to the breeding colony, availability only of peripheral nest sites, late laying, and, because of inexperience, inability to provide sufficient food for nestlings, which as often as not starve before fledging.

Loss of eggs and chicks in some years may be high, largely owing to unfavorable weather, such as heavy rain that inundates nests and prevents adults from finding food or snow that blocks burrows and prevents feeding of the young. Likewise, mortality is high among newly fledged young birds. Most birds found 'wrecked' on beaches after early winter storms are young of the year.

Little is known of the causes of limitation of population numbers in antarctic and subantarctic seabirds. Population studies at breeding stations are in most cases just beginning to provide meaningful information on survival (see the section on banding). Most mortality, however, takes place at sea during winter dispersal. Its extent can only be inferred from returns of banded birds at breeding colonies, and individual causes are unknown.

Ectoparasites

Ectoparasites of antarctic and subantarctic birds have been studied reasonably well, and the mites, hippoboscid flies, ticks, lice, and fleas to be expected on each species are listed in the text. The ectoparasite names were compiled from a variety of sources, and some synonyms, especially of ticks, may be included in the lists. Ticks and fleas are most easily found in nests, whereas mites and lice live on the skin or feathers at all stages. Most parasites, except ticks, are host specific or at most occur on only a few host species. Information on collecting bird ectoparasites may be found in the leaflet by *Watson and Amerson* [1967]. Bird endoparasites in the Antarctic and sub-Antarctic have not been studied as intensively, and no effort has been made to list them.

Arboviruses and other disease organisms that are transmitted by blood-sucking ectoparasites have not been studied much in antarctic and subantarctic birds. At least one rickettsial infection has been found in an emperor penguin, and Kerguelen penguins show general antibody reactions to virus [*Chippaux et al.*, 1972]. Infectious respiratory bacteria have been found in antarctic birds [*Margni and Castrelos*, 1963]. In recognition of the constant threat of bird disease introduction to Antarctica all importation of live domestic poultry is forbidden.

Habitat

Marine habitats frequented by antarctic and subantarctic seabirds may be described in two ways: in terms of life zones or zonal water masses and, within the zones, by the distances that birds range away from the coast. Habitat preference may be an aid in identification of some species. Life zones are discussed on pages 44-52, and Table 1 gives the bird species to be expected in each zone. Most species feed at a characteristic distance from the shoreline. Birds may be found on beaches (littoral), in water within sight of the shore (coastal), within 50 km of the shore (offshore), or far out at sea (pelagic). For instance, sheathbills are littoral, and gulls and cormorants are coastal, whereas albatrosses and gadfly petrels are both offshore and pelagic. During breeding all colonial antarctic and subantarctic birds spend some time near shore but may actually feed in deep water at some distance from the nests. Habitat preference may vary at different times of the year, and immature birds and nonbreeding adults may occur in different habitats from those in which breeding adults feed.

Land birds are generally found in habitats where they can find food, cover, or nest sites: ducks near freshwater streams or meltwater ponds, rails near heavy tussock grass or other vegetation, and passerines along the beach or in tussock, scrub, or woods.

Distribution

The distribution of each species is shown on south polar projection maps or is discussed in the text. Breeding localities are designated by large black circles; questionable breeding localities by half black circles. Breeding localities of different subspecies or closely related forms are shown by white stars in the circles or by black squares. Detailed breeding locality information and documentation on most species is available in the map folio by *Watson et al.* [1971], but later additions are documented by references. The at-sea range is shown on the maps by stippling, and vagrant records outside the usual range are shown by black stars.

Migration movements for each species, if they are known, are discussed in the text. There is marked seasonality in seabird distribu-

TABLE 1. Zonal Distribution of Seabirds of the Antarctic and Sub-Antarctic

Species	Antarctic Life Zone		Sub-Antarctic Life Zone			Subtropic Life Zone
	Continental	Maritime	Transitional	Cold	Temperate	
Emperor penguin	X					
King penguin		*	SGKHM	X		
Adélie penguin	X	X				
Chinstrap penguin	X†	X	SGH			
Gentoo penguin		X	SGKHM	X		
Rockhopper penguin			KHM	X	X	
Macaroni penguin		X	SGKHM	X		
Wandering albatross			SGKM	X	X	
Black-browed albatross			SGKHM	X		
Gray-headed albatross			SGK?M	X		
Yellow-nosed albatross				X	X	
Sooty albatross				X	X	
Light-mantled sooty albatross			SGKHM	X		
Northern giant fulmar			SGKM	X	X	

Southern giant fulmar	X	X	SGKM	X		
Southern fulmar	X	X				
Antarctic petrel	X					
Cape pigeon	X	X	SGKHM	X	X	
Snow petrel	X	X	SG			
Narrow-billed prion			K	X		
Antarctic prion	X	X	SGKHM			
Broad-billed prion				X	X	
Fulmar prion			KH	X	X	
Fairy prion			M?	X	X	X
Blue petrel			KM	X		
Great-winged petrel			K	X	X	
White-headed petrel			KM	X		
Atlantic petrel					X	
Kerguelen petrel			K	X	X	
Soft-plumaged petrel				X	X	
Mottled petrel				X	X	
White-chinned petrel			SGKM?	X	X	
Gray petrel			KM?	X	X	
Sooty shearwater			M	X	X	X
Flesh-footed shearwater					X	X
Greater shearwater					X	

TABLE 1. (continued)

Species	Antarctic Life Zone		Sub-Antarctic Life Zone			Subtropic Life Zone
	Continental	Maritime	Transitional	Cold	Temperate	
Little shearwater					X	X
Wilson's storm petrel	X	X	SGKH	X		
Black-bellied storm petrel		X	SGK	X	X‡	
Gray-backed storm petrel			SGKM?	X		
White-faced storm petrel					X?	X
South Georgia diving petrel			SGKHM?	X		
Kerguelen diving petrel			SGKHM	X	X§	X§
Blue-eyed shag		X	SGH	X	X	
King shag			KM	X	X	
American sheathbill		X	SG			
Lesser sheathbill			KH	X		

South polar skua	X	X				
Brown skua	X? ‖	X	SGKHM	X	X	
Southern black-backed gull		X	SGKHM	X	X	X
Antarctic tern		X	SGKHM	X	X	X
Kerguelen tern			K	X		
Arctic tern						
Brown noddy					X	X

Known to breed in zone or subzone, X; South Georgia, SG; Iles Kerguelen, K; Heard Island, H; Macquarie Island, M; breeding needs confirmation, ?; and occurs at sea in zone, rule.

*Breeding in maritime Antarctic South Shetland Islands during early nineteenth century [*Eights*, 1833] cannot now be substantiated.

†Presumed recent expansion into continental Antarctic Balleny Islands is discussed by *Sladen* [1964].

‡The systematic status of the white-bellied form on Gough Island is unclear; birds of the Tristan da Cunha group belong to the subtropical white-bellied storm petrel, *Fregetta grallaria*.

§The *Pelecanoides urinatrix* populations on or about the Falkland Islands, Tristan da Cunha group, Gough Island, Ile Amsterdam, Ile Saint-Paul, and New Zealand temperate sub-Antarctic Islands are sometimes treated as a separate species.

‖ Balleny Islands only.

tion, particularly in the far south. Antarctic waters teem with birds in summer, when food is plentiful and easily available, but other than incubating emperor penguins, which do not feed while they are ashore anyway, few birds remain in Antarctica in winter. Reduced light, low temperatures, and heavy sea ice make it impossible for most seabirds to secure what little food is present there from April until October. Even such pack ice species as the Adélie penguin and snow and Antarctic petrels move northward to the more open pack and somewhat milder conditions for winter. Seabirds concentrate in waters near land when they are feeding chicks during the spring and summer flush of food, but they disperse widely over the open ocean when food becomes scarcer in fall and winter. Young birds without reproductive responsibilities range great distances all year. Immature albatrosses and giant fulmars may spend 3 or more years circling the world before making their first landfall and seeking nest sites.

In winter most southern species move somewhat northward to open waters between the Antarctic and Subtropical convergences. The southern fulmar, the Cape pigeon, the black-bellied storm petrel, and the two skuas occasionally reach the subtropics, whereas the mottled petrel, the sooty and greater shearwaters, and Wilson's storm petrel undertake long-distance migrations across the equator to subarctic waters. Only one boreal species, the Arctic tern, regularly migrates into antarctic waters during the austral summer and in doing so covers about 35,000 km round trip.

Birds that wander or are blown far beyond the limits of their usual ranges turn up frequently on subantarctic islands and occasionally in Antarctica. Severe storms can cause such long-distance vagrancy, but wandering is common in migrant seabirds and land birds. Some terrestrial and freshwater species are particularly prone to high-latitude vagrancy, among them, herons, ducks, rails, shorebirds, swifts, and swallows. Nearness to land, large size, hospitable climate and habitats, and downwind location from larger landmasses favor landfalls on South Georgia, the Tristan da Cunha group, Iles Kerguelen, and Macquarie Island.

In the species accounts, vagrant individuals of extralimital species that have been recorded in the Antarctic and sub-Antarctic are fully documented. In spite of the efforts of watchful ornithologists, however, many vagrants that come to ground on uninhabited islands, or in Antarctica, undoubtedly perish unsighted and unrecorded.

ANTARCTIC AND SUB-ANTARCTIC LIFE ZONES

Antarctic seabirds are distributed in circumpolar life zones that coincide roughly with zones used by meteorologists, oceanographers, and botanists for classifying climate, seawater masses, and terrestrial plants (see pages 21, 5-7, and 16-19). The boundary lines for seabird life zones are less well defined, however, because although birds are restricted by breeding requirements on land, they are highly mobile at sea. A species may breed on an island in one zone and feed largely in an adjacent zone; another species may occur in a southern zone in summer and migrate to a more northern zone in winter. Nevertheless, some seabirds are so closely identified with particular zonal water masses all year that they can be used as indicator species. For instance, the emperor and Adélie penguins and Antarctic and snow petrels are found exclusively in the Antarctic zone, and their presence practically guarantees the proximity of pack ice.

The Antarctic convergence and northern limit of permanent pack ice are significant zonal boundaries. Less well defined boundaries are the Subtropical convergence and the 10°C mean annual surface water isotherm. The Antarctic convergence separates the Antarctic life zone to the south from the sub-Antarctic life zone to the north, whereas the Subtropical life zone extends north of the Subtropical convergence, beyond the scope of this handbook. The northern limits of permanent pack ice and the East Wind drift, which are essentially coextensive, subdivide the Antarctic into continental and maritime Antarctic subzones. The sub-Antarctic can also be subdivided into transitional, cold, and temperate sub-Antarctic subzones in the south near the convergence and about midway at the 10°C isotherm.

Environmental and physical conditions that birds encounter on islands where they can breed do not necessarily coincide with the birds' extreme physiological tolerances at sea. For instance, because gadfly petrels must have soft soil for burrowing, they breed on islands in the sub-Antarctic life zone, but some species regularly feed over cold water at the edge of the pack ice in the Antarctic life zone. Accordingly, a zonal classification of birds based solely on breeding areas is less satisfactory than one based on at-sea occurrence. Both are shown in Table 1.

The Antarctic continent, its coastal islands, and Peter I, Balleny, and Scott islands are considered to lie within the continental Antarctic subzone; the Antarctic Peninsula and its archipelago, southern islands of the Scotia ridge (South Shetland, South Orkney, and all but possibly the northernmost South Sandwich islands), and Bouvetøya are in the maritime Antarctic subzone. Because of the greater variety of breeding habitats, South Georgia and Heard Island, which lie south of the convergence, share more species of breeding birds with Iles Kerguelen and Macquarie Island, which lie on or immediately north of the convergence, than they do with other islands

TABLE 2. Zonal Differentiation of Birds of the Antarctic and Sub-Antarctic

Species	Antarctic Life Zone	Sub-Antarctic Life Zone: Transitional and Cold	Sub-Antarctic Life Zone: Temperate
Large penguins, *Aptenodytes*	emperor *A. forsteri*	king *A. patagonicus*	
Gentoo penguin, *Pygoscelis papua*	*P. p. ellsworthi*	*P. p. papua*	
Eudyptes penguins		rockhopper *E. c. crestatus*	rockhopper *E. c. moseleyi*
		macaroni *E. chrysolophus*	
Wandering albatross, *Diomedea exulans*		*D. e. exulans*	*D. e. dabbenena*
Mollymauk albatrosses, *Diomedea*		black-browed *D. melanophris*	
		gray-headed *D. chrysostoma*	
			yellow-nosed* *D. chlororhynchos*

Sooty albatrosses, *Phoebetria*	light-mantled *P. palpebrata*	sooty *P. fusca*
Giant fulmars, *Macronectes*	southern *M. giganteus*	northern *M. halli*
Cape pigeon, *Daption capense*	*D. c. capense*	*D. c. australe*
Prions, *Pachyptila*	Antarctic *P. desolata* narrow-billed *P. belcheri* broad-billed *P. v. salvini* fulmar *P. crassirostris*	broad-billed *P. v. vittata* fairy *P. turtur*
White-chinned petrel, *Procellaria aequinoctialis*	*P. a. aequinoctialis*	*P. a. conspicillata*

TABLE 2. (continued)

<table>
<tr><th rowspan="2">Species</th><th rowspan="2">Antarctic Life Zone</th><th colspan="2">Sub-Antarctic Life Zone</th></tr>
<tr><th>Transitional and Cold</th><th>Temperate</th></tr>
<tr><td>Wilson's storm petrel, Oceanites oceanicus</td><td>O. o. exasperatus</td><td>O. o. oceanicus</td><td></td></tr>
<tr><td>Black-bellied storm petrel, Fregetta tropica</td><td colspan="2">F. t. tropica</td><td>F. t. melanoleuca</td></tr>
<tr><td>Kerguelen diving petrel, Pelecanoides (urinatrix)</td><td></td><td>P. (u.) exsul</td><td>P. u. dacunhae</td></tr>
<tr><td>Shags, Phalacrocorax</td><td colspan="2">blue-eyed
P. atriceps</td><td></td></tr>
<tr><td></td><td></td><td colspan="2">king
P. albiventer</td></tr>
<tr><td>Skuas, Catharacta</td><td>south polar
C. maccormicki</td><td colspan="2">brown
C. lonnbergi</td></tr>
</table>

Entries centered between columns indicate that species or subspecies occurs in both zones or subzones.
*In the cold sub-Antarctic subzone occurs only on Prince Edward Island.

that are geographically in the maritime Antarctic subzone. South Georgia, Iles Kerguelen, Heard Island, and Macquarie Island as well as nearby water may best be considered a transitional sub-Antarctic subzone intermediate between the maritime Antarctic and cold sub-Antarctic subzones. Marion Island and Iles Crozet, plus most of the so-called New Zealand sub-Antarctic Islands (Campbell, Auckland, Antipodes, and Bounty islands), Cape Horn, and the Falkland Islands, are in the cold sub-Antarctic subzone, whereas the southern coasts of South America, the Tristan da Cunha group, Gough Island, Ile Amsterdam, Ile Saint-Paul, Snares, Stewart, and Chatham islands, and South Island of New Zealand are in the temperate sub-Antarctic subzone.

Zonal speciation has occurred in several genera with the result that distinct species or subspecies inhabit the Antarctic and sub-Antarctic life zones or the transitional, cold, and temperate sub-Antarctic subzones (Table 2).

With the still enigmatic exception of the snow petrel, not one of the species that inhabit the Antarctic life zone shows regional differentiation within the zone. This systematic unity is a reflection of the uniform environment, circumpolar movement of individuals in many species, and relative continuity of breeding stations around the continent. In contrast, regional or meridional differentiation has occurred in several species between the widely separated sectors or islands within the sub-Antarctic life zone (Table 3).

The few species of land birds that breed on some of the northern islands suggest the avifaunal affinities of the islands. The undifferentiated population of speckled teal of South Georgia is a very recent arrival from South America or the Falkland Islands. The South Georgia pipit and the local subspecies of yellow-billed pintail likewise have their closest relatives in the same area. The Kerguelen pintail is a miniature insular subspecies of the holarctic pintail duck that reaches southern India and tropical Africa on migration. The freshwater Kerguelen tern is closely related to the marine Antarctic tern. The monotypic American sheathbill, which breeds on the Antarctic Peninsula and Scotia ridge islands and migrates to South America and the Falkland Islands, has only one close relative, the resident lesser sheathbill, which has differentiated into subspecies on each of the four Indian Ocean subantarctic island groups. The extinct endemic subspecies of banded rail and red-fronted parakeet and the surviving populations of gray duck, starling, and redpoll on Macquarie Island were derived from other nearby New Zealand sub-Antarctic Islands where the same or related subspecies occur. The starling and redpoll were introduced to New Zealand from Europe, whereas the weka, whose insular occurrence is of no zoogeographic

TABLE 3. Regional Differentiation in Sub-Antarctic Life Zone

Species	Subspecies		
	Atlantic Sector	Indian Sector	New Zealand Sector
Macaroni and 'royal' penguins, *Eudyptes chrysolophus*		*chrysolophus*	*schlegeli*
Black-browed albatross, *Diomedea melanophris*		*melanophris*	*impavida*
Antarctic prion, *Pachyptila desolata*	*banksi*	*desolata*	*altera*
Great-winged petrel, *Pterodroma macroptera*		*macroptera*	*gouldi*
Atlantic and white-headed petrels, *Pterodroma incerta* and *P. lessoni*	*incerta*		*lessoni*

Greater, flesh-footed, and sooty shearwaters, *Puffinus gravis, P. carneipes,* and *P. griseus*	*gravis* *griseus*			*carneipes*	 *griseus*
Blue-eyed shag, *Phalacrocorax atriceps*	*atriceps* *georgianus* *bransfieldensis**		*nivalis*		
King shag, *Phalacrocorax albiventer*	*albiventer*		*melanogenis* *verrucosus*		*purpurascens*
American and lesser sheathbills, *Chionis alba* and *C. minor*	*alba*		4 island subspecies of *minor*		
Antarctic and Kerguelen terns, *Sterna vittata* and *S. virgata*	*georgiae* *gaini**	 *tristanensis*	*vittata* *virgata* *mercuri*		*bethunei* *macquariensis*

Entries centered between columns indicate that species or subspecies occurs in both sectors.
*Occurs in maritime Antarctic subzone.

significance, was introduced to Macquarie Island from Stewart Island.

The rails, thrush, and buntings of the Tristan da Cunha group and Gough Island have differentiated considerably from their presumed ancestors and have even subspeciated within the islands. The Gough moorhen, of which another subspecies used to live on Tristan Island, could have been derived from either the African or the South American form of the cosmopolitan common gallinule or moorhen. The origin of the Inaccessible Island flightless rail is obscure. The thrush is most probably related to an African species, but the three buntings are all derived from South American species [*A. L. Rand*, 1955]. Tristan and Wilkins' buntings probably represent two colonizations. The Gough bunting is least differentiated from its presumed ancestor.

BANDING

Metal bands (also called 'rings') bearing a return address and serial number have been used to mark birds in studies of long-distance movements, orientation, population ecology, and breeding behavior in Antarctica and the subantarctic islands. Individual birds may thus be recognized, and their movements, survival, and behavior studied. Particularly informative have been remote recoveries of albatrosses and southern giant fulmars that demonstrate an eastward circumpolar movement, especially in birds of prebreeding age [*Gibson*, 1963, 1967; *Stonehouse*, 1958; *Tickell and Gibson*, 1968; *Tickell and Scotland*, 1961]. The 35,000-km annual round trip of the Arctic tern between its arctic breeding grounds and its antarctic off-season feeding grounds was also confirmed by banding. Information on fidelity to natal colony, age of first return to colony, age at first breeding, and longevity is accessible only through recognition of individually marked birds of known age and origin, whereas other investigations of breeding behavior and physiology are facilitated by knowledge of the age and sex of individual birds [e.g., *LeResche and Sladen*, 1970]. In birds that commence breeding only after an extended period of immaturity, banding studies are necessarily of long duration. At some colonies, especially on Ross Island near McMurdo Station and on Signy Island, South Orkney Islands, significant numbers of known age banded birds are already available for research [*Sladen et al.*, 1968].

Most of the markers used in Antarctica and the subantarctic islands are tarsus or flipper bands made of aluminum or monel metal, but some flipper bands for emperor penguins have been made of a metal strip bonded between two Teflon (fluorinated ethylene propylene) strips [*Penney and Sladen*, 1966]. A useful innovation in working with relatively tame birds, especially skuas, has been a tall band with vertically stamped numerals. These tarsal bands and penguin flipper bands have numbers large enough to be read with binoculars, eliminating the necessity for capturing a bird in order to read its band [*Sladen et al.*, 1968].

Adults are routinely banded on the right tarsus, and chicks (i.e., birds of known age) on the left tarsus. All penguin bands are designed for attachment to the left flipper, where the number can be read upright. Adults and chicks, on their breeding grounds, may be color banded on the other tarsus to show locality. During the International Geophysical Year, skuas were color banded in the Antarctic in a great variety of combinations. These have now been reduced to seven major colors and may also be used on other species in the following sectors:

0° to 50°E	white	160°W to 80°W	black
50°E to 105°E	blue	80°W to 50°W	red
105°E to 160°E	green	50°W to 0°	gray
160°E to 160°W	yellow		

Banding in Antarctica and on subantarctic islands is coordinated through the Scientific Committee for Antarctic Research, Biology Working Group on Antarctic Bird Banding, with representatives from each of the participating nations. The following national bands are known to have been used extensively in Antarctica and the subantarctic islands: United States (U.S. Antarctic Research Program through U.S. Fish and Wildlife Service and Smithsonian Institution), Great Britain (Falkland Islands Dependencies Survey and British Antarctic Survey through British Trust for Ornithology and British Museum), South Africa (South African National Antarctic Expedition through Percy Fitzpatrick Institute), France (Expéditions Polaires Françaises through Paris Museum), Australia (Australian National Antarctic Research Expedition through Commonwealth Scientific and Industrial Research Organization), and New Zealand (New Zealand Department of Scientific & Industrial Research through Dominion Museum, now called National Museum of New Zealand, and New Zealand Wildlife Service). Bands bearing return addresses in Argentina, Japan, Norway, Chile, and the USSR may also be encountered. Permits for bird banding are issued by each country's National Committee for Antarctic Research.

Any bands recovered on dead birds should be reported or, better still, sent to the address shown on the band. Report serial number, band placement, species, locality and date found, and circumstances of finding. Bands on live birds should be left intact. After recording pertinent data and which leg the band is on, release the bird intact.

Rates of recovery of birds away from the place of banding are generally low. Such species as small petrels and penguins are rarely recovered in numbers, although rates of recovery in conspicuous species under intensive cooperative study in widespread areas, particularly albatrosses and giant fulmars, occasionally reach 8%. On the other hand, much higher rates, more than 90% in some species, have been recorded in birds returning to the site of banding at penguin rookeries.

Some other methods of marking birds have been used in Antarctica and the subantarctic islands. Penguins chicks may have the webs of one or both feet punched to indicate year born. Note the feet (left, right, or both) and webs (inner, outer, or middle) that show punch holes as well as the number of holes on each web. Birds may also be dyed with bright conspicuous colors or may carry colored leg streamers or neckbands to show home range. Any such peculiar markings should be recorded.

RECORDING OBSERVATIONS AT SEA

Records of seabirds seen during open ocean crossings are valuable in providing distributional and ecological data. Identification, numbers, position, and date are sufficient basic information for most distribution and abundance studies. Local environmental data and behavioral observations enhance the basic information and may provide further insight into the ecology of the birds observed. For many species, even distribution is poorly known. For nearly all, routes and timing of migration or other seasonal movements are either unknown or only inferred from a few sightings. Methods of feeding, as well as the food taken, need further study. Such observational data at sea are particularly useful when they are combined with observations of the same species at nearby breeding colonies.

A method of keeping a running seabird log that has proved satisfactory consists of two portions, a species log and an environmental log. The species log should include vertical columns for time (which may later be converted to position), identification, number, flight direction, and remarks (age, sex, behavior, molt, and associated organisms). Entries for each sighting are made horizontally in the appropriate columns. The environmental log is maintained as a separate table with entries every 2 or 4 hours or when conditions change. The environmental data are easily obtained on research vessels and are generally available from the bridge on other ships. *Routh* [1949] has discussed observation- and data-recording methods in antarctic waters.

Observations from even a single cruise are useful, but if they can be combined with those from different years and seasons, their value is greatly enhanced. Several institutions keep seabird logs of many observers in archives. It is also possible to computerize individual observations in order to facilitate retrieval and analysis [*King et al.*, 1967].

The following information should be entered in the logs.

Species Log

Time and position. The local time of each observation is preferable. By extrapolation this may later be converted to position, for normally it is not possible to obtain accurate fixes for each observation.

Identification. This should include the degree of certainty of identification. If you are in doubt, give as many characters as possible (see the section on identification). Also note whether the bird is an adult or immature and whether it is molting.

Number. Record the true count if it is possible; estimate large flocks of passing birds.

Flight direction. Determine this from the ship's compass or known heading.

Behavior. Record flight pattern, height of flight above water, ship following, flocking, calls, and method of feeding.

Associated organisms. The presence of small fish or other food, seals, whales, and other birds may prove useful.

Environmental Log

Positions. Fixes are mainly available at sunrise, noon, and sunset in clear weather and more frequently near shore or if electronic navigation equipment is operating. Extrapolation will be necessary to determine the position of individual observations. The distance from the nearest shore and depth on inshore observations are helpful.

Weather conditions. Include temperature, precipitation, cloud cover and type, wind velocity and direction, and visibility.

Surface water conditions. Record at least temperature and wave height; if salinity, clarity, color, plankton abundance, and local productivity are available, give them also.

Ice conditions. Record the amount of coverage (usually expressed in eighths or 'oktas'), the kind of ice, and the distance to the edge of the pack [*U.S. Naval Oceanographic Office*, 1968].

Day length. Record the time of local sunrise and sunset.

PRESERVING AND SHIPPING SPECIMENS

Occasionally birds fly aboard ship, become entangled in nets, or are injured in other ways and thereby come into hand. Storm and diving petrels, which are attracted to lights, are particularly likely to be found stranded on deck during cloudy nights. If such specimens are properly preserved, they are immediately useful for verifying preliminary at-sea identifications and ultimately for systematic, anatomical, or food habit studies. Most museums are pleased to receive specimens and will provide identifications.

Adequate means for preparing and storing study skins are seldom available at sea. On the other hand, birds may be preserved satisfactorily with the following minimal equipment and supplies either by freezing or by pickling:

Freezing
- Labels of sturdy paper
- Pen and indelible ink or pencil
- Polyethylene bags
- String or thread for attaching labels and closing bags

Pickling
- Freezing supplies (see above)
- Pickling container with a wide mouth (galvanized or plastic garbage pail is good)
- Syringe with assorted needles
- Preservative (one or more of the following)
  - Formalin (37% formaldehyde), dilute to about 4% with 9 parts of salt water or freshwater
  - Alcohol (either ethyl, i.e., grain, or methyl, i.e., wood), dilute to 70% by mixing 7 parts of alcohol with 3 parts of water
  - Vinegar, use full strength
  - Salt solution, add table salt to water until no more will dissolve
- Household detergent or soap, either bar or flakes
- Cloth, porous cheesecloth is good

Data for each specimen should be recorded on a sturdy paper label written in indelible ink or soft dark pencil and tied onto one of the legs. The label should include date, locality (latitude and longitude are sufficient at sea), and collector's name or any other relevant specimen information such as food regurgitated (useful if it is preserved with the bird), weight, and colors of unfeathered areas (eyes, bill, bare facial skin, and feet). Since colors of the unfeathered parts may change rapidly after death, record or photograph them in life or immediately after death.

Frozen specimens are best protected against dehydration by airtight polyethylene bags tied with string or sealed by heat. They should be deepfrozen, not just refrigerated, as quickly as possible. A ship's food freezer is ideal, but stewards are frequently loath to 'contaminate' frozen food supplies with specimens.

Birds may be pickled in 10% formalin solution, alcohol, strong vinegar, or saturated brine in a large covered container. In vinegar or brine, plumage colors may change, but since color characters are relatively unimportant for identifying most seabird species, this change presents no problem. Ideally, specimens should be injected with preservative from a syringe. The abdominal cavity and major muscle masses (breast, wings, and legs) should be pumped full of liquid until the bird begins to swell slightly. It is safer to inject too much preservative than to inject too little. If no syringe is available, cut open the abdominal cavity to expose the viscera. Immerse the bird in preservative (wear gloves for formalin), and squeeze it gently to insure penetration of fluid into the throat and body cavity. In very large birds continue the superficial abdominal incision in the skin up one side of the breast to the shoulder, and loosen the skin from the underlying muscle before immersing the bird in preservative. A little alcohol, liquid household detergent, or soap added to formalin or brine solutions will break surface tension on feathers and therefore permit faster wetting and better preservation.

Specimens should be left in preservative for at least 1 week. If space is a problem, they may then be removed from the pickling container and placed in a sealed plastic bag with a little liquid preservative. Specimens wrapped in cloth stay wetter, and the claws and bills will not puncture the plastic bags.

For shipment, pickled specimens should be placed wet in sealed plastic bags, and the bags packed in a sturdy metal can or drum or in a wooden box. Frozen specimens must not thaw in transit. Wrap each bagged frozen specimen in tinfoil, newspaper, or cloth to provide insulation. Place wrapped specimens in a crush-proof container, also insulated with newspaper, sawdust, excelsior, or styrofoam, and pack with dry ice if deepfreeze accommodations are not available for transit. Packed frozen specimens should be sent by airfreight or airmail, special delivery. If more than 48 hours will elapse between packing and delivery, request that the airline reice the shipment en route. Be sure that the airline notifies the recipient prior to arrival of the shipment.

Collection and Importation Permits

The area south of 60°S is subject to provisions of the Antarctic Treaty [*Conference on Antarctica*, 1959], and permits are needed to take specimens of birds and other wildlife. Scientific collecting permits, subject to international restrictions, are issued by each country's National Committee for Antarctic Research. Specimens imported into the United States and many other countries also require import permits and may have to pass through designated ports of entry. Check with customs or the recipient before shipment to determine the pertinent import regulations.

CONSERVATION

The avifauna of Antarctica and the subantarctic islands is in delicate ecological balance. Human disturbance could well tip the scale in critical cases. The inherent instability of antarctic and subantarctic ecosystems when they are subjected to human disturbance results from (1) the small number of species, (2) their low potential for population recovery, (3) the severe summer environment, and (4) the absence of natural terrestrial predation.

Even without outside disturbance, climatic extremes or disease may cause natural fluctuations in reproductive success. In a simplified ecosystem of few species that breed in large concentrations such fluctuations may have major long-lived effects, especially on populations of albatrosses and petrels, which lay only one egg a year and have delayed reproductive maturity. The extent and effects of natural population fluctuations in the Antarctic are largely unknown, but egg and chick mortality is high during prolonged summer storms.

Antarctic and subantarctic birds have evolved in isolation, and their continued survival depends on maintained isolation from nonindigenous predation and competition. Few species show innate wariness toward introduced terrestrial predators or man, although they have evolved behavioral or other adaptations that reduce predation by their natural enemies, skuas and leopard seals. Introduced dogs, cats, rats, and wekas have wrought havoc among nesting birds on the subantarctic islands, and by devastating tussock grass on Iles Kerguelen and Macquarie Island, rabbits have caused erosion of turf, which petrels use for nest burrows. No introduced animals are known to have become established on the peninsula and continent, but abandoned sled dogs have survived over winter there and presumably could become established.

There are other human threats to bird survival in the far south. The limited sites that permit successful bird reproduction are often those that are most attractive to man. Research stations constructed on or near penguin rookeries have drastically reduced populations. Penguin rookeries and other bird colonies that are easily accessible are disturbed by curious visitors. Research activities of scientists can reduce breeding success in birds that they study. Low-flying helicopters and discharge of firearms near colonies may scare penguins and cormorants from their nests long enough for skuas and sheathbills to snatch unattended eggs or young chicks. The human disturbance, of course, adds to the natural burden of predation. Some well-intentioned people, under the impression that predation by skuas on penguins interferes with the 'balance of nature,' advocate killing skuas to 'protect' penguins.

Aside from causing nest abandonment and providing opportunities for predation, disturbance of incubating, brooding, or molting birds may have other more subtle effects. Excitement results in increased metabolic rate, especially in fasting individuals that may have lowered their body temperature to decrease energy consumption. In extreme cold, excited birds may use up reserves that are critical for survival. Disturbance in some bird colonies, especially penguin rookeries, may also scare off young birds prospecting for future nest sites late in the season. Recruitment of new breeders may, thereby, be adversely affected.

Add to all this disturbance the present threat of pesticide and other contamination that hangs over all carnivorous birds. Even in the tissues of such antarctic species as Adélie and chinstrap penguins, blue-eyed shags, and brown skuas, low concentrations of DDT and other chlorinated hydrocarbons have already been found, presumably transported thousands of kilometers by sea or air currents or by other animals [*Risebrough and Carmignani*, 1972].

In the past, penguins were exploited for oil production when whales became scarce. King penguins were extirpated from some breeding sites, and populations of other penguins were severely reduced. Eggs, fledglings, and adults of other species of antarctic and subantarctic birds have been used as food for humans or their dog teams, but presumably this occasional exploitation has not resulted in total extirpation. In the areas covered in this handbook, only Tristan and Macquarie islands are known to have lost endemic populations. A moorhen and a bunting disappeared from Tristan Island, and local ground-nesting races of a widespread rail and a parakeet fell victim to introduced predators on Macquarie Island around the turn of the century.

In recognition of the need for international conservation of wildlife and plants the 12 nations that signed the Antarctic Treaty adopted Agreed Measures for the Conservation of the Antarctic Fauna and

Flora [*Third Antarctic Treaty Consultative Meeting,* 1964]. The measures only apply south of 60°S latitude, however, and leave conservation on the subantarctic islands in national hands; birds and other animals in the pack ice and on the high seas remain unprotected.

The following selected articles of the agreed measures pertain to birds and should be observed by all professional ornithologists and amateur bird watchers alike, especially those visiting the area for the first time.

Article VI: Protection of Native Fauna

1. Each Participating Government shall prohibit within the Treaty Area the killing, wounding, capturing or molesting of any native mammal or native bird, or any attempt at any such act, except in accordance with a permit.

2. Such permits shall be drawn in terms as specific as possible and issued only for the following purposes:

(a) to provide indispensable food for men or dogs in the Treaty Area in limited quantities, and in conformity with the purposes and principles of these Agreed Measures;

(b) to provide specimens for scientific study or scientific information;

(c) to provide specimens for museums, zoological gardens, or other educational or cultural institutions or uses.

3. Permits for Specially Protected Areas shall be issued only in accordance with the provisions of Article VIII.

4. Participating Governments shall limit the issue of such permits so as to ensure as far as possible that:

(a) no more native mammals or birds are killed or taken in any year than can normally be replaced by natural reproduction in the following breeding season;

(b) the variety of species and the balance of the natural ecological systems existing within the Treaty Area are maintained.

5. The species of native mammals and birds listed in Annex A of these Measures shall be designated 'Specially Protected Species,' and shall be accorded special protection by Participating Governments. [No birds are presently listed in Annex A.]

6. A Participating Government shall not authorize an appropriate authority to issue a permit with respect to a Specially Protected Species except in accordance with paragraph 7 of this Article.

7. A permit may be issued under this Article with respect to a Specially Protected Species, provided that:

(a) it is issued for a compelling scientific purpose, and

(b) the actions permitted thereunder will not jeopardize

the existing natural ecological system or the survival of that species.

Article VII: Harmful Interference

1. Each Participating Government shall take appropriate measures to minimize harmful interference within the Treaty Area with the normal living conditions of any native mammal or bird, or any attempt at such harmful interference, except as permitted under Article VI.

2. The following acts and activities shall be considered as harmful interference:

(a) allowing dogs to run free,

(b) flying helicopters or other aircraft in a manner which would unnecessarily disturb bird and seal concentrations, or landing close to such concentrations (e.g. within 200 metres),

(c) driving vehicles unnecessarily close to concentrations of birds and seals (e.g. within 200 metres),

(d) use of explosives close to concentrations of birds and seals,

(e) discharge of firearms close to bird and seal concentrations (e.g. within 300 metres),

(f) any disturbance of bird and seal colonies during the breeding period by persistent attention from persons on foot.

However, the above activities, with the exception of those mentioned in (a) and (e) may be permitted to the minimum extent necessary for the establishment, supply and operation of stations.

3. Each Participating Government shall take all reasonable steps towards the alleviation of pollution of the waters adjacent to the coast and ice shelves.

Article VIII: Specially Protected Areas

1. The areas of outstanding scientific interest listed in Annex B shall be designated 'Specially Protected Areas' and shall be accorded special protection by the Participating Governments in order to preserve their unique natural ecological system. [Fifteen specially protected areas on the continent and peninsula and on Balleny, South Shetland, and South Orkney islands are listed in Appendix III of Annex B. Several were selected on the basis of their bird populations.]

Article IX: Introduction of Non-Indigenous Species, Parasites and Diseases

1. Each Participating Government shall prohibit the bringing into the Treaty Area of any species of animal or

plant not indigenous to that Area, except in accordance with a permit.

2. Permits under paragraph 1 of this Article shall be drawn in terms as specific as possible and shall be issued to allow the importation only of the animals and plants listed in Annex C [sledge dogs, domestic animals (except poultry) and plants, and laboratory animals and plants]. When any such animal or plant might cause harmful interference with the natural system if left unsupervised within the Treaty Area, such permits shall require that it be kept under controlled conditions and, after it has served its purpose, it shall be removed from the Treaty Area or destroyed.

PENGUINS: Spheniscidae

Penguins are stocky flightless highly aquatic birds with dark upperparts and white underparts. The wings are reduced to short hard flippers covered with scalelike feathers. The seven penguins that breed in the Antarctic fall into three natural generic groups: the two large *Aptenodytes* species have yellow patches on the neck, the three small black and white *Pygoscelis* species lack bright colors on the head and body, and the two crested *Eudyptes* species have yellowish head plumes. The sexes are alike, but males are on the average larger and have heavier bills than females. Penguins literally fly through the water, using their flippers for propulsion and their feet and tail merely to steer. When they are swimming rapidly at speeds estimated at 40 km/h, they porpoise out of the water to breathe; at rest or when they are swimming slowly, they float on the surface with their head and back exposed and their tail occasionally held erect. Their progression on land is clumsy, consisting of either a waddling upright walk or short hops. In snow they toboggan on their belly, using feet and flippers for rapid propulsion. The stiff tail and the bill are effectively used as props in climbing. Penguins eat euphausiid shrimp and other crustaceans, fish, and squid, which they capture in diving, in some species to great depths. In turn their eggs and young are preyed upon by giant fulmars, sheathbills, and skuas on land, and adults and young are preyed upon by the leopard seal in water. They breed in large rookeries on sheltered beaches, exposed headlands, and ice-free islands or, in the case of the emperor, on fast shelf ice. Penguin colonies are marked by a pronounced yellowish or pinkish color caused by guano deposits and contrast with dark rocks and white snow. Courtship and mutual displays, which in most species are variations on a basic posture of stiffly outstretched flippers and vertically pointing bill (except in the emperor, which displays with head bent forward), are accompanied by raucous calls. The smaller species usually lay two white spherical eggs and incubate by lying horizontally on a pebble- or grass-lined nest; the larger *Aptenodytes* penguins make no nest but incubate by standing upright, holding their single egg between the upper surface of their feet and a fold of vascular skin on the belly. Both sexes incubate, the roles varying in different species. Penguin chicks at hatching have a short first down, which later is forced out on the tips of a darker and longer second down. Although chicks are initially brooded and guarded by the parents, they later gather together in small groups called crèches while both parents seek food at sea. The young reach into the parents' bills to take regurgitated food. Severe storms cause great chick mortality. Adult penguins gain weight prior to molting, during which they do not feed. *Eudyptes* and *Pygoscelis* penguins molt on land or pack ice after breeding, replacing the worn brownish black feathers with a fresh 'penguin blue' coat. The two lighter blue gray *Aptenodytes* species molt in the spring and summer either just before breeding (king) or after the adults stop feeding their chicks (emperor). In addition to the seven penguin species resident in the Antarctic and high sub-Antarctic, two temperate zone species occur as vagrants.

EMPEROR PENGUIN *Aptenodytes forsteri*

Resident

Identification. Plate 1. 44 in. (112 cm)/13 in. (33 cm). The largest of the penguins, having broad yellow and white areas on either side of the head and neck that are continuous with the breast. The remainder of the head and throat are black, contrasting with the light blue gray back. The dark color of the throat is sharply demarcated from the light breast, and a black border, which separates the dark upperparts from the yellowish white underparts, is especially distinct on the sides of the neck. From behind, considerable white shows on the sides of the head and neck. The underside of the flipper is white. The base of the relatively short decurved bill is feathered, only the narrow bright orange, coral pink, purple, or lilac mandibular plate being left exposed. The outer side of the tarsus is feathered. See the king penguin.

When it fledges at 5-6 months, the overall pattern of the markedly smaller juvenile is similar to that of the adult, but the light area on the side of the head and neck is yellowish gray, the throat is grayish white, the crown is gray, and the bill is all black. With age and wear the back becomes brownish, and the mandible dark orange. After molting again at about 18 months, the young bird is indistinguishable from the adult.

The newly hatched chick is thinly covered with down, nearly white on the back and silvery gray on the belly and throat. The head, except for a white mask over the eyes, and stubby tail are black. As the chick matures, the down becomes denser and darker gray.

Habits. The extreme climatic conditions under which the species breeds during the antarctic winter force it to be more socially tolerant and less territorial than other penguins. Incubating birds and chicks huddle together for warmth and protection from bitter winds. Overprotective unemployed adults may kill young birds during squabbles. Although the emperor is generally docile in temperament, it exhibits great strength, especially when it is handled.

Voice and display. The short recognition call of single birds is a powerful guttural 'kra-a-a-a' given with the head and neck extended and the beak open. Exhibition displays take place in March and April after the birds have returned to the rookeries in single-file marches over the ice from the sea. The display begins with slow head wagging from shoulder to shoulder. The 'song' proper, which differs in the two sexes, is often given in duet with the neck crooked forward, the head leaning on the breast, and the beak shut. It consists of several notes in modulated cadence ending on a long note in the male and of a mixture of cooing and cackling with a very brief final note in the female. In mutual displays two birds face each other and after giving their songs raise their heads on high while they stand chest to chest and beat their flippers slowly up and down. The call of the young is a shrill piping whistle.

Food. Mainly fish and squid, also euphausiid and schizopod crustaceans. Underwater dives may last more than 18 min and reach more than 250 m in depth [*Kooyman et al.*, 1971].

Reproduction. Highly colonial. Most of the known colonies are on fast shelf ice; two are on shingle beaches. Colonies are situated in the shelter of icebergs, islands, coastal ice cliffs, or steep hills. Birds return to colonies in late fall and court ashore without feeding for up to 2 months before laying. The female loses about 20% of body weight during courtship and laying. She departs to feed at sea immediately after laying, leaving the male to incubate alone in the darkness and cold of the antarctic winter. No nest is built, the egg being held close to the body on the feet of the adult until hatching. The male loses 35-45% of his weight during a total fasting period of 3-4 months. After an absence of 60-70 days the female, having regained most of her weight, returns to the colony and relieves the male. If she has not returned by hatching time, as may normally be the case, the male feeds the young chick an esophageal secretion rich in protein and lipids. Otherwise the female gives the chick a partly digested regurgitation as its first meal. Both parents thereafter share brooding and feeding duties. The following schedule pertains to most continental rookeries, but the Cape Crozier schedule is about 2 months later.

Arrival: March to early April.

Eggs: Early May to June. Clutch, 1 chalky greenish white egg; 85 × 125 mm and 450 g.

Hatching: Mid-July to August. Incubation period 60-65 days. Chicks are brooded for 40 more days.

Fledging and departure: Chicks molt into juvenile plumage and depart for the open sea in December at the age of 150-170 days.

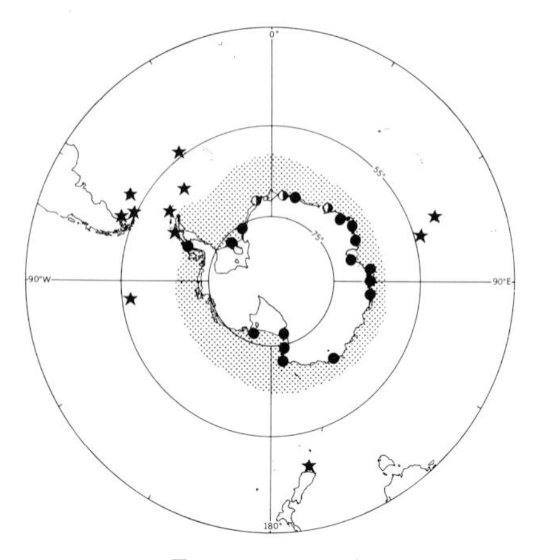

Emperor penguin

Molt. Mid-November to late February. Molt in individual birds lasts about 6 weeks and usually takes place on large floating blocks of sea ice (more data needed).

Predation and mortality. Mortality is very low in adults but normally reaches 25-33% in eggs and chicks. Blizzards may kill exposed chicks directly, and prolonged bad weather, which hampers adults in obtaining food, may indirectly result in the death of chicks through undernourishment. Early spring breakup of ice may result in the total loss of the production of a year in a rookery. Giant fulmars may take numbers of chicks when they begin to form crèches in September and October. Leopard seals attack emperors at sea.

Ectoparasites. Ticks, *Ixodes uriae* and *I. auritulus;* feather louse, *Austrogoniodes mawsoni.*

Habitat. Confined to the open pack ice in the Antarctic zone.

Distribution. Breeds at more than 25 known localities on the continent (some dots on the map represent two or more localities) and at the Dion Islands on the southern Antarctic Peninsula. See map for pelagic range.

KING PENGUIN *Aptenodytes patagonicus*

Resident

Identification. Plate 1. 37 in. (94 cm)/13 in. (33 cm). The more brightly colored and smaller of the two *Aptenodytes* penguins with well-defined bright orange spoon-shaped auricular patches on the black head. These patches extend down to the orange breast as narrowing black-bordered bands. The light breast blends with the 'V'-shaped lower end of the black throat. In adults, white feathering is confined to the lower breast and belly, and from behind, only a little yellow shows on the sides of the neck. The underside of the flipper is broadly tipped with blue gray. The conspicuous unfeathered mandibular plate of the long slightly decurved bill is orange or red. The tarsus is entirely unfeathered. The sexes are alike, but the male is on the average larger. See the emperor penguin.

In the juvenile the auricular patches are lemon yellow rather than orange, the black crown feathers are tipped with gray, the throat is whitish, and the inconspicuous mandibular plates are black, streaked with pink. When the king penguin is newly fledged at 9 months, the back is light bluish gray, but it becomes brownish with wear before the bird molts into adult dress at the end of its second year. In older immature birds the mandibular plate is ivory white.

The chick is nearly naked at hatching, only a few strands of gray or black down obscuring the dark gray skin. By 10 days, however, it is covered with a thick pale gray or brownish down. The facial mask and neck are light gray. The bill is black.

Habits. Although this species is less tolerant of crowding than the preceding one, perhaps because of its less rigorous habitat and summer breeding habits, it is highly gregarious.

Voice and display. A resonant musical trumpeting accompanies the advertising displays of both sexes. The call is given with the bill raised vertically, the neck extended, the neck feathers erected, the eyes half closed, and the flippers held out from the sides. This display is followed by stationary head flagging and the male leading the female in an advertising march in which he swings his head from side to side and thus shows the orange neck patches. Mutual displays at the nest involve side by side high pointing, nibbling, slow fencing with the head and neck, and bill clattering. Chicks have a shrill piping whistle, which becomes a clear three-note whistle in yearlings. Single adults on the breeding grounds and at sea or on land emit an assembly call consisting of a single monosyllabic 'aark.'

Food. Cephalopods and fish.

Reproduction. Colonial, sometimes in very large numbers, on gently sloping protected beaches that are free of snow and ice. Nearby snowdrifts or meltwater pools are essential to provide freshwater for chicks. Owing to the 10- to 13-month chick period that lasts through the winter, reproduction in king penguins is apparently on a unique

3-year cycle during which a pair can successfully rear only two chicks. In the first season the pair is on an early schedule, laying in November. If they are successful the first year, they are on a late schedule the second year, and providing the egg is laid by early February, they can rear a second chick. In the third year, laying is so late, however, that the retarded chick, if it hatches at all, cannot survive the winter. The following spring the unsuccessful parents are ready to begin on the early schedule of another 3-year cycle. Winter survival of chicks depends on their attaining sufficient weight by late fall to endure starvation and cold. Parents begin feeding the chicks irregularly in May, when food becomes scarce owing to bad weather. During the ensuing winter, chicks live mostly on stored fat and lose about one third of their maximum May weight. Early chicks attain weights of about 11-12 kg by May, whereas late chicks usually weigh only 8-10 kg. Very late chicks that have not attained this minimum weight by May do not survive the winter. In October, when parents resume intensive feeding, chicks again gain weight before molting to juvenile plumage. The previous discussion and the following schedule are based on *Stonehouse*'s [1960] detailed study on South Georgia. The schedule in Iles Crozet may be up to 3 weeks earlier, a small number of early-hatched chicks perhaps developing fast enough in their first summer to fledge and leave the colony in May [*Voisin*, 1971]. It is unknown whether their parents might consequently be on an annual schedule.

Arrival: Early breeders, unsuccessful in the previous season, return to begin courtship in September; late breeders return in November; and adults that continued to feed chicks sporadically through the winter are already on the breeding grounds in the spring before the others arrive.

Eggs: Late November to early March. Clutch, 1 pear-shaped pale greenish white egg, the base color usually being obscured by a heavy white chalky deposit; averages 74 × 106 mm and 304 g.

Both sexes share incubation duties, although the male spends more time with the egg. The female incubates for a few hours after laying and then departs to feed at sea. She loses about 26% of body weight during the 2-3 weeks of courtship and laying. The male, who fasts during courtship and the first long incubation shift of about 19 days, may lose up to 30% of body weight. The female then returns to incubate for a similar long period. The male's second shift lasts about 12 days, and the egg usually hatches during the female's ensuing shorter shift.

Hatching: Late January to April. Incubation period 52-55 days. Chicks are brooded about 30 days.

Fledging and departure: November to April of the following year depending on the month of hatching. Total chick period is 10-13 months.

Molt. The molt, which takes place on or near the breeding site prior to courtship, averages about 32 days in individual birds. Pairs that were unsuccessful in breeding during the previous season molt from September to November; successful birds, which remain at sea for a month to put on fat after abandoning their young, delay beginning the molt until December or even later. The mandibular plates are replaced 1-2 months after the feathers are renewed. After molting, adults leave the colony to feed at sea for 2-5 weeks.

Predation and mortality. Giant fulmars and, to a lesser extent, skuas take weakened chicks in rookeries. Sheathbills take unattended eggs and distract parents feeding young. The leopard seal is the major predator of adults at sea, and such predation contributes to chick mortality by starvation of parentless birds. Leopard seals are less frequent near breeding colonies in summer, when young birds first enter the water.

Ectoparasites. Tick, *Ixodes uriae;* feather louse, *Austrogoniodes brevipes.*

Habitat. Low-latitude antarctic and subantarctic waters, generally near the breeding islands.

Distribution. Breeds on the Falkland (very few), South Georgia, Prince Edward, Marion, Crozet, Kerguelen, Heard, and Macquarie is-

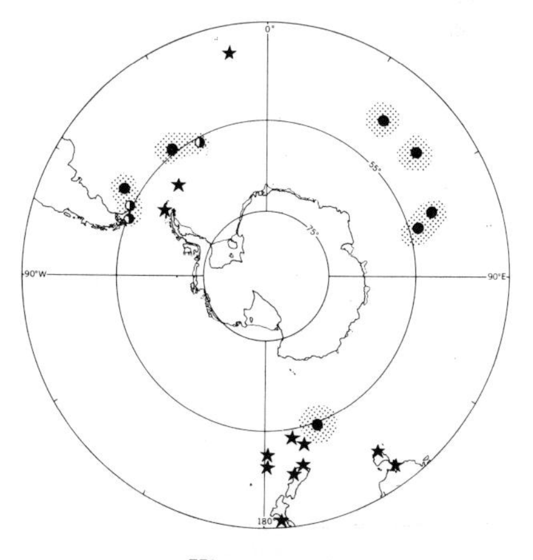

King penguin

lands. Status on South Sandwich Islands is unclear, possibly breeds on Zavodovski Island, and recorded on Saunders Island. Although all populations were reduced by exploitation for oil in the past and the species was possibly extirpated from the South Shetland Islands (if indeed it really did breed there [*Eights*, 1833; *Stonehouse*, 1967]), it is increasing at most colonies at present [*Budd and Downes*, 1965; *Budd*, 1968, 1970; *Conroy and White*, 1973]. Formerly, it bred on Isla de los Estados (Staten Island) and in the Cape Horn region, but its present status is in question.

ADELIE PENGUIN *Pygoscelis adeliae*

Resident

Identification. Plate 1. 28 in. (71 cm)/6.5 in. (17 cm). A small stocky penguin whose conspicuous white eyelids give it a comical button-eyed appearance. The head and throat are black, and the lower margin of the throat is V shaped. The stubby dark brick red bill is feathered over half its length. The feet are dull white or flesh colored on the upper surface and black on the soles.

The juvenile has black eyelids and white, or sometimes grayish, cheeks and throat. The border between the black and white areas of the head lies below the eye. The white throat and partly feathered bill differentiate the juvenile Adélie from the young *Eudyptes* penguins. At 1 year the eyelids become white, and shortly later in February the throat becomes black through molt. See also the chinstrap penguin.

The newly hatched downy chick is silvery gray with a sooty gray head. By 10 days after hatching it assumes the second down, which is much longer and entirely grayish brown. The chick fledges about 50 days after hatching.

Habits. In the three pygoscelid penguins, upright progression on land is a slow walk with the flippers held out from the body and the head bent forward. They walk with high short steps and sway from side to side. Adélie penguins are highly gregarious both on the breeding stations and at sea. They are also inquisitive and unafraid of man but are second only to the chinstrap penguin in aggressive defense of the nest.

Voice and display. In the ecstatic display during pair formation the male waves his head from side to side, slowly stretches his closed bill vertically, and with jerky rhythmic beating of the flippers emits softly at first and then louder a drumlike rolling 'ku-ku-ku-ku-ku-ku.' As the climax is reached, this leads into a reverberating 'kug-gu-gu-gu-gu-ga-aaaa' with the head and neck fully extended; the eyes rolled down and back, the whites thus being exposed; the occipital crest erected; and the bill slightly open. This call is often followed by a soft 'gurr-gurr-gurr-gurr-gurr' as the head is rocked from side to side, the bill being directed alternately toward either axilla, while the wings continue beating. Pairs also indulge in mutual head-waving displays at the nest while they face each other with their bills pointed upward and flippers at their sides. The mutual display call is a raucous 'gug-gug-gug-gaaa' with the bill open. This display and call vary in intensity and loudness. The young whistle shrilly while they beg for food. In the ice floes at sea the usual call is a guttural 'aark.'

Food. Adélies eat small shoaling organisms found near the surface. These are predominantly euphausiid shrimp, but locally, fish, other crustaceans (amphipods), and cephalopods make up a significant proportion of the food.

Reproduction. Colonial, in immense numbers, usually on broad flat protected areas above the beach up to 200 m high. To reach most colonies early in the season, Adélies must travel 30-100 km over ice. Some colonies may be over 300 km from open water in October. The nest is a simple platform of stones, some of which are stolen from neighboring nests throughout breeding. Early in the season the nest and eggs may be nearly submerged by melting ice, and incubating birds may occasionally be buried in falling snow.

Arrival: Early to mid-October.

Eggs: Early to mid-November. Clutch, 2, sometimes 1 in inexperienced breeders, ovoid greenish white eggs with a thin chalky outer deposit; 61-77.5 × 47-59.5 mm and 75-148 g. The second egg is usually slightly smaller than the first and is laid about 3 days later.

Like the emperor penguin the male Adélie has a remarkably long fast period during reproduction. After courtship and laying, which last 17-24 days, the female departs for the sea, having lost about 20% of her weight. The male fasts for another 14-17 days during his first incubation shift and loses about 30% of his weight. The female then relieves the male for about 14 days. He returns from feeding at sea for a shorter final incubation fast before the eggs begin hatching.

Hatching: Mid-December to late December. Incubation period 33-37 days; the second egg hatches about 1 day later than the first. The chicks are brooded for 22-25 days.

Fledging and departure: Late January to mid-February, after 50-55 days.

Molt. January to March, lasting about 2 weeks in an individual. Molt takes place mostly on sea ice and less frequently on land.

Predation and mortality. Mortality in adults is low, but it ranges

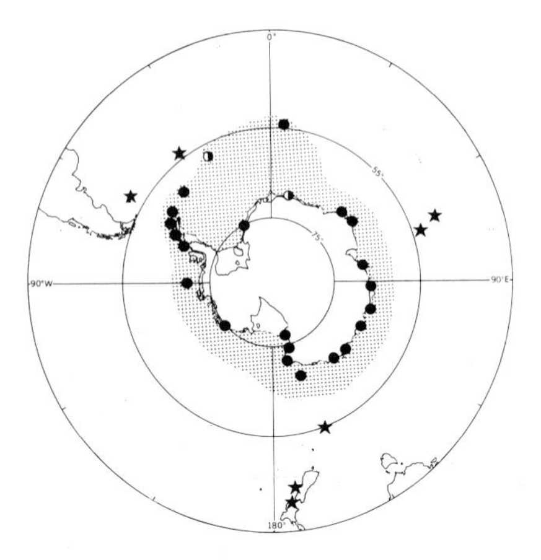

Adélie penguin

from 50 to 70% in eggs and chicks. Skuas and, to a lesser extent, giant fulmars take eggs and weakened isolated chicks. Sheathbills rob unattended nests of eggs and after deliberately distracting adult penguins while they are attempting to feed chicks snatch up any spilled food. Leopard seals prey on Adélies offshore about the breeding colonies.

Ectoparasites. Tick, *Ixodes uriae;* nasal mite, *Rhinonyssus schelli;* feather lice, *Austrogoniodes antarcticus* and *A. bifasciatus.*

Habitat. The open pack ice, the birds coming to land during the winter only when gales break up sea ice and leave open water near the coast.

Distribution. Breeds on South Orkney, South Shetland, Bouvet, Balleny, and Peter I islands and at numerous localities on the Antarctic Peninsula and continent including the adjacent islands. Status on South Sandwich Islands, where young have been recorded (P. C. Harper, unpublished data, 1966), is unclear. See map for pelagic range. Vagrant records for Australia are suspect [*Serventy et al.*, 1971].

CHINSTRAP PENGUIN *Pygoscelis antarctica*

Resident

Identification. Plate 1. 30 in. (76 cm)/7.5 in. (19 cm). A white-cheeked black-crowned penguin with a conspicuous narrow black band extending from ear to ear across the white throat. This species differs from the immature Adélie at all ages in having a narrow black band across the throat and the border between the black crown and the white cheek lying above the eye. The slim black bill is unfeathered to the base; the feet are flesh colored or dull white. See the immature Magellanic penguin.

The juvenile is similar to the adult but is smaller, has a weaker bill, and shows some dark spotting on the chin and cheeks.

The newly hatched chick is entirely silvery white, quite unlike the dark-crowned chick of the gentoo and the darker Adélie chick. The longer second down is gray, slightly lighter on the head and cheeks and darker on the chin and throat. The underparts vary from nearly pure white to pale gray.

Habits. Gregarious but very aggressive, attacking potential enemies with the bill and flippers. The raised bristling hackles and deliberate bold approach to an intruder are characteristic.

Voice and display. Most of the displays are similar to those of the Adélie, but the ecstatic call 'ah, kauk, kauk, . . .' is lower in pitch and given with the bill open. When the chinstrap penguin is approached closely, it utters a low muttered growl and hissing accompanied by foot stamping. The chinstrap is the least studied of the *Pygoscelis* penguins.

Food. Chiefly euphausiid shrimp and other crustaceans, also fish.

Reproduction. Colonial, breeding in large dense rookeries. Millions of birds breed on some Scotia ridge islands. The largest colonies are located on flat coastal areas. Some sites on high plateaus that are not frequented by other *Pygoscelis* penguins may be shared with *Eudyptes* penguins. The nest is a low platform of pebbles and an occasional bone.

Arrival: Early to mid-November.

Eggs: Late November to early December. Clutch, 2, sometimes 1, pale greenish white eggs with a chalky coating; 49-53 × 64-75 mm and 91-118 g.

Hatching: Early January. Incubation period averages 37 days.

Fledging and departure: Early February to mid-March. No precise information on the duration of brood or chick periods. All birds have left by late March to late April.

Molt. Early February to March on land.

Predation and mortality. Skuas, sheathbills, and giant fulmars take eggs and young that are weakened or injured. Leopard seals capture birds offshore near the rookeries.

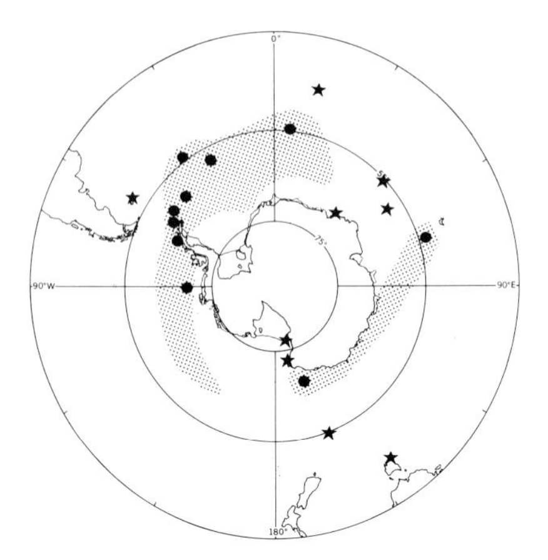

Chinstrap penguin

Ectoparasites. Tick, *Ixodes uriae;* feather lice, *Austrogoniodes gressitti* and *A. macquariensis.*

Habitat. Antarctic waters, less frequently in pack ice than the Adélie.

Distribution. Breeds on South Georgia, South Sandwich, South Orkney, and South Shetland islands; on the Antarctic Peninsula south to Anvers Island; and on Bouvet, Heard, Balleny, and Peter I islands. The species is apparently increasing in numbers and spreading from its center of distributin in the American sector. Breeding may have been established only recently at Heard, Balleny, and Peter I islands, where small numbers occur among Adélie or gentoo rookeries, and success in breeding may be very low (not yet fully documented on the Balleny Islands [*Sladen,* 1964]). See map for pelagic distribution.

GENTOO PENGUIN *Pygoscelis papua*

Resident

Identification. Plate 1. 30 in. (76 cm)/9 in. (23 cm). The only pygoscelid penguin with brightly colored bill and feet and a triangular white patch above each eye. These conspicuous patches are variable but usually extend from eye to eye and meet across the crown. The eyelids are white. Scattered white feathers occur on the black head and neck, which are markedly darker than the back. The long bill is carrot red with a black culminicorn and tip; the feet are yellow orange. Gentoos from the South Shetland Islands and Antarctic

Peninsula (*P. p. ellsworthi*) are smaller in bill, flipper, and foot measurements than more northern populations, and other insular populations may be distinguishable [*Stonehouse*, 1970*a*].

The juvenile differs from the adult in having a narrower head patch with some white flecking near the edge of each eye and a mottled grayish chin and throat. The eyelids are dark, and the bill is more dully colored and weaker than that of the adult.

The newly hatched chick is covered with short silvery gray down on the back; the crown, face, and cheeks are slaty gray; and the underparts vary from grayish white to nearly brown. The second down is essentially like the first in pattern but is deep gray above and immaculate white below. The flippers are more broadly edged with white than those of any other penguin chick, and the white head patches of the adult are suggested by a pale gray pattern.

Habits. The gentoo is the most timid of the penguins, sometimes preferring to retreat rather than to defend even its young chicks. They avoid rough terrain for walking.

Voice and display. A repeated low-pitched double-noted braying or trumpeting ecstatic call 'ah, aha, aha, aha-e' given with the head pointed upward and the bill open. The mutual display consists of bowing accompanied by a wheezing hiss.

Food. Mainly fish, also euphausiid shrimp and cephalopods. The gentoo feeds on nototheniids, scopelids, and other bottom fish and can dive to depths of at least 100 m [*Conroy and Twelves*, 1972].

Reproduction. Colonial, in small to large groups, most northern birds nesting on open tussock and *Azorella*-covered coastal flats and slopes rather than on beaches. Some gentoos select low rocky headlands and exposed inland hills, but the landing places are generally sand beaches. Many colonies are marked with whitish guano deposits, an indication of fish in the diet, rather than pink deposits like those of the colonies of the other two krill-feeding *Pygoscelis* penguins. The nests, which are more widely separated and contain more material than those of the Adélie and chinstrap penguins, are constructed of stones, bones, seaweed, moss, tussock grass, and other vegetation depending on local availability. The breeding schedule of this widely ranging species is long and variable, being earliest on the northernmost breeding stations or those with permanent snow-free areas. Success in rearing replacement clutches is common.

Arrival: Late August at northern stations to late October at more southerly stations.

Eggs: Mid-August to mid-November at northern stations and November to early January elsewhere. Clutch, 2, rarely 1 or 3, spherical white eggs with a greenish cast and a chalky outer coating; 56.2-61.0 × 66.5-73.0 mm and 117-153 g. The second egg is slightly smaller than the first and is laid 3 days later.

Hatching: Early September to late December in the north and January to mid-February elsewhere. Incubation period 35-39 days; brood period 4 additional weeks.

Fledging and departure: Between December and late February, from 2 to 3 months after hatching. Departure of adults takes place from March to June.

Remarks: The breeding cycle on Marion Island and probably on Iles Crozet occupies up to 8 months, and egg laying may take place in the winter. Adults begin nest building as early as March, eggs are found from June through October, and young hatch in late July to November.

Molt. December to March, following fledging of the young.

Predation and mortality. Skuas, giant fulmars, sheathbills, and, to a lesser extent, gulls take eggs and poorly guarded young chicks. The leopard seal is the major predator of gentoos at sea.

Ectoparasites. Feather lice, *Austrogoniodes gressitti, A. keleri,* and *A. macquariensis;* flea, *Parapsyllus longicornis.*

Habitat. Subantarctic and antarctic waters, usually avoiding the pack ice. Except for populations in the southern part of the American sector, gentoos usually stay near the breeding island all year.

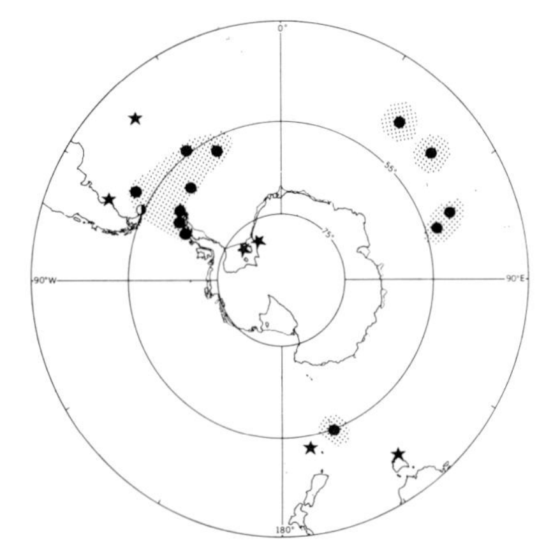

Gentoo penguin

Distribution. In the Antarctic the gentoo penguin breeds on South Georgia, South Sandwich, South Orkney, and South Shetland islands; on the Antarctic Peninsula south to about 65°S; and on Prince Edward, Marion, Crozet, Kerguelen, Heard, and Macquarie islands. Farther north it breeds on the Falkland Islands, but the Staten Island breeding record needs confirmation. Most birds in the Antarctic Peninsula populations probably migrate north after breeding, but Indian Ocean and Macquarie Island populations are sedentary around the breeding islands. See map.

ROCKHOPPER PENGUIN *Eudyptes crestatus*

Resident

Identification. Plate 2. 24 in. (61 cm)/7 in. (18 cm). A small crested penguin with a black forehead. The narrow pale yellow eyebrows begin just behind the bill and do not meet in a patch on the forehead. The pale crest plumes, which bristle out from the sides of the head, vary geographically in length and fullness and together with the elongated black crown feathers make the head appear shaggy, especially when the crests are erected. The black throat border is nearly straight, not markedly V shaped like that of the macaroni. In all *Eudyptes* penguins the eyes are dull garnet red, the bill reddish brown with a fleshy margin, and the feet whitish flesh with black soles. Males are larger than females, and their bills are usually more massive. Northern birds, *E. c. moseleyi,* have fuller crests, darker underwing tips (Figure 5), and a larger average size than southern birds, *E. c. crestatus* (see the section on distribution). The margin of bare flesh between the bill and the feathered cheek is black in birds from Staten, Falkland, and Gough islands, Ile Amsterdam, and Ile Saint-Paul but pink in individuals from Heard, Campbell, and Macquarie islands [*Carins,* 1974]. Information is needed on its color in other populations to determine whether it can be considered an additional subspecific character.

The juvenile differs from the adult in having the head plumes suggested by an inconspicuous line of creamy feathers over the eyes and the throat and chin mottled with gray. The bill is weaker and browner than that of the adult. See the immature macaroni penguin. The unfeathered bills of immatures of all three species of *Eudyptes* are noticeably longer than the bill of the somewhat similar, but white-throated, young Adélie. In 2-year-old birds the crests are shorter and more erect than they are in adults.

The downy chick is blackish brown on the head, throat, and back and white on the belly. The lower margin of the dark throat is nearly straight. See the macaroni penguin chick.

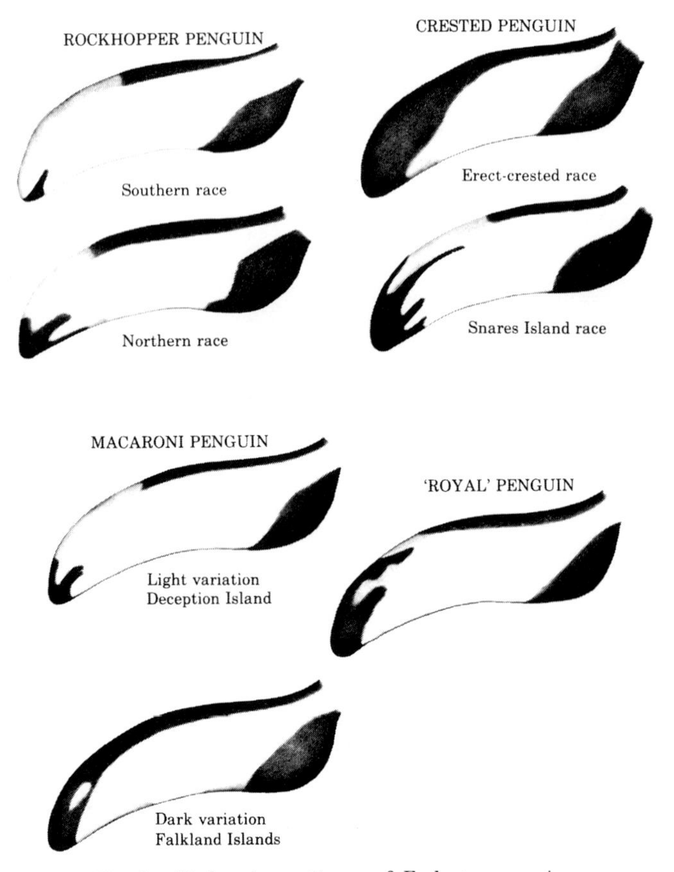

Fig. 5. Underwing patterns of *Eudyptes* penguins.

Habits. True to its name, this species usually progresses on land in a series of stiff hops with both feet together 'like a man running in a sack race.' This gait and its sharp toenails make it the most agile penguin species on steep and rocky ground. It can climb uphill about 30 cm in a single hop. Unlike other penguins, which always dive into the water, the rockhopper generally jumps in feetfirst.

Voice and display. In ecstatic display at the nest two birds stand side by side, point their bills at each other, and call 'caa, caa.' They then raise their bills high and beat their flippers slowly in time to a trumpeting 'hurrrah-hurrrah-hurrrah.' Mutual displays include the male bowing, raising his head vertically, and shaking it from side to side while he raises his flippers and delivers pulsating raucous cries with the bill open. The female bows and quivers her head in response, then rises to face the male, and calls as she reaches toward his head. Chicks have a loud peeping cry.

Food. Mainly crustaceans, including amphipods, copepods, and isopods, at most localities, but squid are important to the diet in the Tristan da Cunha group.

Reproduction. Colonial, nesting near the sea among large boulders, on sheltered rock ledges, in caves, and, on Ile Amsterdam, in burrows a meter or more long. The colonies, which are small in the south, are frequently on steep cliffs inaccessible to other penguin species. Landing places are rocks rather than beaches. The nest is a depression in the ground, lined with small stones and occasionally with quantities of tussock grass. The rockhopper is remarkably regular in its breeding schedule, returning to the rookery on almost the same date each year. The following schedule applies to the southern rookeries. On Tristan da Cunha group, Gough Island, Ile Amsterdam, and Ile Saint-Paul the schedule may be 2 months earlier. Both sexes incubate, the female taking the first shift.

Arrival: Mid-October to early November.

Eggs: Late October to mid-November. Clutch, 2 (3 on Tristan da Cunha group and Gough islands) white or greenish white eggs. The first egg is usually much smaller (averages 64 × 46 mm and 76 g) than the second (averages 72 × 53 mm and 111 g), and if it is incubated to hatching, the resulting chick seldom survives longer than 2 days.

Hatching: Mid-December through January. Incubation period 32-34 days. The male broods the chick an additional 19-25 days.

Fledging and departure: February to March, after 67-72 days. Departure takes place from March to June.

Molt. At most localities, adults molt from mid-March to April; yearlings from mid-January to mid-March. The schedule on Tristan da Cunha group is 2 months earlier.

Predation and mortality. Skuas and sheathbills take both eggs and young; wekas eat the eggs at Macquarie Island.

Ectoparasites. Tick, *Ixodes uriae;* feather lice, *Austrogoniodes conci, A. hamiltoni, A. macquariensis, A. keleri,* and *A. cristati;* fleas, *Parapsyllus magellanicus* and *P. longicornis.*

Habitat. Subantarctic and antarctic waters (at Heard Island).

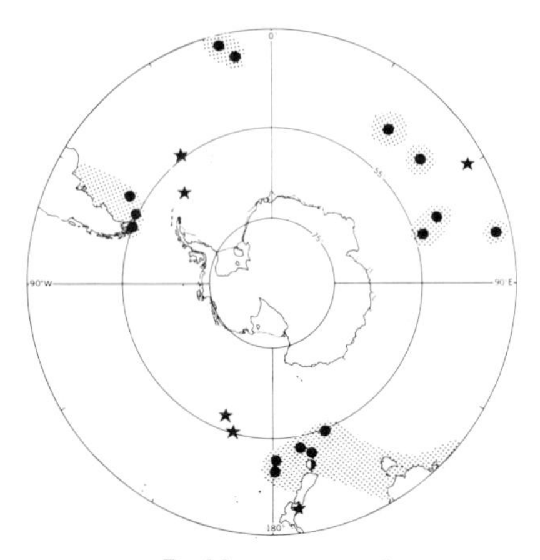

Rockhopper penguin

Rockhoppers are highly pelagic: the long period spent at sea is indicated by barnacles sometimes found adhering to their feet.

Distribution. In the south, *E. c. crestatus* breeds on Prince Edward, Marion, Crozet, Kerguelen, Heard, and Macquarie islands; the same subspecies breeds farther north on islands near Cape Horn, possibly including Staten Island, and on the Falkland Islands. *E. c. moseleyi* breeds on the Tristan da Cunha group, Gough, Amsterdam, Saint-Paul, Campbell, Auckland, Antipodes, and Bounty islands and possibly on the Snares Islands. Little is known of the pelagic range of island populations. See map.

MACARONI PENGUIN *Eudyptes chrysolophus*
(includes royal penguin) (*Eudyptes schlegeli*)

Resident

Identification. Plate 2. 28 in. (71 cm)/7.5 in. (19 cm). A medium-sized crested penguin whose orange yellow eyebrow feathering originates from a broad patch on the forehead. The forehead and crown feathers are tipped with black, the forehead thus tending to be streaked. The elongated lateral crest feathers droop closely along the sides of the head rather than bristling outward as they do in the rockhopper. The transverse black border of the throat forms a V on the upper breast, and adults have a white patch at the base of the tail. The reddish brown bill is much stouter and more heavily sculptured than that of the rockhopper. The 'royal penguin' of Macquarie Island, *E. c. schlegeli*, is a large-billed heavy-bodied bird with white or light gray and very rarely black cheeks and throat. The underwing pattern shows more black than that in *E. c. chrysolophus* (Figure 5). Albinism is not infrequent in this well-differentiated subspecies, which some authors regard as a separate species.

Head plumes in the juvenile are reduced to scattered short yellow feathers behind the eyes and on the forehead and do not form a well-defined eyebrow streak as they do in the immature rockhopper and crested penguins. The throat is dark gray, and the white tail patch is absent. The eyes and relatively weaker bill are brown. Yearling birds remain smaller and slimmer than adults and have rather small heads and bills. The eyes may be as bright a reddish brown as those of the adult, but the bill stays dull brownish.

The downy chick is similar to that of the rockhopper, but the border of the throat extends ventrally in a V as it does in the adult.

Habits. A noisy aggressive penguin, which in locomotion is similar to the pygoscelid penguins rather than to the rockhopper.

Voice and display. A loud trumpeting accompanies mutual ecstatic displays at the nest site. Either the bill is pointed vertically, or the head and bill are stretched forward as the wings beat in time with the calls. Chattering accompanies quivering displays at the nest and often leads to mutual or lone bowing and head swinging associated with deep throbbing calls or raucous cries. At sea the macaroni emits a harsh barking, which is lower pitched, more nasal, and less sharp than that of the chinstrap penguin.

Food. Mainly euphausiid shrimp and other crustaceans, also cephalopods.

Reproduction. Colonial, usually in vast rookeries on gentle open sloping ground with a gravel or sandy substrate. Access to the colonies follows small permanent streams, especially on Macquarie Island. On Heard Island, small numbers of rockhoppers may nest on the periphery of macaroni colonies, and in the South Shetland Islands, where macaronis are uncommon, they share colonies with *Pygoscelis* penguins, including steep rocky coastal slopes with chinstraps. The nest is a shallow depression lined with stones, sand, or grass. Laying can begin as early as 5 years of age; most 7-year-olds have returned to the colonies, but first breeding can be delayed until 11 years of age in some individuals. Birds come ashore progressively earlier each year and stay longer with increasing age. The schedule, which is highly synchronized within colonies, is 1 month earlier on Macquarie Island than on Heard Island.

Arrival: Mid-September to early October on Macquarie Island

and late October to early November on Heard Island. Males precede females by about 8 days and select nest sites.

Eggs: Mid-October to late October on Macquarie Island and mid-November to late November on Heard Island. Clutch, 2 chalky white eggs. The smaller first egg, which averages 70.5 × 49.1 mm and 93.9 g, is usually discarded or lost when the larger second egg, which averages 80.9 × 58.6 mm and 154.4 g, is laid 4-6 days later. The measurements are from Heard Island.

Hatching: The female incubates the first 12-14 days, the male incubates the second 12-14 days, and both are present for hatching, which takes place in mid-November to mid-December on Macquarie Island and mid-December to early January on Heard Island. The incubation period is 32-37 days. The male broods the chick an additional 2-3 weeks while the female feeds it.

Fledging and departure: Chicks fledge in January (Macquarie Island) and February (Heard Island), after about 65 days. They leave Macquarie Island in late January and Heard Island in late February and are followed soon after by the adults, which remain at sea for up to 5 weeks fattening in preparation for molt back on the breeding grounds in March. The colonies on Heard Island are totally deserted by late April, and those on Macquarie Island by mid-May.

Molt. Year-old birds begin molting on the breeding grounds as early as late January; older birds begin molt somewhat later. Successful breeders do not begin molt until early March on Macquarie Island or late March on Heard Island. Molt lasts 24-29 days.

Predation and mortality. In addition to heavy accidental loss of smaller first eggs, mortality is high during breeding. Skuas, gulls, sheathbills, and wekas take unattended eggs and small chicks; larger chicks and fledglings fall prey to giant fulmars. Egg and chick mortality results in about only one third of the pairs in a colony successfully rearing one chick. Elephant seals cause mortality when they are hauled out near breeding colonies, but direct predation by fur and leopard seals on adults and young birds at sea is probably insignificant.

Ectoparasites. Tick, *Ixodes uriae;* feather lice, *Austrogoniodes bicornutus, A. cristati, A. hamiltoni, A. macquariensis,* and *A. keleri;* fleas, *Parapsyllus magellanicus* and *P. longicornis.*

Habitat. Warmer antarctic waters, north of the pack ice. Often observed far from land.

Distribution. Breeds on South Georgia, South Sandwich, South Orkney, and South Shetland islands; on the Antarctic Peninsula; and

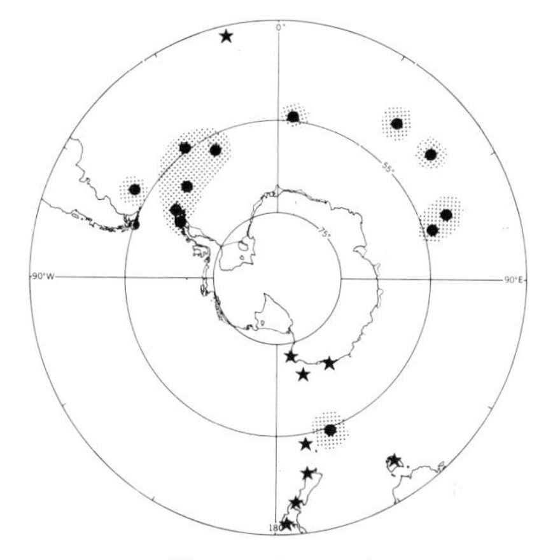

Macaroni penguin

on Bouvet, Prince Edward, Marion, Crozet, Kerguelen, Heard, and Macquarie islands. Suggested South American record of breeding on Isla Noir (Isla Negra [*Reynolds*, 1935]) needs confirmation, but small numbers breed in the Falkland Islands (Kidney Island). *E. c. chrysolophus* is the breeding bird in all localities except Macquarie Island, where *E. c. schlegeli* breeds. Immatures resembling *E. c. chrysolophus* were found at Cape Hallett in February 1964 and on the Balleny Islands in March 1964 [*Hatherton et al.*, 1965]. Macquarie Island birds have been reported once on the Adélie Coast [*Jouanin and Prévost*, 1953], and birds identified as *E. c. schlegeli* have been seen on Marion Island, 20 in mid-December 1960 [*Voous*, 1963] and one in 1965 [*van Zinderen Bakker*, 1971*a*], once in Iles Crozet in mid-February 1971 [*Despin et al.*, 1972], and once in 1968 on Iles Kerguelen [*Prévost and Mougin*, 1970, p. 36]. These records, remote from known breeding areas, may be either extreme vagrants or individual mutants from nearer colonies. At sea, range is poorly known. See map.

CRESTED PENGUIN (includes erect-crested and Snares crested penguins) — *Eudyptes pachyrhynchus* (*E. sclateri, E. robustus*)

Vagrant

Identification. Plate 2. 29 in. (74 cm)/7.5 in. (19 cm). A crested penguin similar in size and bill proportions to those of the macaroni but with the black forehead, pale yellow eyebrows, and straight throat border of the rockhopper. The eyebrows are broad, forming a conspicuous light stripe. The crests are shorter and project posteriorly instead of bristling laterally like those of the rockhopper. The head is jet black, much darker than it is in either of the other two *Eudyptes* species, and the underside of the flipper shows a heavier black anterior border and more black distally (see Figure 5). Two very distinct subspecies occur on islands south of New Zealand. The 'erect-crested penguin,' *E. p. sclateri*, has shorter more upturned crests and heavier dark markings on the underwing than the 'Snares crested penguin,' *E. p. robustus* (see the section on distribution).

The juvenile is smaller and has less conspicuous crests and a weaker bill than the adult. The pattern on the underside of the flipper and the whitish throat are useful for separating this species from the young rockhopper.

Distribution. Breeds on New Zealand and several of its outlying islands. Vagrant individuals of the two southernmost populations, the Snares (breeds on Snares Islands) and erect-crested penguins (breeds on Bounty, Antipodes, Campbell, and Auckland islands), have been recorded on Macquarie Island several times [*Keith and Hines*, 1958; *Warham*, 1969], and a single erect-crested penguin is reported to have interbred with a rockhopper in the Falkland Islands [*Napier*, 1968].

MAGELLANIC PENGUIN *Spheniscus magellanicus*

Vagrant

Identification. Plate 1. 30 in. (76 cm)/7 in. (18 cm). A medium-sized temperate zone penguin with conspicuous black and white bands on the face, neck, and upper breast. A broad curving white superciliary stripe outlines the black face. Two black bands cross the throat and upper breast, the narrower lower one continuing down the flanks to the legs. The bill is brownish black with yellowish markings; the feet are brownish black mottled with pale gray. The tail lacks the long stiff quills of the *Pygoscelis* and *Eudyptes* species.

The juvenile, which is paler and grayer blue, has only a single indistinct breast band and a dirty gray cheek. The young Magellanic penguin resembles the chinstrap or young Adélie, but both antarctic species have immaculate white underparts, and the white cheeks are sharply set off from the black crown.

Distribution. Breeds in southern South America and the Falkland Islands. Occurs at sea in the Drake Passage and as a very rare vagrant on South Georgia [*Tickell*, 1965] and possibly in the Tristan da Cunha group [*Elliott*, 1957]. There is one extreme vagrant record for New Zealand [*Robertson et al.*, 1972].

ALBATROSSES: Diomedeidae

Albatrosses, the largest flying seabirds, have long narrow wings, which may exceed 11 feet (3.35 m) in the larger species. The nostrils open in nasal tubes located on either side of the culmen. The southern species may be divided into three natural groups: the very large wandering and royal albatrosses are mostly white as adults; the smaller mollymauks have a dark mantle and wings and a short rounded dark tail; the two entirely dark sooty albatrosses have long wedge-shaped tails. The mollymauks present problems in at-sea identification, but underwing pattern, head color, and bill characters are useful in specific identification. The giant fulmars are superficially similar to albatrosses in size, shape, and general habits, but their nostrils are united in a single tube on top of the bill as they are in other members of the family Procellariidae. Albatrosses depend on strong winds for their effortless gliding flight. They follow a regular pattern in the air: rising into the wind, coasting across it, and then losing altitude but gaining speed while they dip to leeward and bank to turn and rise into the wind once more. In such a manner these birds can follow a ship for hours or days, never flapping their wings but making slight adjustments at the wrist and elbow to change effective wing area. Albatrosses cannot take off from the water unless they are headed into the wind, and even then they must paddle clumsily along the surface with much laborious wing flapping to become airborne. They also walk clumsily on land with a distinctive side to side swaying motion of their body. Although they alight on the water to feed on galley refuse during the day, their staple diet is squid, presumably caught at the surface at night. The molt is gradual, and individual wing feathers may not be replaced annually (needs confirmation). The nest in the southern hemisphere species is a truncated cone of seaweed, grass, and mud usually placed on an exposed windswept hillside. A single large white egg is laid. Both adults incubate; the downy young remain in the nest for many months before fledging and are fed by regurgitation. Chicks repel intruders and predators by ejecting squirts of stomach oil and partly digested food. Molt in adults probably takes place at sea. Five species breed on islands near the Antarctic convergence and range south to the pack ice, and three other species breed farther north and occur less frequently in the far south.

WANDERING ALBATROSS *Diomedea exulans*

Resident

Identification. Plate 3. 42-48 in. (107-122 cm)/120-138 in. (305-351 cm). A large albatross that in all stages has mostly white underwings. The bill is pinkish white with a yellow tip and no dark markings on the cutting edge. Nostril openings are elliptical and directed slightly upward. Eyes are brown; eyelids are pale gray or pink; feet are flesh with a faint bluish tinge. The great variation in the color of body and upperwing surface is largely due to age. The best at-sea character for identification of most adults of this species and the royal albatross is the white body contrasting with the dark upperwings. Juvenile wanderers are largely brown, the face, throat, and occasionally the belly being white. Northern birds (*D. e. dabbenena*), which nest on Tristan da Cunha group, Gough Island, and Ile Amsterdam, breed in this largely dark plumage. As they mature, wanderers pass through various stages during which the chest, mantle, head, and wings gradually whiten. Birds of intermediate age ('leopard' stage) are white with dark wings and white 'elbow' patches. Definitive adult males are entirely white except for the primaries and perhaps some dark spots on the tail feathers (*'chionoptera'* stage). Females, which are on the average slightly smaller, take longer to assume a definitive white plumage and usually retain light transverse vermiculations on the back and a dark patch on the crown. With extreme age, however, even these traces may be lost. The royal albatross, which is also large and mostly white at all ages, may not be distinguished readily at sea from the adult wanderer. It differs mainly in bill characters and eyelid color. Any large albatross with dark markings on the body, however, must be a young wanderer. Therefore at-sea records should include plumage notes for species verifications. See also the giant fulmars.

The downy chick at hatching is dirty white, but the longer second down is light gray. The bill is fleshy white with a yellowish tip. The feet are bluish gray at first but later become whitish flesh with slightly darker webs; the iris is dark gray.

Flight and habits. In strong wind the wanderer flies in long sweeping glides on fully outstretched wings. It seldom resorts to prolonged wing flapping except during calms, when it is more frequently seen resting on the water.

Voice and display. A harsh croaking when it is competing for food at sea. Squealing, gobbling, braying, and bill clapping accompany courtship displays on land. Adults bow, bill fence, and 'dance' with their wings and tails spread in mutual displays.

Food. Squid and fish, caught mainly at night while the bird sits on the surface. Follows ships tenaciously to obtain galley refuse.

Reproduction. Semicolonial, in small dispersed groups with nests from 18 to 23 m apart. Breeding sites are usually on exposed coastal tussock-covered ridges or hilltops. The nest of mud and tussock grass is 0.30-0.90 cm high and is reused and maintained by the occupying birds. Although both adults may be absent from the breeding grounds for extended periods during the winter, they return at intervals to feed the young until they are fully fledged. Because of its 13-month breeding cycle the wandering albatross breeds every other year if it is successful, but if a chick is lost early in the season, a pair may breed again the following year. The following schedule refers only to birds breeding in the antarctic area. There is no information on the schedule of Ile Amsterdam birds.

Arrival: November to December.

Eggs: Mid-December to early January. Clutch, 1 white egg with reddish spots, more numerous on the larger end; 120-140 × 75-83 mm and 429-487 g.

Hatching: Early to mid-March. Incubation period averages 78 days. Chicks are brooded an additional 4-5 weeks.

Fledging and departure: Mid-December to early February of the following year, after an average of 278 days in the nest.

Molt. Schedule may be biennial in successful breeders (needs confirmation).

Predation and mortality. Skuas are infrequent predators on unattended eggs and young, and some chicks die during bad weather because of starvation. Combined mortality in eggs and chicks is 30-40%. Adults may survive over 50 years.

Ectoparasites. Tick, *Ixodes uriae;* feather lice, *Austromenopon affine,*

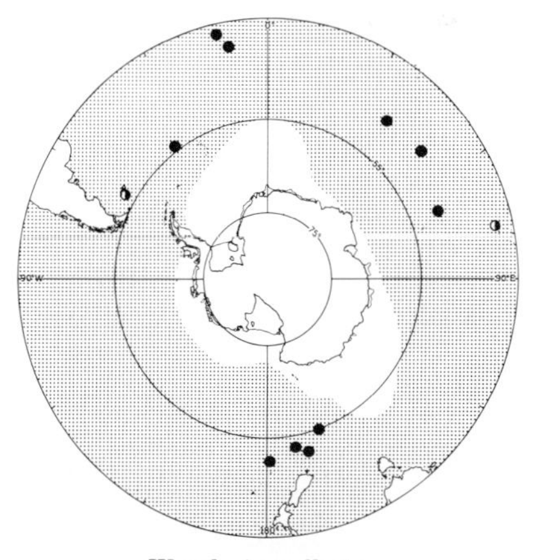

Wandering albatross

Perineus concinnoides, Paraclisis hyalina, Naubates fuliginosus, Harrisoniella hopkinsi, Docophoroides brevis, and *Episbates pederiformis.*

Habitat. Highly pelagic in antarctic and subantarctic waters but rarely occurs near pack ice.

Distribution. In the Antarctic the wanderer breeds on South Georgia, Prince Edward, Marion, Crozet, Kerguelen, and Macquarie islands. Farther north it also breeds on the Tristan da Cunha group, Gough, Auckland, Campbell, and Antipodes islands. Its status on the Falkland Islands, Ile Amsterdam, and Ile Saint-Paul needs clarification (bred on Ile Amsterdam in 1951, but no later records [*Segonzac,* 1972]). See map for the pelagic range of birds following the West Wind drift around the world; young birds disperse north of 40°S.

ROYAL ALBATROSS *Diomedea epomophora*

Vagrant

Identification. Plate 3. 42-48 in. (107-122 cm)/120-138 in. (305-351 cm). Very similar to the adult wandering albatross in size and appearance but lacks the dark immature stages and has a mostly white tail. At all ages the body and underwings are white, the upperwing being dark and variably freckled with white. A few dark spots may be present on the tail. At close range the black eyelids (very old birds may lose dark pigment) and the dark margins along the cutting edges of both pale pinkish yellow mandibles are diagnostic. The nasal tubes are more protuberant in the royal albatross than in the wandering albatross, and the round nostril openings are directed forward rather than upward. Flight, habits, and food are like those of the wandering albatross.

Distribution. Breeds on the South Island of New Zealand and on Campbell, Auckland, and Chatham islands. Although Pacific Ocean pelagic records are rare owing to the difficulty of separating this species from the adult wandering albatross, it regularly occurs in offshore temperate waters on both coasts of South America. Vagrants should be looked for in the Drake Passage and have also possibly been recorded from the vicinity of Macquarie Island [*Gillham,* 1967, p. 208]. There are no African nor Indian Ocean records.

BLACK-BROWED ALBATROSS *Diomedea melanophris*

Resident

Identification. Plate 4. 35 in. (89 cm)/90 in. (229 cm). A white-headed mollymauk with a reddish-tipped yellow bill generally resembling a large dark-backed gull. The axial portion of the underwing is white and broadly margined with black, somewhat more so on the leading edge. A dark patch about each dark brown eye contrasts with the white head to give the bird a 'frowning' look (*D. m. melanophris*). Campbell and Antipodes island birds (*D. m. impavida*) have honey-colored eyes, very dark eye patches, and broader dark margins on the underwing. In both subspecies the culminicorn is expanded where it joins the forehead feathering. The feet are pale blue gray. Both sexes are alike.

The juvenile has dark markings on the neck and a dark underwing and bill. Gray margins on the feathers of the nape (and occasionally the crown) form an incomplete collar, which becomes less pronounced with wear. The lower neck, however, remains gray even after the head whitens at 3-5 years of age. The underwing is mainly dark with a narrow grayish axial streak. When the fledgling leaves the nest, the bill is entirely black, or dark brown with a black nail, and less robust than that of the adult. As the bird matures in 3-5 years, the bill gradually lightens to yellow, the tip remaining darker. The young black-browed albatross is almost impossible to distinguish at a distance from the immature gray-headed albatross, but its whiter head, darker underwing, and lighter bill may be helpful in identification. Bill structure, which is diagnostic at close range, is impossible to observe at sea.

The chick's grayish white first down is replaced by a longer light gray second down. Incipient adult characters in the bill provide one means for differentiation from the gray-headed chick, but see that species and compare the breeding schedules (young black-browed chicks will generally be larger than gray-headed chicks, since they hatch earlier).

Flight and habits. The general flight pattern is similar to that of the wandering albatross but with more flapping and less soaring. Gregarious, in flocks of 30 or more.

Voice and display. At the nest, loud braying accompanies bill fencing and mutual bowing, nibbling, and preening. Although adults are usually silent at sea, they croak noisily when they are squabbling over food.

Food. Feeds while it is on the surface, taking fish and squid and, to a lesser extent, crustaceans (mainly euphausiids) and salps. Follows ships for galley refuse.

Reproduction. Colonial, usually breeding on narrow terraces of tussock-covered slopes and cliffs overlooking the sea. The nest is cone shaped, is composed of grass cemented with mud, and is from a few centimeters to 60 cm high with a saucer-shaped depression on top. Birds may use the same nest year after year (needs confirmation). Breeding probably does not begin until after 7 years of age but is probably annual thereafter.

Arrival: Mid-September to early October.

Eggs: Early to mid-October. Clutch, 1 dull white egg with scattered dark reddish spots on the large end; averages 65 × 105 mm and 257 g.

Hatching: December. Incubation period about 56 days (needs confirmation). Chick is brooded 1 additional month.

Fledging and departure: Mid-April, 4-5 months after hatching. Departure takes place from April to early June.

Molt. Complete molt takes place from January to July (more information needed).

Predation and mortality. Adult female remains on guard over the nest during incubation and early chick stage. No predation is known.

Ectoparasites. Tick, *Ixodes uriae;* feather lice, *Perineus circumfasciatus, Paraclisis diomedeae, Harrisoniella ferox, Docophoroides harrisoni, D. simplex,* and *Naubates fuliginosus;* flea, *Parapsyllus longicornis.*

Habitat. Antarctic and colder subantarctic waters, usually north of the pack ice. Occurs both offshore and far out to sea but tends to favor the seas near continental landmasses.

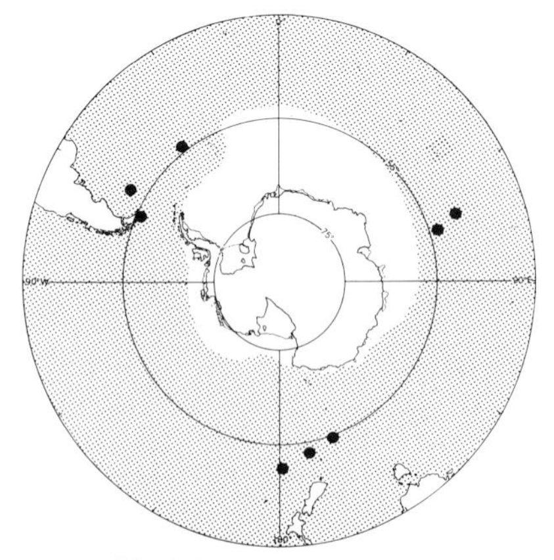

Black-browed albatross

Distribution. In the Antarctic the black-browed albatross breeds on South Georgia, Kerguelen, Heard, and Macquarie islands. There is no good evidence of breeding on Marion, Prince Edward, or Crozet islands. Farther north it also breeds on the Falkland Islands; Islas Ildefonso, Evout, and Diego Ramírez; and Staten, Campbell, and Antipodes islands. See map for pelagic range.

GRAY-HEADED ALBATROSS *Diomedea chrysostoma*

Resident

Identification. Plate 4. 33 in. (84 cm)/85 in. (216 cm). This mollymauk is similar to the black-browed albatross in underwing pattern but differs in having a bluish gray head and neck and a dark bill. The white axial portion of the underwing tends to be more extensive than it is in the black-browed albatross. The gray head contrasts with the white breast and accentuates a small white crescent on the lower eyelid. Frequently, the forehead is paler than the rest of the head but does not form a well-defined pale patch as it does in the white-capped albatross. The black bill has a yellow culminicorn and yellow line on the ramicorn, which, however, are not easy to observe at sea. The narrow culminicorn is bordered by a fleshy black membrane and is rounded but not expanded where it meets the forehead feathering. The feet are bluish flesh.

The juvenile is very similar to the young black-browed albatross, but the head is usually darker gray, even darker than that of the adult gray-headed albatross, and contrasts more with the white breast. The underwing generally shows some white instead of being all dark. The white crescent below the eye is suggested in the juvenile gray-headed albatross. The all-black bill has the same structure as that of the adult. In older immatures, 2-4 years, the face or sometimes the entire head is white with a gray collar. The gray head of the adult begins to appear at 5 years. The culminicorn begins turning yellowish at 2 years but even at 5 years is paler than that of the adult.

The downy chick may be distinguished from the young black-browed albatross on the basis of bill characters, lighter gray down, and usually smaller size due to difference in breeding schedule.

Flight and habits. Similar to the flight and habits of the black-browed albatross.

Voice and display. Essentially like the voice and display of the black-browed albatross.

Food. Similar to that of the black-browed albatross but also in-

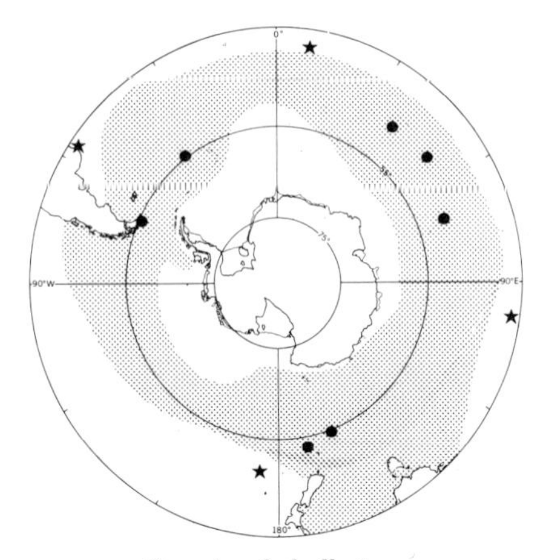

Gray-headed albatross

cludes lampreys. The fact that lampreys are fed to the young strongly suggests a difference in the feeding methods of the two species. The gray-headed albatross is not attracted to ships as readily as the black-browed albatross.

Reproduction. Colonial, often associated with the black-browed albatross. Nest sites and structure and timing of reproductive cycle are similar in the two species except on South Georgia, where the cycle of the gray headed albatross is about 1 month longer. Although the peaks of egg laying and hatching are only about 1 week apart, the peak of departure is late April for the black-browed albatross and mid-May for the gray-headed albatross. If a chick is successfully reared in one season, its parents do not breed the next year.

Arrival: Mid-September.

Eggs: Mid-October. Clutch, 1 white egg with a ring of small reddish spots around the larger end; averages 74 × 101 mm.

Hatching: Mid-December. Incubation period 9-10 weeks (needs confirmation). Both adults incubate, the male taking the first period.

Fledging and departure: Mid-April. No precise information on brood and chick periods. Departure takes place from May to early June.

Molt. Takes place after breeding; more information needed.

Predation and mortality. Like those of the black-browed albatross.

Ectoparasites. Tick, *Ixodes uriae;* feather mite, *Brephosceles diomedi;* feather lice, *Docophoroides simplex* and *D. harrisoni.*

Habitat. Similar to that of the black-browed albatross, but its tendency to occur in higher latitudes where landmasses are absent makes it predominantly a bird of the open oceans.

Distribution. Only known to breed on Diego Ramírez off Cape Horn, South Georgia, Prince Edward, Marion, Crozet, Kerguelen, Macquarie, and Campbell islands. See map for pelagic range.

YELLOW-NOSED ALBATROSS *Diomedea chlororhynchos*

Resident

Identification. Plate 4. 32 in. (81 cm)/80 in. (203 cm). This, the smallest, most slender, and longest tailed of the mollymauks, has a white head and largely white underwings. The slender black bill appears very long and has a yellow culminicorn and pinkish tip. The pointed proximal end of the culminicorn does not meet the forehead feathering and is bordered by a black membrane. The head and hind neck are faintly tinged with gray in freshly molted birds. The dark area about the eyes is prominent. The underwing has narrow dark margins, slightly broader on the leading edge. The legs and feet are pinkish blue.

The juvenile differs from the adult in having an entirely glossy black bill, less prominent dark eye patches, a paler mantle and tail, and more extensive dark margins on the underwing.

The chick at hatching is covered with a thick light gray down, which later becomes nearly white. The bill is black, and the feet are pale bluish gray.

Flight and habits. Similar to the flight and habits of the black-browed albatross but the yellow-nosed albatross is shier and less inclined to follow ships.

Voice and display. In courtship the male walks ponderously around the female on the nest. Both raise heads and utter a high-pitched clattering cry, which has also been called braying. Other displays include slow deliberate bowing, tail fanning, lifting folded wings, preening, and touching bill tips. The chick twitters or squawks when it is soliciting food.

Food. Mostly cephalopods, also fish and large shrimp.

Reproduction. On Tristan, Inaccessible, and Gough islands, nests are solitary, but at favored locations on Nightingale and Prince Edward islands some birds nest in groups of several hundred pairs. The nest is a truncated mud cone, 30-60 cm high, placed on a ledge on a cliff, slope, or plateau or even in a boggy flat valley. The site is some-

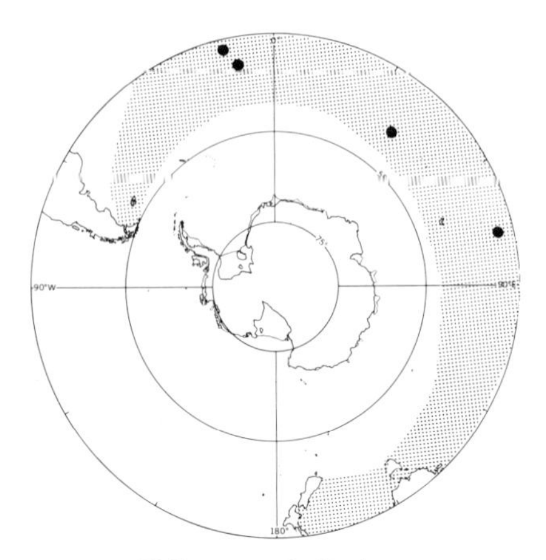

Yellow-nosed albatross

times sheltered by lush vegetation including *Phylica* trees, ferns, or tussock or is in a protected gulley. The schedule is well known on the Tristan da Cunha group and has not been studied much or is unknown elsewhere.

Arrival: August to September.

Eggs: Early September (Nightingale Island) and early October elsewhere. Clutch, 1 white egg with tiny reddish spots; averages 94.3 × 62.5 mm and 206 g.

Hatching: December. Incubation period about 78 days; brood period 2-3 weeks.

Fledging and departure: Late March to mid-April, at about 130 days. Adults leave in late March; young in mid-April to mid-May.

Molt. One bird reported with outer primaries regrowing in March.

Predation and mortality. On Tristan da Cunha group, giant fulmars and skuas attack unguarded young.

Ectoparasites. Ticks, *Ixodes diomedeae, I. percavatus, I. zumpti, I. auritulus,* and *I. rothschildi;* feather mite, *Brephosceles marginiventris;* feather lice, *Docophoroides simplex* and *Paraclisis diomedeae.*

Habitat. Largely pelagic in subtropical and warmer subantarctic waters.

Distribution. Breeds on the Tristan da Cunha group, Gough, Prince Edward (but not Marion), Saint-Paul (probably only occasionally), and Amsterdam islands. Occurs at sea in the South Atlantic and Indian oceans between 20° and 50°S and is numerous at sea near Iles

Kerguelen. In the Pacific, it reaches New Zealand, but there are no west coast South American records.

WHITE-CAPPED ALBATROSS *Diomedea cauta*

Vagrant

Identification. Plate 4. 36 in. (91 cm)/96 in. (244 cm). The largest of the mollymauks with a virtually white underwing and a light colored bill. The wings are broader in proportion to length than those of other mollymauks. In *D. c. cauta* (breeds on the Auckland Islands and Tasmania), the only subspecies likely to be seen in antarctic waters, the head and neck are mostly white with a faint wash of gray on the cheeks, especially in females. The bill is pale greenish or yellowish gray. In two other smaller temperate water subspecies, either the head is gray with a white cap and the bill gray with an ivory culminicorn (*D. c. salvini,* breeds Bounty and Snares islands), or the head is gray, the crown paler, and the bill entirely bright yellow (*D. c. eremita,* breeds Chatham Islands). Juveniles resemble adults but have a uniformly gray bill and more gray on the head and neck.

Flight and habits. In both, the white-capped albatross resembles the greater albatrosses more than it does the smaller mollymauks.

Distribution. Breeds on islands off Tasmania and New Zealand (Chatham, Bounty, Auckland, and Snares islands) and ranges widely, but mainly offshore, in subantarctic waters of the Atlantic and Indian oceans. *D. c. cauta* has been recorded occasionally about Macquarie Island (R. A. Falla, personal communication, 1967) and Ile Saint-Paul [*Segonzac*, 1972].

SOOTY ALBATROSS *Phoebetria fusca*

Resident

Identification. Plate 3. 34 in. (86 cm)/80 in. (203 cm). A graceful all-dark albatross with long slender wings, wedge-shaped tail, and black bill. Although the head is somewhat darker than the brownish gray body, it does not contrast sharply with it. In worn plumage the tips of the back and breast feathers become pale, almost sandy buff, but never ashy gray like those of the light-mantled sooty albatross. At close range a yellow groove (sulcus) shows on the side of the lower mandible, and an incomplete ring of white feathers partially surrounds the dark eyes. The culmen is nearly straight, and the feet are pale gray.

Juveniles resemble adults but have dark rather than white shafts on the wing and tail quills. The feathers of the head and neck are particularly subject to wear, a pale buffy collar and whitish nape thus being produced. Reports of sulcus color in juveniles vary from dull yellow to gray and even violet. See the light-mantled sooty albatross and the giant fulmars.

There is little information available concerning the appearance of the downy chick, but presumably, it is similar to that of the light-mantled sooty albatross.

Flight and habits. This and the light-mantled sooty albatross are by far the most graceful of the albatrosses, flying with precise effortless gliding. They appear to be very inquisitive and follow ships closely.

Voice and display. On the breeding grounds a loud two-note call, the first note shrill and the second deeper, is given both in the air and on the ground. It has been described as a wild drawn-out scream descending in scale and has been likened to the bleating of a young goat on Tristan Island, where the call has occasioned the local name 'pee-oo' for the bird. It is lower pitched and less hoarse than the call of the light-mantled sooty albatross. Both sooty albatrosses are silent at sea.

Food. Squid, fish, crustaceans, and carrion, including remains of prions, diving petrels, and penguins, presumably found floating at sea.

Reproduction. Loosely colonial, usually nesting in small groups and only rarely singly among heavy vegetation, including tussock clumps, on steep coastal cliffs. The low conical nests, about 15 cm high, are composed of mud, grass, and moss. The following schedule is for

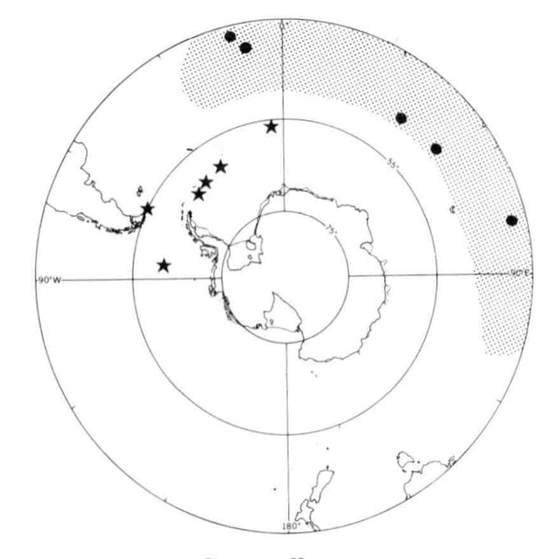

Sooty albatross

Marion Island mostly; the cycles on the Tristan da Cunha group and Gough Island are up to 1 month earlier.

Arrival: Mid-July; occupy nests in late August.

Eggs: Late September (Tristan da Cunha group) to early November (Marion Island). Clutch, 1 pale grayish white egg covered with gray brown specks, more numerous and larger on the broad end; averages 102.8 × 65.3 mm and 248 g.

Hatching: Mid-December to late December. Incubation period about 62 days. In one case, possibly up to 10 weeks (needs confirmation).

Fledging and departure: April to May, at 5 months of age. Departure takes place by late June.

Remarks: On the two islands where both sooty albatrosses nest, the breeding times are separate. In Iles Crozet the sooty albatross breeds 2 weeks earlier than the light-mantled sooty albatross, whereas on Marion Island it nests 1 month later. The two nest in mixed colonies on Marion Island but almost invariably in separate colonies in Iles Crozet.

Molt. No information is available.

Predation and mortality. Skuas take eggs and chicks on Gough Island.

Ectoparasites. Tick, *Ixodes zumpti;* feather louse, *Paraclisis diomedeae.*

Habitat. Mostly in temperate and subantarctic waters. Highly pelagic.

Distribution. In the Antarctic the sooty albatross breeds along with the light-mantled sooty albatross on Prince Edward, Marion, and Crozet islands; farther north it breeds alone on the Tristan da Cunha group, Gough, Amsterdam, and Saint-Paul islands. At sea it ranges between 30° and 50°S in the Atlantic and Indian oceans. Southernmost records are in the northern Weddell Sea [*Novatti*, 1962, Table 2] and 58°S, southwest of Bouvetøya [*Bierman and Voous*, 1950, p. 29]. Its status off southern Australia needs study. The only Pacific records are at 61°S, 90°W [*Holgersen*, 1957, p. 18].

LIGHT-MANTLED SOOTY ALBATROSS *Phoebetria palpebrata*

Resident

Identification. Plate 3. 34 in. (86 cm)/84 in. (213 cm). Differs from the sooty albatross in having pale ashy gray mantle and underparts, which contrast with the brownish black head and dark gray wings. The sulcus on the side of the lower mandible is blue and is on the average a little narrower than that of the sooty. The culmen is slightly concave. The feet are light bluish gray.

The juvenile is similar to the adult in plumage characters, but the shafts of the wing and tail quills are dark rather than white, and the sulcus is pale yellow or gray. The first feather coat is subject to extreme wear with the result that the faded mantle and breast appear barred with pale buff. This species can be confused with the worn adult and immature sooty albatross, but the underparts are usually paler and grayer. Some young *Phoebetria*, however, may be impossible to identify even in the hand. See the giant fulmars.

The chick at hatching is covered with a fine silvery gray down, which is lightest on the sides of the head. The first down on the head is replaced by a velvety dark gray second down on the crown and chin and a contrasting white eye mask. The second body down is longer than the first. The bill is black.

Flight and habits. Similar to the flight and habits of the sooty albatross.

Voice and display. On the breeding grounds a two-note call 'pee-ow,' the first, a low shrill expiratory note made with the bill open, immediately followed by a lower quieter inspiratory note made with the bill closed and pointing downward. Field notes comparing the calls of the two sooty albatrosses would be useful.

Food. Squid, euphausiids and other crustaceans, fish, and carrion. Digested remains of prions have been found in stomachs. Usually, the light-mantled sooty albatross alights on the water to feed, but it may also plunge from the air for food below the surface. Follows ships.

Reproduction. Usually nests singly, rarely semicolonially with up to four or five pairs together. Nest placement and structure are like those of the sooty albatross, but this species prefers inland breeding sites rather than coastal cliffs.

Arrival: September in Iles Crozet and early October elsewhere.

Eggs: Mid-October to early November. Clutch, 1 white egg with reddish spots toward the larger end; averages 103 × 67 mm and 257 g.

Hatching: Late December to January. Incubation period 63-70 days; brood period 20 days.

Fledging and departure: Late April, in 135-150 days. Departure takes place by late June.

Molt. Adults have been found with their outer primaries in molt in October and March.

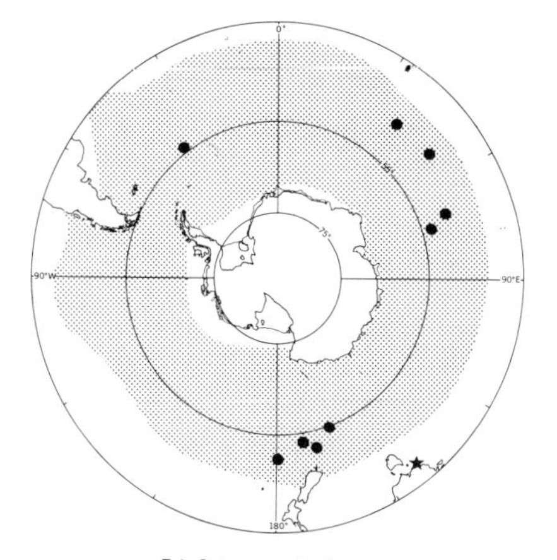

Light-mantled sooty

Predation and mortality. Unattended eggs and young fall prey to skuas and sheathbills, about 30% mortality of eggs and 30% mortality of chicks thus being caused. Adults are long lived.

Ectoparasites. Tick, *Ixodes uriae;* feather lice, *Perineus circumfasciatus* and *Paraclisis diomedeae;* flea, *Parapsyllus magellanicus.*

Habitat. Antarctic waters, its southern limit seems to be heavy pack ice at the edge of the Antarctic continent. Highly pelagic.

Distribution. Breeds on South Georgia, Prince Edward, Marion, Crozet, Kerguelen, Heard, Macquarie, Auckland, Campbell, and Antipodes islands. See map for pelagic range. Vagrants occur as far north as 33°S.

FULMARS, PRIONS, GADFLY PETRELS, AND SHEARWATERS: Procellariidae

The procellariids are a diverse group of long-winged pelagic petrels. The nostrils are situated on the top of the bill united in a single nasal tube but separated by a visible median septum. The procellariids fall into four groups based on anatomical and ecological differences. fulmarine petrels, prions, gadfly petrels, and shearwaters. Flight patterns, feeding habits, and bill shape are useful for separating the groups at sea and in the hand. Characters for at-sea species identification include contrasting patterns of the body, head, and wing and color of the bill. Most species, except the more migratory shearwaters, have a gradual molt that generally takes place after breeding. They retain the power of flight during molt, although the wings may appear very ragged and flight is labored. All procellariids lay a single relatively large white egg in a variety of colonial nest sites. Both parents share incubation duties and feed the chick by regurgitation. The downy young remain in the nest 2 or more months and have 2 generations of down before fledging. They and the adults repel predators and intruders by squirting stomach oil and undigested food from the open bill.

FULMARS

Fulmarine petrels are small to very large pelagic birds with relatively long broad wings. Color ranges from all white through pied to all dark. Males are on the average larger and heavier billed than females. Flight patterns are varied: slow and powerful beats in the giant fulmars and moderately fast to rapid beating in the other species. None fly as erratically as the gadfly petrels, and all are good gliders. They feed from the surface on squid and crustaceans. Giant fulmars and the Cape pigeon are well-known ship-following scavengers and are among the most familiar of seabirds. Giant fulmars nest on open level ground; the other species occupy rock crevices and cliff ledges. Colonies are usually coastal, but Antarctic and snow petrels may use snow-free cliffs far inland. Six fulmarine petrels breed in the Antarctic: two frequent the pack ice, and the other four migrate or wander north almost to the equator.

SOUTHERN GIANT FULMAR *Macronectes giganteus*
NORTHERN GIANT FULMAR *Macronectes halli*

Resident

Identification. Plate 3. 30-36 in. (76-91 cm)/80-95 in. (203-241 cm). Very large brownish gray or white fulmarine petrels, the size of a mollymauk, with massive pale bills. The prominent nasal tube extends more than half the length of the bill. Colors of plumage, bill, and eye vary geographically and with age. The southern population, *M. giganteus*, has both dark and light phases, and in both the bill is pale yellowish green, greenest near the tip. Dark-phase adults are mostly brownish gray with dirty white on the head, neck, and sometimes the breast but with little tendency toward dark mottling of the head. The head, neck, and upper breast become progressively whiter with increasing age, and the leading edge of the wing is mottled with pale feathers in very old birds. The eyes are either dark brown or somewhat less frequently pale gray in breeding adults. All-white birds have only a few scattered dark feathers, and over 10% of the breeding birds in some localities may be in this phase. The percentage of white individuals seen in the open pack ice may be even higher (needs confirmation).

Males are on the average larger than females in weight, wing length, and tarsus and tail measurements, and there is no overlap in bill length. Males have exposed bills that measure over 97 mm long, and those of females measure less than 97 mm [*Conroy*, 1972].

A darker, browner, and more heavily marked population, *M. halli*, of which no white phase is known, breeds farther north on subantarctic islands. Although the face, throat, and upper breast become paler and mottled gray in older birds, fully white feathering is confined to the face near the bill. The crown, back of the head, and wings are always dark. The bill is yellowish olive, dull pink, or reddish near the tip of the culmen and dark brown on the inside margin of the hook. The eyes are mostly pale gray, sometimes appearing almost white. A few birds breed when the eyes are still gray brown.

Although there is much overlap in size, northern birds are on the average slightly smaller in wing and tail length (and probably in weight) but are larger in bill, tarsus, and toe measurements. The systematic status of the two populations remains in doubt, since some Atlantic Ocean islands (Gough Island and the Falkland Islands) apparently have populations with intermediate characters. They are best considered two full species, however, because both breed on Macquarie Island, Marion Island, and Iles Crozet, where they are on different reproductive schedules (see the sections on reproduction and distribution).

When juvenile dark-phase birds in both populations are first fledged, they are entirely glossy sooty black, but the plumage soon fades to dull brown. This dark plumage may be retained for up to 4 years. The bill is horn colored, the tip being greenish when the juvenile is first fledged. All but the nostril tube becomes pale yellow green as plumage becomes browner. The bill color in northern juveniles needs verification. The eyes are dark brown. White-phase birds, however, are white from fledging. See the immature wandering albatross, the sooty albatrosses, and the white-chinned petrel.

The first down of northern chicks (*M. halli*) is white below and gray on the back, darkest on the crown and nape, where it forms a hood contrasting with the white face. The second down is uniform iron gray. The first down of all chicks of the southern form (*M. giganteus*) is white with some grayish markings on the back. The second down of the white phase is also white, but that of the dark phase is gray, distinctly lighter than the second down of the northern form. Chicks of both forms have straw colored bills.

Because it has only recently been suggested that there might be two species of giant fulmars, there is little comparative biological information except from Iles Crozet and Macquarie Island [*Bourne and Warham*, 1966; *Voisin*, 1968]. The following details therefore generally represent a composite picture.

Flight and habits. The giant fulmars are similar in gliding and soaring flight to the smaller albatrosses but are decidedly less graceful stiff-winged fliers. The heavy head and long neck contribute to an overall impression of clumsiness. Like albatrosses the giant fulmars cannot take off easily, requiring a stiff wind or a clear runway to become airborne. If these heavy-bodied petrels are pursued on land, they can run surprisingly fast and may vomit their stomach contents to lighten themselves before takeoff. After feeding on carrion or fat and bloody prey they bathe vigorously in the sea. They are solitary at sea, but groups converge around ships. On Iles Crozet, *M. halli* is less timid than *M. giganteus*. In the Australian and New Zealand sectors, northern giant fulmars are strongly attracted to stationary and moving ships, and southern giant fulmars pay little attention to ships [*Johnstone*, 1974]. Comparative information is needed from the South Atlantic and south Indian oceans.

Voice and display. Courtship calls include a hoarse drawn-out croaking 'like the neighing of a colt' and catlike 'mewing' while two birds face each other or sit side by side and touch bills, nibble faces, or wave heads. The voice of *M. giganteus* may be sharper but less loud than that of *M. halli*. Nauseating retching noises, loud hissing, vicious bill snapping, and squirting stomach oil by both adults and nestlings discourage nest intruders. They are generally silent at sea except for croaking sounds when they are quarreling over scraps.

Food. During the summer, dead and moribund penguins and petrels together with eggs and chicks provide the majority of food for these predatory scavengers. They also gather at seal pupping grounds in October to feed on the mammals' placentas and carcasses of dead young. Some southern giant fulmars remain through the winter on the continent to take advantage of the hibernal breeding habits of emperor penguins. While giant fulmars are swimming on the water like albatrosses, they capture cephalopods, fish, and crustaceans, which probably constitute the bulk of their natural food at sea, but they also regularly follow ships for galley refuse. Near Iles Crozet, *M. halli* regularly fishes in coastal waters, whereas *M. giganteus* rarely does.

Reproduction. The main ecological differences presently known between the two giant fulmars involve their breeding habits. On Macquarie Island and perhaps elsewhere the northern form nests alone in secluded sheltered areas with heavy vegetation or in broken coastal terrain, which affords concealment. The southern form nests sociably in conspicuous loose colonies of up to 300 pairs on open coastal plateaus or headlands. On Iles Crozet both species tend to nest in colonies, although a few northern form birds nest in isolation. Some colonies are mixed, northern birds occupying more sheltered sites and southern birds more exposed ones. Nest size and construction depend on the availability of vegetation. In the far south the nests are merely low mounds of small stones, but moss and grass are used in building bulkier nests farther north. Where the two forms occur together on Marion Island, Iles Crozet, Macquarie Island, and possibly Iles Kerguelen, the breeding cycle of the northern solitary form is 5-8 weeks earlier than that of the southern colonial form. The prolonged breeding schedule on South Georgia and possibly on Heard Island suggests the presence of both species. The following dates pertain to both species, the earliest *M. halli* dates being from Iles Crozet; presumably dates on Tristan and Gough islands are similar or even somewhat earlier, but there are few data. Regular breeding may commence as early as the sixth year, but such young birds are rarely successful. Even younger birds may occasionally breed, but most of the 3-, 4-, and 5-year-old birds at colonies are prebreeders that arrive after established breeders have laid.

Arrival: Some adults remain near the breeding grounds throughout the year; reestablishment of nesting territories begins in July (*M. halli*), August, or September (*M. giganteus*). After reoccupying nest sites, adults spend about 2 weeks at sea feeding before returning to lay.

Eggs: Mid-August to early September (*M. halli*) or early October to mid-November (*M. giganteus*). Clutch, 1 elongate dull white egg; 60-79 × 98-115 mm and 220-283 g. The very rare two-egg clutches recorded are probably due to the use of the same nest by two females or to disturbance.

Hatching: Mid-October to early November (*M. halli*) and late November to early January (*M. giganteus*). Incubation period 55-65 days, *M. halli* possibly averaging a few days less than *M. giganteus*. Both sexes incubate, the male taking the first long shift. Adults guard the chick in the nest for 2-3 weeks until it assumes the warmer second down and can defend itself.

Fledging and departure: Early February (*M. halli*) to April (*M. giganteus*), at an age of about 115 days. Departure takes place from February (*M. halli*) to May (*M. giganteus*).

Molt. Northern form birds begin wing molt in November on Iles Crozet and complete molt in March. Body and wing molt is in progress in early October on Macquarie Island and lasts at least until February. On South Georgia, wing molt begins in early December and ends in March. On the South Orkney Islands, body molt begins in late November during incubation and is completed in March. Primary molt begins in February and probably lasts 3 months. Prebreeders and unsuccessful breeders may molt primaries somewhat earlier than birds feeding chicks.

Predation and mortality. Egg loss may average 70% in early-breeding northern birds in Iles Crozet. Breeding success is generally higher in southern populations, however, where except for a few instances of egg and chick cannibalism, there is little evidence of natural predation. Skuas, sheathbills, and wekas take eggs and young but only from disturbed or unguarded nests. Nests in sheltered locations may be buried in snow. Leopard seals may catch adults in the water (only one recorded occurrence).

Ectoparasites. Tick, *Ixodes uriae;* feather lice, *Austromenopon ossifragae, Paraclisis obscura, Perineus circumfasciatus, Docophoroides murphyi, Trabeculus heteracanthus,* and *Saemundssonia gaini;* fleas, *Parapsyllus cardinis, P. magellanicus,* and *Notiopsylla kerguelensis.*

Habitat. Pelagic in antarctic and subantarctic waters, the northern form occurring almost exclusively north of the convergence. Adults of the southern form frequent antarctic waters all year, but immatures are common farther north nearly to the tropics. This zonal segregation of the two species is most apparent from November to May in the Australian and New Zealand sectors [*Johnstone*, 1974].

Distribution. Northern giant fulmar breeds on Gough, Prince Edward, Marion, Crozet, Kerguelen, Macquarie, and New Zealand sub-Antarctic islands (Chatham, Stewart, Auckland, Antipodes, and Campbell islands but most probably not on the Snares Islands [*Conroy*, 1972]). Southern giant fulmar breeds on the Falkland, possibly Diego Ramírez (C. C. Olrog, personal communication, April 1974), South Georgia, South Sandwich, South Orkney, and South Shetland islands; at four recorded localities on the Antarctic Peninsula; on Bouvet (needs confirmation), Prince Edward, Marion, Crozet, Kerguelen, Heard, and Macquarie islands; and on the Adélie Coast, Windmill Islands, and Enderby Land on the continent. A few northern birds breed at South Georgia, and some southern birds with green-tipped bills may breed on the Chatham Islands (G. W. Johnstone, personal communication, January 1975). Northern birds

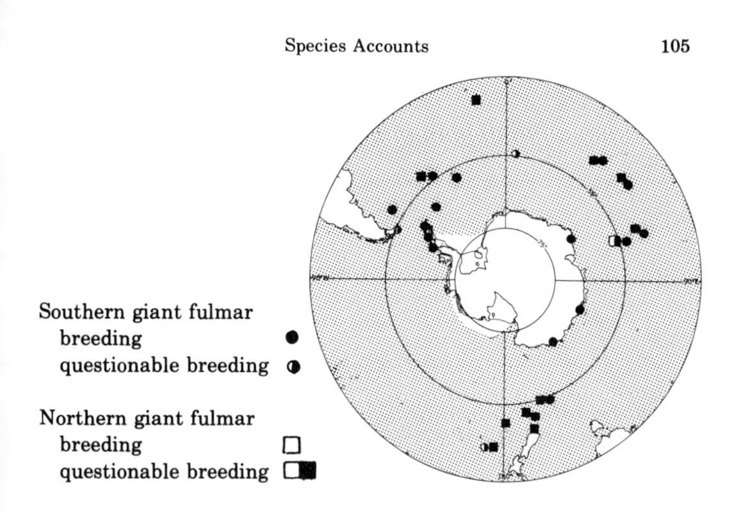

may also breed on Heard Island, where both light and dark birds occur [*Bourne and Warham*, 1966; *Downes et al.*, 1959]. Prebreeding birds of both species range widely at sea. Young of the southern form follow the circumpolar West Wind drift, regularly reaching north to the Tropic of Capricorn in all oceans. Adults of the northern form probably do not occur far south of the Antarctic convergence, and long-distance movements of young birds, other than a few recoveries of Macquarie Island birds from South America, are little known owing to limited banding. The pelagic distributions of the two forms are not separated on the map.

SOUTHERN FULMAR *Fulmarus glacialoides*

Resident

Identification. Plates 5 and 6. 18 in. (46 cm)/50 in. (127 cm). A pale gray petrel superficially resembling a gull but with conspicuous white wing flashes formed by the white inner webs of the slate gray primaries. Underwings and underparts are white, the latter being lightly washed with gray on the sides of the breast and flanks. The bill is pink with a black tip and bluish markings on the lower mandible and nasal tubes; the feet are flesh. See the white-headed petrel and the gray petrel.

The first-year bird has a weaker and more slender bill than the adult.

The chick in first down is mostly white with a bluish gray wash on the mantle. The second down is somewhat darker, light gray on the upperparts and flanks shading to white on the forehead and underparts. The bill is mostly pink like that of the adult.

Flight and habits. Alternates fast wingbeats and long glides; soars with a light rocking motion. Wings are extended fully in gliding and appear exceptionally long and narrow. Gregarious at sea, small flocks resting or feeding together.

Voice and display. The courtship call is a soft droning or guttural croaking given while birds sit side by side waving heads, nibbling, and mutually preening. When they are disturbed at the nest or quarreling over food at sea, they also utter a shrill rattling cackle.

Food. Mainly euphausiid shrimp and other crustaceans, some cephalopods and fish. Follows ships to feed on refuse.

Reproduction. Colonial, nesting on rocky ledges of steep coastal cliffs. The shallow nest lined with stone chips is usually somewhat sheltered from the wind. Colonies are highly synchronous in schedule and may number several hundred birds.

Arrival: October.

Eggs: Late November to mid-December. Clutch, 1 white elongated egg; averages 75 × 50 mm and 103 g.

Hatching: Mid-January to late January. Incubation period averages 46 days. The chick is brooded or guarded an additional 3 weeks.

Fledging and departure: Mid-March, when they depart with adults at about 51 days after hatching.

Molt. Breeding adults probably molt after chicks fledge, but nonbreeders may molt as early as late December (needs confirmation).

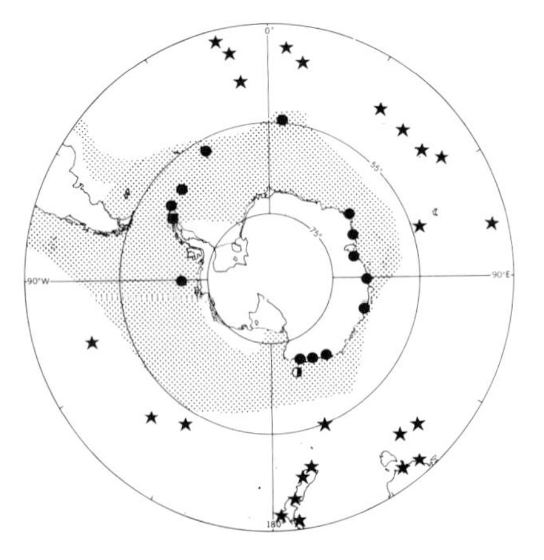

Southern fulmar

Predation and mortality. Skuas may take some abandoned eggs and chicks, but extreme weather conditions are much more important causes of egg and chick mortality of 40 and 25%, respectively.

Ectoparasites. Feather lice, *Ancistrona* sp., *Perineus nigrolimbatus, Procellariphaga brevifimbriata, Harrisoniella chilensis,* and *Saemundssonia bicolor;* flea, *Glaciopsyllus antarcticus.*

Habitat. Cold antarctic waters during the breeding season but avoiding pack ice. Nonbreeding birds commonly wander into the subtropics but only where they follow cold currents along the western coasts of southern continents. Highly pelagic.

Distribution. Breeds on South Sandwich, South Orkney, and South Shetland (Clarence and Gibbs) islands; at Cape Roquemaurel on the Antarctic Peninsula; on Bouvet, Peter I, and probably Balleny islands; and at several localities on the continent. See map for at-sea range.

ANTARCTIC PETREL *Thalassoica antarctica*

Resident

Identification. Plates 5 and 6. 17 in. (43 cm)/41 in. (104 cm). A brown and white fulmarine petrel with conspicuous dark-edged white wing bars and a dark-tipped white tail. The head, throat, back, and leading edge of the wings are dark; the underparts and underwings are mainly white, the latter being margined with brown. The bill is brown; the feet are flesh. The dark areas of birds in fresh plumage appear almost black, but the color fades rapidly to light grayish brown. The dark tips to the feathers of the throat and nape may also wear, so that these areas become white, especially in first-year birds. See the Cape pigeon.

In fresh plumage the juvenile is slightly darker than the adult and has an almost black bill.

The chick in first down is pale gray on the upperparts, foreneck, and breast and white on the head and belly. Information is needed on the appearance of the second down.

Flight and habits. Similar to the flight and habits of the southern fulmar, but the Antarctic petrel flaps more with shorter, stronger, and more rapid wingbeats. It usually flies high over the water, occasionally at great heights. Gregarious, large flocks gathering on ice floes.

Voice and display. Courtship consists of two birds sitting face to face in the nest and mutually nibbling head and bill. One bird lowers beak to breast and slowly raises and extends head until beak lies horizontally on back. Head wagging is less intense than it is in the southern fulmar. Displays are accompanied by churring, soft clucking, and cackling calls. Loud calls are stronger and more resonant than similar calls of the southern fulmar. Displays at the nest may be broken by short flights out to sea.

Food. Euphausiid shrimp, pteropods, cephalopods, fish, and medusae. The Antarctic petrel feeds while it is swimming but can dive well, both from the surface and from the air.

Reproduction. Colonial, from 1000 to 1,000,000 birds nesting together on low snow-free cliffs exposed to wind. Offshore island colonies may have less than 100 pairs. One known colony is 250 km from the coast on an isolated inland nunatak [*Konovalov and Shulyatin*, 1964]. The nests, which are on rock ledges or in shallow crevices, are sometimes lined with stone chips.

Arrival: Early to mid-October.

Eggs: Late November to early December. Clutch, 1 white egg; averages 70 × 48 mm and 90 g.

Hatching: Mid-January. Incubation period more than 45 days.

Fledging and departure: Late February to mid-March. No precise information on brood or chick periods. No data on departure, but adults frequent edge of fast ice near colonies in winter.

Molt. January to March. Up to one half of the flight feathers may be in molt simultaneously.

Predation and mortality. Adverse weather causes egg losses of about 25%. No information exists on chick or adult mortality.

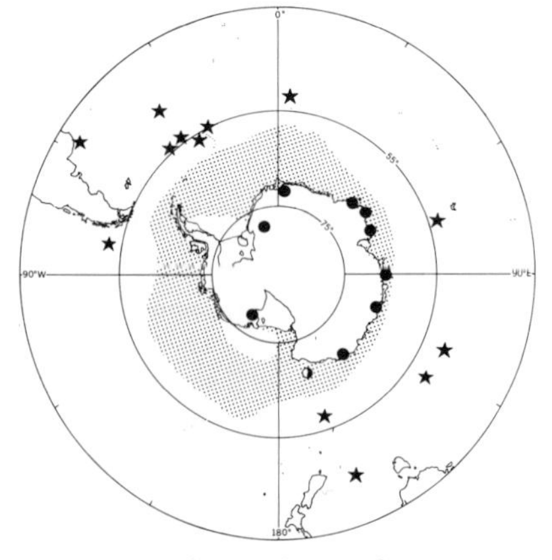

Antarctic petrel

Parasites. Feather lice, *Ancistrona* sp., *Austromenopon oschei, Pseudonirmus lugubris,* and *Saemundssonia nivea.*

Habitat. Most frequent in open water with scattered icebergs and ice floes, also found in open pack ice.

Distribution. Breeds at nine known widely scattered localities on the continent and possibly on the Balleny Islands. There are probably other unknown colonies judging by its abundance, especially in the Weddell Sea. See map for pelagic range. One vagrant has been recorded in South Africa, and another at 33°S in the eastern Indian Ocean.

CAPE PIGEON *Daption capense*

Resident

Identification. Plate 5. 16 in. (41 cm)/36 in. (91 cm). A chunky fulmarine petrel with a black and white checkerboard pattern on its back and two white patches on each wing. The head and nape are black, and the tail is white with a black tip. The underparts and underwings are white, the latter having narrow black margins. The bill and feet are black with some white mottling on the toes. The intensity of the black and the amount of white on back, wings, and chin vary geographically, with wear, and possibly with age. The New Zealand subspecies (*D. c. australe*) is smaller and much darker on the mantle than even freshly molted young antarctic birds (needs confirmation); older antarctic birds tend to be whiter (Figure 6).

The bill in first-year birds is narrower, but bill size is highly variable and is not a good character for aging individual birds.

The downy chick is slaty gray, somewhat lighter on the underparts. The second down is slightly paler.

Flight and habits. Similar to the flight and habits of the Antarctic petrel but with more flapping and shorter glides. Cape pigeons are gregarious and gather in large flocks to feed.

Voice and display. Raucous chattering 'cac-cac, cac-cac, . . . ' in increasing tempo. The Cape pigeon is extremely vocal both when it is at the nest and when it is feeding at sea. Greeting displays at the nest consist of neck and wing stretching and mutual head swaying.

Food. Predominately euphausiids and other crustaceans, cephalopods, and, to a lesser extent, fish and carrion. Food habits vary locally. The Cape pigeon readily follows ships to take refuse. While it is feeding on the water, it pecks like a hen, using its feet as paddles to bring plankton to the surface. It may take some food by

Fig. 6. Back patterns of two races of the Cape pigeon, *Daption c. capense* (above) and *D. c. australe* (below).

hydroplaning (see the section on the food of the Antarctic prion) and, on rare occasions, makes shallow dives.

Reproduction. Colonial, from less than 10 to 2000 pairs nesting together. The site is generally protected from the prevailing winds except on the continent, where exposed rocks are snow free early in the season. Nests are located on rock ledges or in shallow crevices, occasionally just above the tidal zone, and consist of stone-lined scrapes. Breeding commences in the fourth year or later.

Arrival: Mid-October. In northern colonies some adults return as early as August and may be present in April, but the rest of the schedule is essentially the same as it is in more southerly localities.

Eggs: Late November to early December. Clutch, 1 white egg;

averages 62 × 43 mm and 67 g. Reported two-egg clutches are from two females sharing the same nest.

Hatching: Early to mid-January. Incubation period 41-50 days; brood period about 19 days.

Fledging and departure: The young leave with adults in early to mid-March. The chick period lasts about 7 weeks.

Molt. December to mid-April. Unsuccessful and prebreeding birds begin molt earliest; successful breeders molt body feathers during incubation and begin molting flight feathers as young near fledging. Tail molt begins as primary molt ends, most of the feathers being shed simultaneously.

Predation and mortality. Starvation and exposure to cold are the most important causes of mortality, especially in the far south, but eggs may be deserted by inexperienced birds or eaten by skuas. Egg mortality is 24-40%; chick mortality is 10-20%. Mean annual adult survival may be near 95%.

Ectoparasites. Tick, *Ixodes uriae;* feather lice, *Austromenopon daptionis, Ancistrona procellariae, Naubates testaceus, Pseudonirmus gurlti,* and *Saemundssonia stammeri;* flea, *Parapsyllus magellanicus.*

Habitat. Generally pelagic in subantarctic and antarctic water but avoiding the pack ice. The Cape pigeon is widespread at sea both in the far south during the breeding season and in the north to 35°S at other times, even following cooler currents into tropical waters.

Distribution. In the Antarctic the Cape pigeon breeds on South

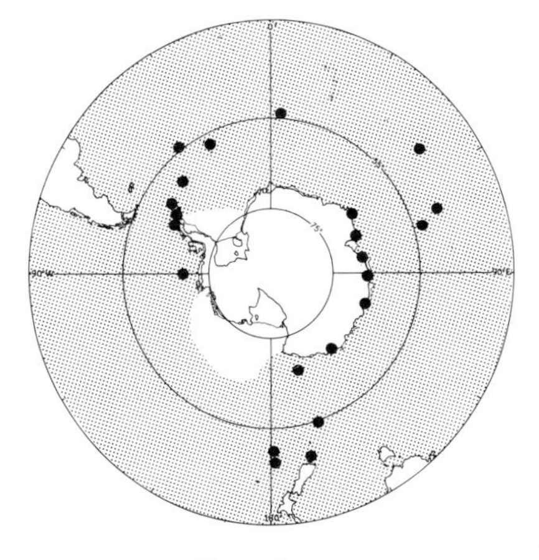

Cape pigeon

Georgia, South Sandwich, South Orkney, South Shetland, Bouvet, Crozet (Ile de la Possession only), Kerguelen, Heard, Macquarie, Balleny, and Peter I islands and at eleven known localities on the Antarctic Peninsula and continent. Farther north it also breeds on the Bounty, Antipodes, and Snares islands off New Zealand. See map for pelagic range.

SNOW PETREL *Pagodroma nivea*

Resident

Identification. Plate 6. 13.5 in. (34 cm)/31 in. (79 cm). An all-white small fulmarine petrel with conspicuous dark eyes, small black bill, and bluish gray feet. In poor light the white plumage appears gray, especially on the underparts. Beneath the white plumage the down is dark gray. Birds from two nesting localities, Géologie Archipelago on Adélie Coast and the Balleny Islands (*P. n. confusa*, also called *P. n. major* by some authors), are heavier (average 388 versus 265 g) and have longer wings (291 versus 262 mm), longer tarsi (37.7 versus 33 mm), and longer (23.5 versus 21 mm) and stouter bills (see Figure 7) than those from all other populations studied (*P. n. nivea*) [*Prévost*, 1969; *Isenmann*, 1970*a*]. More measurements of known breeding birds are needed. Possibly, both large and small forms breed at Géologie Archipelago, in which case they may have to be considered separate species.

Young birds in the first plumage have faint gray vermiculation on their back and a gray wash on the tips of their primaries, but these characters are not discernible at sea and may be present in some adults, especially females.

The chick is lavender gray on the upperparts, head, and foreneck; ivory or grayish white on the abdomen; and pure white on the forehead. The second down is slightly lighter in color than the first.

Flight and habits. Very erratic almost batlike flight with short

Fig. 7. Heads of two races of snow petrel, *Pagodroma n. nivea* (right) and *P. n. confusa* (left).

rapid wingbeats and infrequent gliding. The snow petrel tends to fly about 10 m high over the water and may fly very high over land. It is frequently seen hovering low over the water but is rarely observed swimming. It is more agile on land than other small procellariids. Flocks are characteristically seen sitting on ridges of icebergs.

Voice and display. The copulation call is a mechanical 'chirring' note. In nest defense a guttural and disagreeable 'teck-teck-teck' similar to that of a tern is uttered. From undisturbed nests, another 'more melodious' call has been described as being between the croaking of a fulmar and the screeching of a gull.

Food. At sea the snow petrel eats mainly fish with some cephalopods, other mollusks, and euphausiids. Apparently, the snow petrel also feeds on seal placentas and the carcasses of dead seals, whales, and penguins and occasionally eats refuse on land. Comparison of food, especially sizes of fish, of *P. n. confusa* and *P. n. nivea* is needed. Feeding methods at sea have not been described.

Reproduction. Colonial, nesting in small to large cliff colonies usually near the sea, but snow petrels also breed on inland cliffs almost 325 km from the coast. The nest is a simple pebble-lined scrape usually in a deep rock crevice with overhanging protection. Some nests at Géologie Archipelago are fully exposed from above. The nest often contains dehydrated eggs and mummified corpses of chicks of previous seasons. The average schedule below may be advanced 2 weeks in the South Orkney Islands and 1 month later in inland mountain colonies on the continent.

Arrival: In some areas, birds are present all year, but the main influx is from mid-September to early November.

Eggs: Late November to mid-December. Clutch, 1 white egg; averages 59.4 × 41.9 mm and 57.4 g (Adélie Coast) or 55.5 × 39.4 mm and 47.4 g (Davis Station).

Hatching: Early to mid-January. Incubation period 41-49 days, averaging 45 days for the large form on the Adélie Coast and possibly 2 days less for the small form elsewhere. Young are brooded an additional 8 days.

Fledging and departure: By mid-March, at an age of 42-50 days. Departure takes place from late February to mid-May, latest on the Antarctic Peninsula. The chick remains in the nest about 7 weeks.

Molt. November to April, during the breeding cycle. Body molt begins during incubation; flight feathers molt as young near fledging. Unsuccessful and prebreeders molt earliest.

Predation and mortality. Skuas are important predators, but severe weather conditions, especially heavy snow that blocks nest

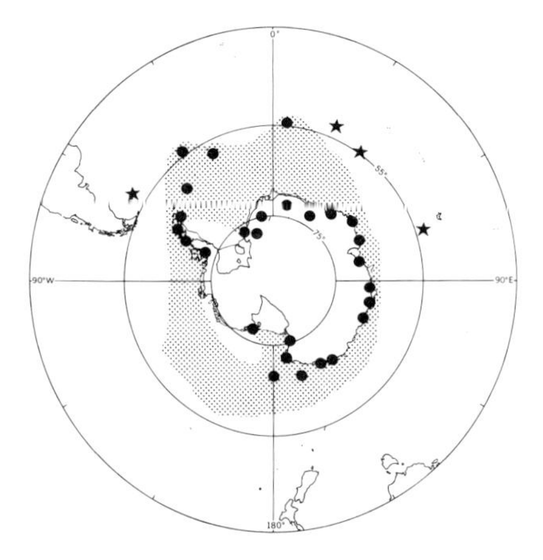

Snow petrel

entrances, may cause adults to abandon eggs and allow chicks to starve. Egg mortality is about 50%, and chick mortality is 10-15%, but mean annual adult survival is 93-96%.

Ectoparasites. Feather mite, *Zachvatkinia hydrobatidii;* feather lice, *Ancistrona* sp., *Austromenopon*? *daptionis, Pseudonirmus charcoti,* and *Saemundssonia nivea;* flea, *Glaciopsyllus antarcticus.*

Habitat. Almost entirely restricted to the colder antarctic waters with pack ice or icebergs and ice floes.

Distribution. Breeds on South Georgia, South Sandwich, South Orkney, Bouvet, Balleny, and Scott islands and at numerous localities on the Antarctic Peninsula and continent. See map for pelagic range.

PRIONS: *Pachyptila* spp.

Prions, also known as whalebirds, are small bluish gray petrels with white underparts and underwings and prominent black tipping on the wedge-shaped tail and undertail coverts. Dark coverts and primaries form an open 'M' on the upper surface of the spread wings in flight. The bill and feet are blue. The species are so similar on the wing, differing mainly in the shape of the bill and the head pattern, that they are almost impossible to identify with certainty. Most at-sea observations should be recorded only as 'prions,' and any specimens should be submitted to a museum for specific identification. Bill structure is usually diagnostic for identifying specimens in the hand, but immatures have weaker and narrower bills, so that they may superficially resemble adults of a smaller-billed species. In the large species the sides of the bill have fine palatal lamellae, which are used in straining their crustacean food from the water. Other species characters include overall size, shade of back color, width of dark tail band, and relative prominence of white eyebrow and narrow dark eye mask. In general, larger species are darker and have proportionately larger heads. Prions are remarkable for their buoyant erratic flight, 'hydroplane' feeding, and highly social flocking. Five species breed in burrows on low-latitude antarctic and subantarctic islands, but only two range south of 60°S with any regularity.

NARROW-BILLED PRION *Pachyptila belcheri*

Resident

Identification. Plate 7. 10.5 in. (27 cm)/22 in. (56 cm). This, the palest of the prions, has a distinct head pattern and proportionately the narrowest bill. The pure white lores and broad white eyebrows contrast with the narrow bluish gray stripe below and behind the eyes. The open M on the wings and the tip of the tail are grayish black. The bill is narrow and elongated, 3 times longer than it is wide at the base with a considerable expanse of culmen between the tip and the nostrils. The sides of the bill are straight when the bill is viewed from above. The palatal lamellae are only represented by weak ridges and are not visible laterally when the bill is closed. All prions might be confused with the blue petrel, but in that species the crown is dark gray, the forehead is white, and the rounded tail has a narrow white tip.

For a short period after fledging the juvenile differs from the adult on the basis of its weaker narrower bill and fresh unworn rich blue plumage. Adults at this period are beginning to molt their worn feathers.

The grayish blue downy chick is presumably like that of the Antarctic prion but has not been described in detail.

Flight and habits. Extremely agile and erratic. The narrow-billed prion twists from side to side in impetuous weaving flight with no apparent pattern, even in calm weather. It generally stays low over the water surface in smaller flocks than those of the Antarctic prion.

Voice and display. Coos and a loud harsh grating alarm note have been reported for birds captured at sea.

Food. Feeds mainly at night, taking amphipods, some small squid, and possibly other crustaceans. The narrow-billed prion picks food from the surface while it glides on stiffly outstretched wings. It hovers in calm weather, when plankton is abundant, and may alight on the water to feed.

Reproduction. Colonial, nesting in burrows near the shore in soft soil or peat. The burrow, from 0.5 to 3.5 m long, may be shared by two or more pairs, each in its own enlarged nesting chamber lined with tussock grass and feathers. Little information is available on the breeding cycle, and the following schedule is based on one set of observations each from the Falkland Islands and Iles Kerguelen.

Arrival: Late September.

Eggs: Early November. Clutch, 1 dull white egg; averages 48 × 35 mm.

Hatching: No information on dates or incubation period.

Fledging and departure: No information on fledging. Departure takes place from late February to early March.

Molt. Immediately following the breeding season. Molting birds recorded in the Pacific Ocean in September to December are probably immatures.

Predation and mortality. Presumably skuas, but more information is needed.

Ectoparasites. Feather louse, *Naubates prioni;* flea, *Parapsyllus magellanicus.*

Habitat. Highly pelagic in sub-Antarctic and Antarctic zones south to the vicinity of the pack ice.

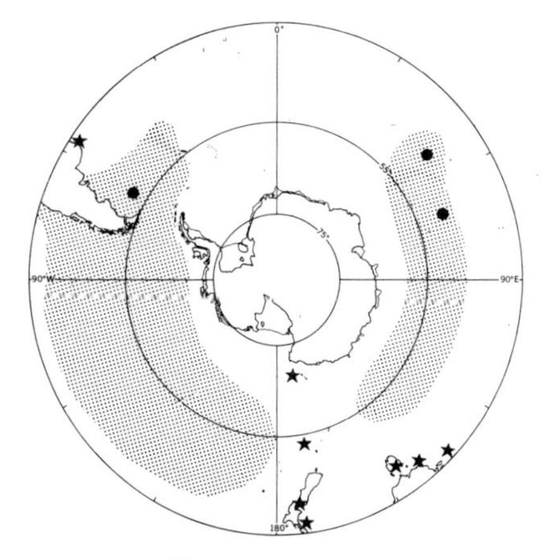

Narrow-billed prion

Distribution. Definitely known to breed on the Falkland, Crozet (Ile de l'Est only [*Despin et al.*, 1972]), and Kerguelen islands, but large numbers of birds seen at sea off Tierra del Fuego suggest that Staten Island and other islands near Cape Horn may also support breeding populations. It might also breed on South Georgia, where remains have been found in skua nests. See map for pelagic range. The at-sea range shown for the Indian Ocean population is mostly hypothetical, but the Pacific Ocean range is based on observations, photographs, and specimens [*Harper*, 1972]. Stragglers occur in South Africa, western Australia, and New Zealand during the winter months.

ANTARCTIC PRION *Pachyptila desolata*

Resident

Identification. Plate 7. 11.5 in. (29 cm)/24 in. (61 cm). Slightly larger and darker than the narrow-billed prion but with a darker more obscure facial pattern and a larger more sculptured bill. The prominent subocular eye stripe is sooty black, the white eyebrow is reduced, and the lores are freckled with black. In flight a gray patch shows conspicuously at either side of the upper breast. The open M on the wings and the black tip on the tail are noticeably darker and a little broader than they are in the narrow-billed prion. The bill is twice as long as it is wide at the base, and its sides are essentially straight. The palatal lamellae, although they are more developed than they are in the narrow-billed prion, are still not visible laterally. See the broad-billed prion and blue petrel.

The juvenile has a weaker and narrower bill than the adult and may be confused with the narrow-billed prion.

The chick in first down is smoky gray, slightly lighter on the belly; the face and throat are practically bare. The second down is a little bluer above and almost white on the belly.

Flight and habits. Similar in flight to the narrow-billed prion but with slightly slower and more purposeful wing action, especially in calm weather. The head is tucked into the shoulders, the neck thus appearing short and thickset. The wings are held forward, so that the length of the tail is accentuated. Highly gregarious, flocks of thousands being common.

Voice and display. A throaty cooing heard in flight and when the bird is in its burrow.

Food. Euphausiids and other crustaceans, pteropods, small cephalopods, and polychaete worms. The Antarctic prion feeds during the day by hydroplaning. The bird rests its breast on the water and with its wings outstretched runs along the surface keeping its

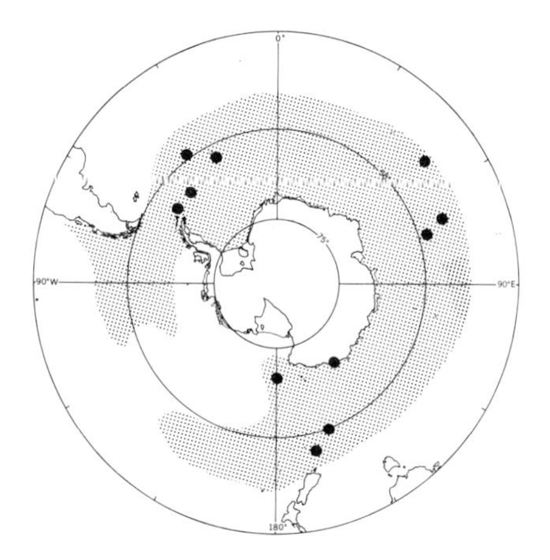

Antarctic prion

bill, or even its whole head, submerged to scoop in the plankton-rich water. Excess water is strained out through the bill lamellae. This species also takes larger individual prey from the surface in flight or while it is swimming and occasionally makes shallow dives.

Reproduction. Colonial, nesting on exposed rock faces of cliffs, in cavities under boulders, or in fairly short twisting burrows under grass tussocks or *Azorella* cushions on steep slopes. Several pairs may occupy *Azorella*-lined nest chambers in the same burrow system.

Arrival: Late October to early November. Experienced birds are first to arrive. Females leave to feed at sea 1-2 weeks before laying.

Eggs: Throughout December. Clutch, 1 white egg; averages 47 × 35 mm.

Hatching: Late January to mid-February. Incubation period about 45 days. Both sexes incubate, the male taking the first shift. Brood period 3-5 days.

Fledging and departure: March, 45-55 days after hatching. Departure begins in mid-March, adults and immatures leaving about the same time.

Remarks: Two forms may breed on Iles Kerguelen. Slightly smaller billed individuals were found on North Island, Iles Kerguelen, in November 1929 breeding somewhat earlier than larger-billed birds elsewhere in the archipelago [*Falla*, 1937]. The situation has not yet been studied fully.

Molt. Takes place at sea following departure from breeding grounds.

Predation and mortality. Skuas capture adult prions about the breeding grounds, particularly early in the season, and skuas and gulls may dig out burrows for eggs and young. Rats kill chicks in the nest. Cats prey on prions on Macquarie Island.

Ectoparasites. Ticks, *Ixodes uriae* and *I. pterodromae;* feather lice, *Naubates prioni, Halipeurus turtur, Saemundssonia desolata,* and *Longimenopon galeatum;* fleas, *Parapsyllus cardinis, P. magellanicus,* and *Notiopsylla kerguelensis.*

Habitat. Highly pelagic in antarctic and subantarctic waters.

Distribution. Breeds on South Georgia, South Sandwich, South Orkney, South Shetland (Elephant Island [*Furse and Bruce,* 1971]), Crozet (Ile de l'Est only [*Despin et al.,* 1972]), Kerguelen (see the section on reproduction), Heard, Macquarie, Auckland, and Scott [*Harper,* 1972] islands and on the continent at Cape Denison, King George V Coast (population possibly extirpated). The Antarctic prion has been recorded ashore on Marion Island, but its status is unknown. See map for pelagic range. The species is apparently absent from part of the Pacific Ocean. Storm-driven birds occur on Australian and New Zealand beaches in winter (not shown on map).

BROAD-BILLED PRION *Pachyptila vittata*
(includes lesser broad-billed or Salvin's prion) (*P. salvini*)

Resident

Identification. Plate 7. 11.5-12 in. (29-30 cm)/25-26 in. (64-66 cm). A large richly pigmented prion very similar in plumage characters to the Antarctic prion but with a much more flattened and broader bill. The culmen is less than twice as long as it is broad at the base and is perceptibly bowed out at the sides. The well-developed palatal lamellae are visible laterally when the bill is closed. The bill is wholly blue in *P. v. salvini,* which breeds on Marion Island and Iles Crozet. In the larger *P. v. vittata* (folded wing length 192-214 mm versus the 182- to 199-mm folded wing length of *P. v. salvini*), which breeds on Tristan da Cunha group, Gough Island, Ile Amsterdam, Ile Saint-Paul, and the New Zealand sub-Antarctic Islands, the bill is steel gray and much deeper and broader, being almost as broad as it is long. The prominent lamellae are lemon yellow; the distensible throat pouch is mauve. The head appears grotesquely large, and the overall color is darker than that of all other prions. See the blue petrel.

The juvenile has a weaker and narrower bill than the adult.

The downy chick is similar to that of the Antarctic prion.

Flight and habits. Not as agile in flight as the Antarctic prion and the narrow-billed prion and weaves less. Gregarious.

Voice and display. No information for *P. v. salvini,* but the courtship call of *P. v. vittata* is 'rerky-rickik-kikkik.'

Food. Mainly cephalopods and pteropods and probably some small crustaceans, but more information is needed. Feeding is largely by hydroplaning as it is for the Antarctic prion.

Reproduction. Colonial. Vast numbers burrow in a variety of coastal sites including *Azorella*-covered slopes, flat lava fields, and caves, i.e., wherever suitable cover and soil are available. Two or more pairs may occupy the same burrow. On Marion Island this species occupies inland sites, whereas fairy prions breed along the coast. On Iles Crozet, nests are above the upper limits of vegetation, generally higher than 185 m. More information is needed on breeding cycles.

Arrival: Early July on the Tristan da Cunha group, September on Iles Crozet, and October on Marion Island.

Eggs: August on the Tristan da Cunha group, no dates on Iles Crozet and Marion Island, and said to lay in both September and November on Ile Saint-Paul (needs confirmation). Clutch, 1 white egg; averages 50.3 × 35.9 mm and 32 g (Marion Island) or 50.5 × 36.7 mm (Ile Saint-Paul).

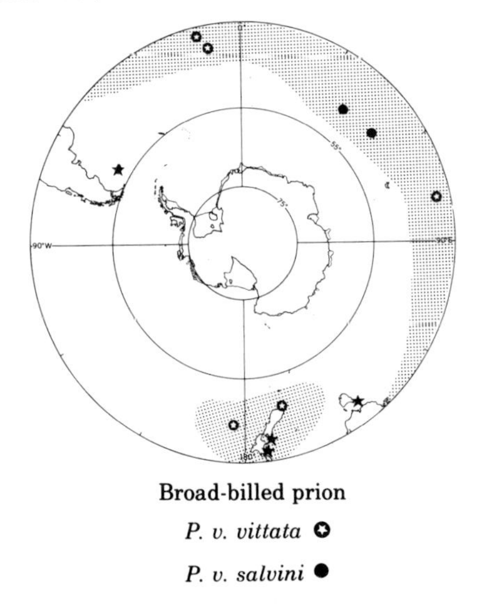

Broad-billed prion

P. v. vittata ✪

P. v. salvini ●

Hatching: Early January on Iles Crozet and no information elsewhere. Incubation period about 56 days (needs confirmation).

Fledging and departure: December on the Tristan da Cunha group and March on Marion Island, at 50-60 days after hatching.

Molt. No information is available for *P. v. salvini.* Molting *P. v. vittata* are found under rocks on the Tristan da Cunha group in February and March.

Predation and mortality. Skuas and gulls dig out burrows and attack chicks. Giant fulmars take adults, which nevertheless may visit burrows during the day. Rats have severely reduced the *P. v. vittata* populations on Ile Amsterdam and Ile Saint-Paul. Immature birds from the Indian Ocean are wrecked on New Zealand beaches during storms from May to September.

Ectoparasites. Feather lice (on *P. v. vittata* only), *Naubates prioni* and *Bedfordiella simsi.*

Habitat. Pelagic in the sub-Antarctic zone.

Distribution. *P. v. salvini* breeds on Prince Edward, Marion, and Crozet islands and occurs off South Africa, Australia, and New Zealand (especially 'wrecks' of young birds) after the breeding season from May to August, but the pelagic range is otherwise not documented. See map. *P. v. vittata* breeds on Tristan da Cunha group, Gough Island, Ile Amsterdam (probably extirpated), Ile Saint-Paul, Chatham and Snares islands, and islets off Stewart Island. It ranges throughout the southern Atlantic Ocean and about New Zealand. The Ile Saint-Paul population (formerly called *P. v. macgillivrayi* but probably not distinct according to P. C. Harper and F. C. Kinsky (personal communication, August 1974)) is probably very small if indeed it still nests successfully at all on the main island. The offshore stacks should be searched. *P. v. vittata* found in western Australia may be from Ile Saint-Paul.

FULMAR PRION *Pachyptila crassirostris*

Resident

Identification. Plate 7. 9.5 in. (24 cm)/20 in. (51 cm). Differs from the narrow-billed, Antarctic, and broad-billed prions in being smaller and in having richer, paler, and bluer upperparts and more black on the wings and tip of the tail. The head is smaller in relation to body size than it is in those larger species and has an indistinct facial pattern and shorter bill. The white eyebrow and gray eye stripe are weakly developed. The M on the wings is broad and very black. The tail feathers are black for nearly one-half their length, and the terminal bar extends fully across the tail rather than merely on the tips of the central feathers as it does in the narrow-billed, Antarctic, and broad-billed prions (Figure 8). The bill is stout with a bulbous nail that is barely separated from the nasal tubes. The palatal lamellae are only suggested by weak ridges.

The juvenile has a weaker and narrower bill than the adult, and its plumage is fresher immediately after fledging.

The downy chick is a sooty gray, darker than the chick of the Antarctic prion.

Flight and habits. The most erratic of the prions with a unique 'loop-the-loop' maneuver. Gregarious.

Voice and display. Harsh cooing and 'gurgling' at the breeding grounds.

Food. Small crustaceans, pteropods, some squid, and rarely fish. The prey may vary seasonally. The fulmar prion feeds like the narrow-billed and fairy prions by picking rather than by hydroplaning as the broader-billed species do.

Reproduction. Colonial, breeding in cracks and crevices in lava cliffs and scree slopes. The nest is merely a few small stone chips.

Arrival: Present all year.

Eggs: November. Clutch, 1 white egg; averages 44.7 × 33.1 mm.

Hatching: Late December. Incubation period unknown.

Fledging and departure: Mid-February. Fulmar prions remain near the islands all year.

Molt. No information is available.

Predation and mortality. No information is available.

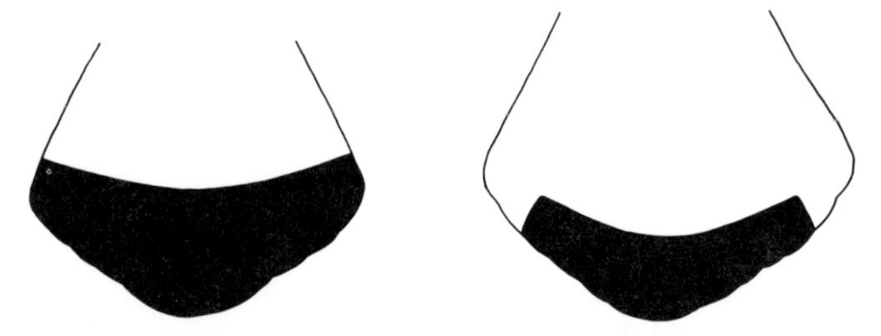

Fig. 8. Tails of prions showing that black is more extensive in fulmar and fairy prions (left) than in narrow-billed, Antarctic, and broad-billed prions (right).

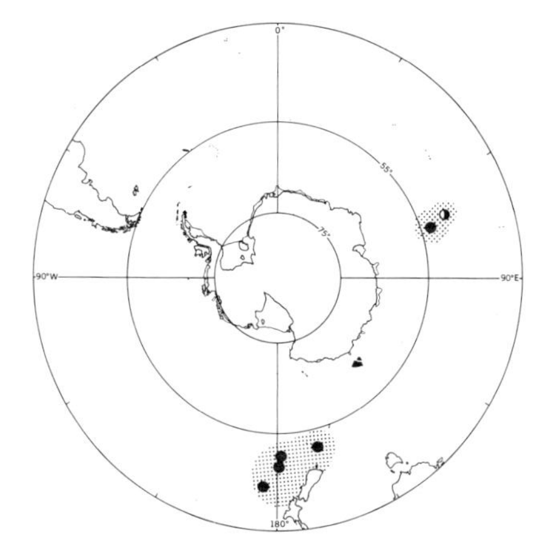

Fulmar prion

Ectoparasites. Tick, *Ixodes auritulus.*

Habitat. Sub-Antarctic and low Antarctic zone, probably remaining near breeding island.

Distribution. Breeds on Heard, possibly Kerguelen, Antipodes, Auckland, Bounty, and Chatham islands; pelagic range is not documented, but vagrant specimens have been recorded from Australia and South Africa. See map. The specific identification of the prion of this small-billed group breeding on Beauchêne Island in the southern Falkland Islands is not yet confirmed. It was originally reported as *P. turtur* [*Strange*, 1968] but may belong to *P. crassirostris.*

Although a specimen of fulmar prion was collected on Iles Kerguelen in 1840 and eventually became the type of *P. c. eatoni* (Mathews), the individual was probably a vagrant from Heard Island, where the subspecies is known to breed [*Falla*, 1937]. There are no other confirmed records of fulmar prions from Iles Kerguelen [*Derenne et al.*, 1974].

FAIRY PRION *Pachyptila turtur*

Resident

Identification. Plate 7. 9 in. (23 cm)/19 in. (48 cm). Very similar to the fulmar prion but slightly smaller with a smaller and narrower bill. The upper surface of the nail forms an angular ridge rather than a bulb like that of the fulmar prion.

The juvenile has a weaker and narrower bill than the adult. The downy chick is clear grayish blue.

Flight and habits. Like the flight and habits of the narrow-billed prion.

Voice and display. Guttural cooing in the burrows.

Food. Small squid and crustaceans,feeding during the day in the same manner as the narrow-billed prion.

Reproduction. Adults have been taken on Marion Island in May and on Macquarie Island in September and October; two nearly fully fledged chicks were found on Marion Island in late February. The presence of soil on the feathers of two Macquarie Island specimens and their enlarged gonads also strongly suggest breeding there. Elsewhere the breeding season extends from September to January. Nests are in burrows and rock crevices along the coast.

Molt. Completed in June and July.

Predation and mortality. No information is available on Macquarie Island, but elsewhere there is evidence of skua and gull predation.

Ectoparasites. Tick, *Ixodes auritulus;* feather lice, *Austromenopon stammeri, Halipeurus turtur,* and *Naubates prioni.*

Habitat. Subtropical and subantarctic waters.

Distribution. In the Antarctic the fairy prion breeds on Marion Is-

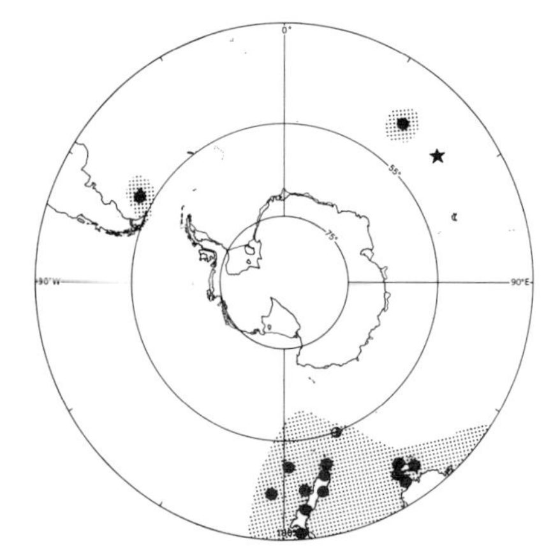

Fairy prion

land [*van Zinderen Bakker*, 1971*a*], possibly on Ile de l'Est in Iles Crozet [*Despin et al.*, 1972], and probably on Macquarie Island [*Keith and Hines*, 1958]. Farther north, either this species or the fulmar prion breeds on the Falkland Islands (Beauchêne Island [*Strange*, 1968]). The fairy prion breeds on islands off Tasmania and New Zealand (including Chatham, Snares, and Antipodes islands). Banding recoveries show that the New Zealand birds range widely. Pelagic range in the Indian Ocean is not documented, but this species has also occurred on the South African coast. See map.

GADFLY PETRELS

Gadfly petrels are small to medium-sized highly pelagic birds superficially resembling shearwaters. The wings are shorter in relation to body length than those of fulmars and shearwaters and broader at the base and more pointed at the tip, the tail is long and wedge shaped, the strongly hooked black bill is short and stout with prominent nostrils in which the nasal openings are directed forward, and the tarsi are rounded. They flap and glide like shearwaters but are more erratic, frequently rising in high swooping arcs above the horizon. They hold their wings noticeably bent at the wrist rather than extended stiffly like those of shearwaters. Gadfly petrels feed largely on squid, but in antarctic waters, crustaceans are probably also important food items. They seldom alight on the water, do not dive, and tend to avoid ships. They nest in burrows excavated in soft ground. Seven gadfly petrels breed on antarctic and subantarctic islands, and all but two range exclusively south of the equator. An eighth species has occurred as a vagrant off the Tristan da Cunha group.

BLUE PETREL *Halobaena caerulea*

Resident

Identification. Plate 7. 12 in. (30 cm)/26 in. (66 cm). A small dark-crowned gray petrel with white underwings and underparts. It resembles a prion except for its conspicuous white-tipped square tail, mottled white forehead, and narrow black bill. The white-tipped tail is unique among petrels. Dark patches at either side of the breast form an incomplete collar. The general color of the upperparts is gray, not blue as it is in the prions, the open M on the upper surface of the wings being indistinct. The feet are pale lilac blue with flesh-colored webs.

The juvenile is slightly brownish and has an ashy forehead and weaker bill than the adult. The feet are pale lilac with dark stains on the outer toe.

The downy chick is slate gray.

Flight and habits. Not as agile or erratic in flight as the prions and glides more, very low over the water. The blue petrel follows ships (unlike prions) and accompanies whales. It swims and dives.

Voice and display. In and around the burrow a pigeonlike cooing that is apparently indistinguishable from that of the Antarctic prion. The blue petrel calls most intensely from September until eggs are laid. Another supposedly distinct flight call needs confirmation.

Food. Mainly euphausiid shrimp and small cephalopods and occasionally small fish.

Reproduction. Colonial, burrowing in soft dry soil under *Acaena* and *Azorella* clumps on coastal slopes. The excavation is between 1 and 2 m long and terminates in a nest chamber lined with plant fibers and leaves. Burrows are frequently associated with those of prions and diving petrels.

Arrival: Early September.

Eggs: Late October to mid-November. Clutch, 1 dull white egg; averages 50 × 36 mm.

Hatching: Late December to early January. Both sexes incubate. Incubation period about 46 days.

Fledging and departure: Late February to March. Departure takes place in late March, at slightly less than 2 months after hatching. Some adults return to the breeding grounds in April to June and occupy burrows without breeding.

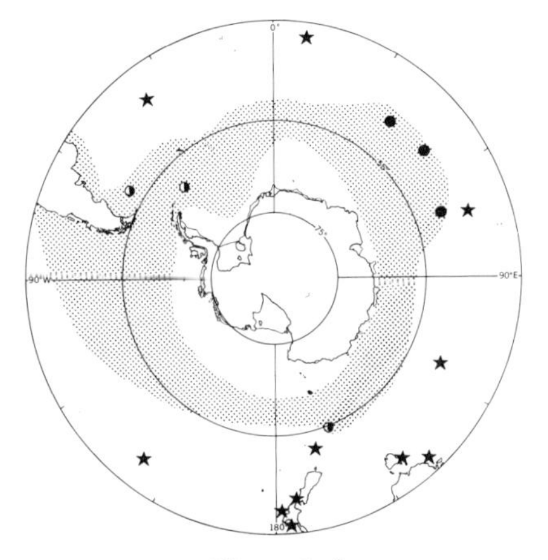

Blue petrel

Molt. Begins in December.

Predation and mortality. Nocturnal on breeding grounds probably as an adaptation against skua predation, which may be severe on Macquarie Island.

Ectoparasites. Feather louse, *Naubates clypeata;* fleas, *Parapsyllus magellanicus* and *Notiopsylla kerguelensis.*

Habitat. Cold subantarctic and antarctic waters, most commonly from +2° to −2°C.

Distribution. At present the blue petrel is known to breed on Prince Edward, Marion, Crozet (Ile de l'Est only), and Kerguelen islands and possibly on Macquarie Island [*Keith and Hines,* 1958; *Merilees,* 1971*a*]. One unconfirmed breeding record for the South Orkney Islands [*Ardley,* 1936] is very doubtful. Although there is no recent information from the Falkland Islands to verify earlier reports of breeding there [*Bennett,* 1926, p. 316], the species probably does breed somewhere in the South Atlantic Ocean and Scotia Sea area.

GREAT-WINGED PETREL *Pterodroma macroptera*

Resident

Identification. Plate 5. 16.5 in. (42 cm)/38 in. (97 cm). A large long-winged brownish black gadfly petrel with a stout strongly hooked black bill and black feet. See the Kerguelen and white-chinned petrels and the sooty shearwater.

The juveniles in Atlantic and Indian ocean populations (*P. m. macroptera*) generally show some pale grayish feathers on the face and chin, but since this character varies geographically in adults (being prominent in the large subspecies *P. m. gouldi* in the western Pacific subtropics and less pronounced or absent in the Atlantic and Indian oceans), it is only of limited use for aging birds in the Atlantic Ocean and near Iles Kerguelen.

The downy chick is sooty gray or brown (more information needed).

Flight and habits. Powerful and swift in the air following the usual gadfly pattern of wheeling in broad arcs. The wings appear particularly long and narrow and are usually somewhat bent in flight.

Voice and display. Generally silent at sea but very noisy at night on breeding grounds. Courtship flights at dusk resemble a 'screaming party of huge swifts.' The aerial calls are a sweet liquid whistle and a squeaky 'kik, kik, kik' occasionally interspersed with a gruff slurred 'quaw-er.' The last call is also commonly uttered on the ground, as is a clear 'eee-aw' courtship call, very like the bray of the donkey, while birds fence with bills, nibble at head, neck, and wings, and mutually preen. When incubating birds and older chicks are alarmed, they break into a repeated sibilant squeaky 'si-si-si . . . .'

Food. Largely pelagic cephalopods and occasionally fish.

Reproduction. Semicolonially or sometimes singly in isolated burrows on oceanic islands or headlands up to 600 m above sea level. The nest sites are either on bare hillsides or on vegetated flat expanses. The burrow varies in length but is about 30-60 cm deep in soft soil with a protective covering of *Azorella* or *Acaena* at the entrance. Farther north in Western Australia they may nest on the surface in rock niches, in tree roots, or under scrub. The nest itself is a bulky structure composed of dry leaves and twigs. Although there are few precise breeding data from either Atlantic or Indian ocean islands, the eggs are presumably laid in midwinter.

Arrival: March to April.

Eggs: Late May to July. Clutch, 1 white egg; averages 65.5 × 50 mm (Indian Ocean) or 69 × 51 mm (Tristan da Cunha group).

Hatching: Late July to August. Incubation period about 53 days; brood period 2-3 days. Both parents incubate eggs and tend chicks.

Fledging and departure: Late November to early December, 128-134 days after hatching.

Molt. Adults are in full molt in April.

Predation and mortality. Skuas, cats, and rats take eggs and chicks.

Ectoparasites. Feather lice, *Naubates clypeatus, N. pterodromi, N.*

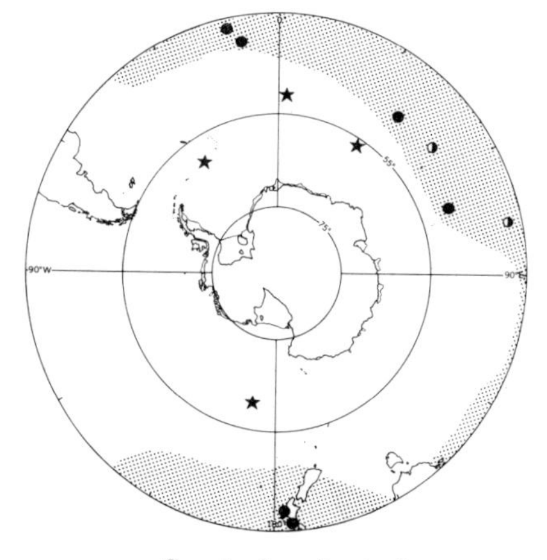

Great-winged petrel

heteroproctus, Halipeurus procellariae, and *Trabeculus schillingi;* flea, *Parapsyllus magellanicus.*

Habitat. Generally occurs well out to sea, where it is largely confined to the subtropical surface water zone, especially in the New Zealand sector. Occasionally, the great-winged petrel strays into the sub-Antarctic and the Antarctic zones during the summer, particularly near Iles Kerguelen.

Distribution. In the Antarctic the great-winged petrel breeds on Marion, Crozet (needs confirmation), and Kerguelen islands. Farther north it breeds on the Tristan da Cunha group, Gough Island, Ile Amsterdam (needs confirmation), and islands off Western Australia and in northern New Zealand. See map for pelagic range; the southernmost vagrant records are highly questionable and might be Kerguelen petrels.

WHITE-HEADED PETREL *Pterodroma lessoni*

Resident

Identification. Plate 5. 18 in. (46 cm) /42 in. (107 cm). A large white-bodied and white-headed gadfly petrel with dark gray underwings and a distinctive black patch around each eye. The upperparts are gray, palest on the neck, mantle, and tail and darker on the lower back and wings. The bill is black; the feet are flesh, brownish on the outer toes and webs. See the southern fulmar and soft-plumaged petrel.

No characters are known for identifying first-year birds.

The chick in first down is uniformly gray with scattered white tufts on the throat. The second down is gray on the back and dull white on the underparts.

Flight and habits. Characterized by 'swift flight and sharp arching of its wings which give it an "M"-like appearance when the back is presented.' The white-headed petrel is usually solitary but may occasionally be seen in small groups. It does not usually follow ships.

Voice and display. Calls are very similar to those of the great-winged petrel. Adults in the burrow emit a soft squeaky 'si-si-si.' The in-flight call given during high-speed aerial chases is a louder more throaty 'wi-wi-wi' variation of the burrow call. This is interspersed with long slurred calls 'oooo-er' and 'oooo-er, kukoowik, kukouwick.' When adults are attacked, they have a loud screaming alarm call. Silent at sea.

Food. Largely squid and, to a lesser extent, fish and crustaceans.

Reproduction. Colonial, on sloping ground at low elevations on Iles Kerguelen and on the open plateau on Macquarie Island. The 1- to 3-m dry burrow is excavated in soft soil under *Azorella* cushions or in *Acaena*-covered rock screes. The mouth of the burrow is frequently revealed by feathers and twigs littering the adjacent ground. The nest consists of a spare to substantial lining of leaves and grasses.

Arrival: Late August to October.

Eggs: Late November to early January. Clutch, 1 ovoid white egg; 69-74 × 49-52.2 mm. The larger measurements are from Iles Kerguelen: the smaller from Macquarie Island.

Hatching: Late January to early February. Incubation period 8-9 weeks; brood period 2-3 days. Both parents incubate the eggs and feed the young.

Fledging and departure: April to May, when the chick is about 15 weeks old.

Molt. Breeding birds replace body plumage before and during incubation. Wing molt takes place at sea after breeding and is completed in September. First- and second-year nonbreeding birds molt wings from October to March.

Predation and mortality. On Macquarie Island, feral cats are the major predators attacking nearly fledged young in the burrows. Wekas may take eggs and chicks from burrows. Rabbits compete for burrow sites. Skuas on both Macquarie Island and Iles Kerguelen attack and kill adults during the day.

Ectoparasites. Feather lice, *Trabeculus schillingi, Halipeurus pro-*

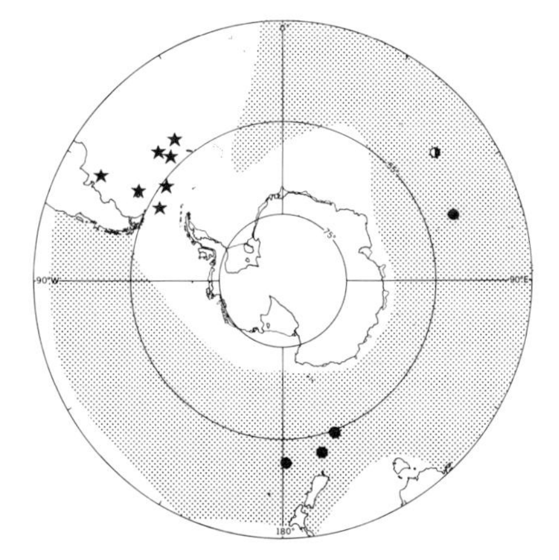

White-headed petrel

cellariae, and *Naubates heteroproctus;* fleas, *Notiopsylla enciari, Parapsyllus magellanicus,* and *P. cardinis.*

Habitat. This species ranges widely in subantarctic waters, and during the summer it can be seen regularly among the icebergs south of the Antarctic convergence. In winter it ventures into subtropical waters.

Distribution. Breeds on Kerguelen, Macquarie, Auckland, and Antipodes islands. See map for pelagic range, which in the South Atlantic Ocean is very poorly known. Its status on Ile de la Possession, Iles Crozet, is in question [*Despin et al.,* 1972].

ATLANTIC PETREL *Pterodroma incerta*

Resident

Identification. Plate 5. 17 in. (43 cm)/41 in. (104 cm). This species is essentially a brown dark-headed Atlantic Ocean representative of the white-headed petrel, with which it has been considered conspecific by some authors. The throat, upper breast, underwing, and undertail coverts are brown. The rest of the underparts are white or faintly washed with gray. In worn plumage the throat may also be white. The bill is black; the feet are flesh, brown on the outer toes and webs. See the gray and Antarctic petrels.

There are no known characters for identifying year-old birds. The downy chick is unknown.

Flight and habits. Similar to the flight and habits of the white-headed petrel, but the Atlantic petrel tends to follow ships.

Voice and display. The flight call is similar to that of the great-winged petrel but is lower pitched and more fluty.

Food. Little is known, but most probably squid is the main food item.

Reproduction. Colonial, burrowing in soft soil and vegetation on exposed ridges from 150 to 600 m high. Breeding takes place during winter.

Arrival: February to mid-March.

Eggs: Mid-June to mid-July. Clutch, 1 white egg; 69.5 × 52 mm (only 1 measurement).

Hatching: No information on dates or on incubation and brood periods.

Fledging and departure: Unknown.

Molt. Probably takes place before breeding.

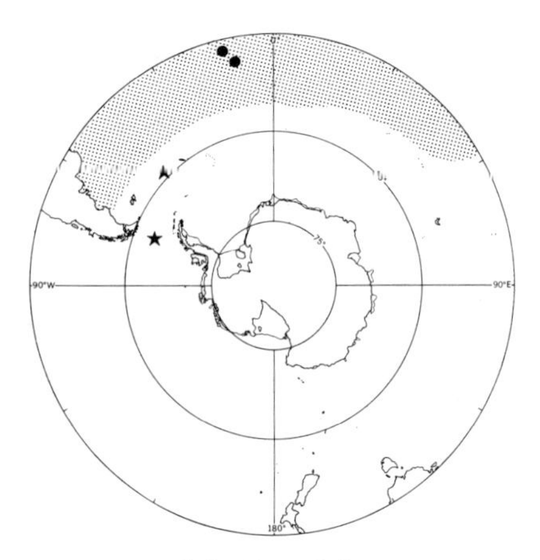

Atlantic petrel

Predation and mortality. No information is available.

Ectoparasites. No information is available.

Habitat. Warmer parts of subantarctic waters.

Distribution. Breeds on the Tristan da Cunha group and Gough Island. The Atlantic petrel ranges over the South Atlantic and western Indian oceans regularly to 50°S.

KERGUELEN PETREL *Pterodroma brevirostris*

Resident

Identification. Plate 6. 13.5 in. (34 cm)/27 in. (69 cm). A medium-sized uniformly gray gadfly petrel that has a glossy silvery sheen in fresh plumage. This sheen is particularly apparent on the undersides and shafts of the primaries and in good light is an excellent in-flight character. The coverts of the underwing and along the leading edge of the upperwing have narrow white edges. The black bill is short and compressed, and the head appears extraordinarily large. Dark tips on the throat feathers may abrade, so that the chin is left pale. The silvery underwing, light primary feather shafts, white-tipped wing coverts, and wholly black feet are useful characters for distinguishing this species from the very rare dark phase of the soft-plumaged petrel.

Unlike other gadfly petrels the immature has a thick bill that is said to become narrower and more compressed in the adult (needs confirmation).

The downy chick is said to be sooty brown.

Flight and habits. Beats its wings very rapidly in gaining altitude and is distinctive in 'floating' in a stiff wind high above the surface of the sea. In calm weather the Kerguelen petrel flies rapidly in a weaving batlike fashion low over the water. The primaries are generally narrower than those of either the soft-plumaged petrel or the mottled petrel and do not taper as abruptly toward the tips. Although this species is attracted to ships, it does not regularly follow in the wake. It is mostly solitary at sea, but small flocks may be seen resting on the water. It dives from the surface.

Voice and display. The high-pitched alarm call is a long drawn-out piercing, but hoarse, screech.

Food. Mainly cephalopods and, to a lesser extent, crustaceans.

Reproduction. Colonial, burrowing into soft, deep, and usually wet soil on a steep incline. The burrow is often branched and curving and may be 1.5-1.8 m long. The nest itself is a truncated cone composed of small twigs of *Acaena* and usually surrounded with water. A drainage channel leads water from around the nest down the sloping burrow. The following schedule is based on very few observations.

Arrival: Probably August, but adults are present all year at Iles Crozet and may be found in burrows in winter.

Eggs: Mid-September to early October. Clutch, 1 white egg; averages 59.5 × 46 mm (Iles Kerguelen) and 56.7 × 44.8 mm and 55.6 g (Marion Island).

Hatching: Late November. Incubation period 47-51 days; brood period 2 days. Both sexes incubate the eggs and feed the young.

Fledging and departure: January on Gough Island and Iles Crozet and February and March on Marion Island, after about 2 months in the nest.

Molt. No information is available.

Predation and mortality. Adults are preyed upon by skuas, and chicks are attacked in the burrows by rats on Iles Crozet, where mortality varies from 60 to 100%.

Ectoparasites. Tick, *Ixodes kerguelenensis;* feather louse, *Bedfordiella unica.*

Habitat. Highly pelagic in subantarctic and antarctic waters north of the pack ice.

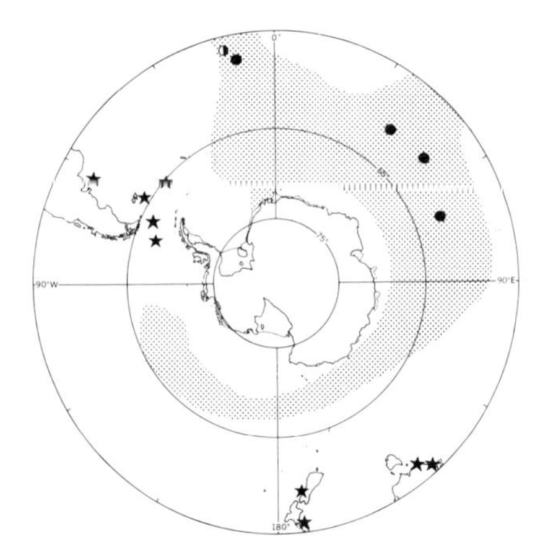

Kerguelen petrel

Distribution. Breeds on Marion, Crozet, Kerguelen, and Gough islands and possibly in the Tristan da Cunha group. Map of pelagic range is based on very few records.

SOFT-PLUMAGED PETREL *Pterodroma mollis*

Resident

Identification. Plate 6. 14 in. (36 cm)/33 in. (84 cm). A medium-sized gray and white gadfly petrel with distinctive dark underwings and a white face. The upper surface from the crown to the tail is gray with darker markings about the eyes. Dark wing coverts produce an open M on the spread wings. In most individuals, dark patches at the sides of the breast form an incomplete band, but the amount of gray elsewhere on the underparts varies. In heavily marked birds the flanks and belly may be flecked with gray; rare individuals have all-gray underparts. This dark-phase '*P. deceptornis*' may be confused with the Kerguelen petrel (or may even be the result of hybridization between the two species), but the soft-plumaged petrel has very dark underwings without any silvery sheen on the relatively broad primaries or their shafts and with no white edges to the upperwing coverts. The bill is black; the feet are flesh with black toes (needs confirmation; there may be variation).

No characters are known for differentiating between first-year birds and adults.

The chick in first down is dark gray above and somewhat lighter on the chin and underparts. The second down is a slightly lighter gray.

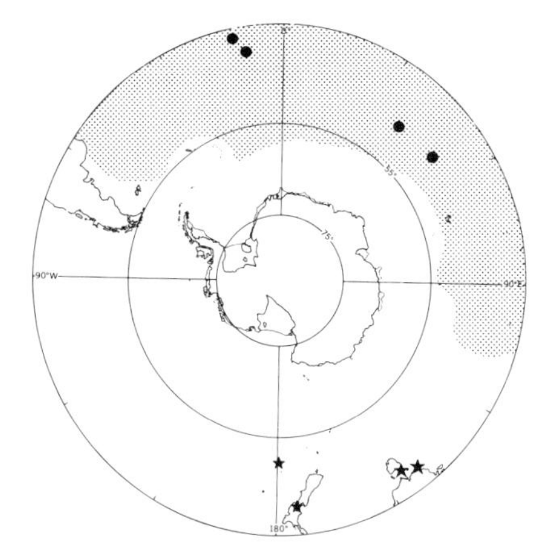

Soft-plumaged petrel

Flight and habits. Fast and impetuous with rapid wingbeats and much gliding. The soft-plumaged petrel swings in great arcs high above the surface of the water, but not as high as the Kerguelen petrel, and does not float in the air. It is generally shy and solitary.

Voice and display. On the breeding grounds the flight call is a flutelike 'uuuuuuu-hi' also described as a shrill whistle 'treee-pi-peee.' Moaning cries are also given in aerial chases. Silent at sea.

Food. Cephalopods and fish.

Reproduction. Colonial, breeding in steep slopes along the perimeters of oceanic islands. The relatively long burrow may have two nest chambers, each occupied by a pair of petrels. Dry grass and *Acaena* stems line the nest. The following schedule is based on scanty information from the Tristan de Cunha group and Gough Island and from four nests on Marion Island:

Arrival: As early as the beginning of August, but most birds probably return and begin courting in the burrows in September.

Eggs: Laying begins in early November and continues to late December. Clutch, 1 white egg; averages 57.9 × 42.5 mm and 52.5 g.

Hatching: Date of hatching and incubation period unknown. Downy chicks are present from late January to mid-April.

Fledging and departure: Fledging takes place in May, and non-breeding birds are still present in late May.

Remarks: The schedule on Marion Island is therefore probably 6-8 weeks later than that of the Kerguelen petrel.

Molt. No information is available for the Marion Island population.

Predation and mortality. No information is available.

Ectoparasites. Tick, *Ixodes kerguelenensis;* feather lice, *Trabeculus schillingi, Halipeurus procellariae,* and *Naubates pterodromi.*

Habitat. Mostly subtropical and subantarctic waters but occasionally entering the Antarctic zone during the summer.

Distribution. Breeds on Tristan da Cunha group, Gough, Marion, Crozet (Ile de la Possession only), and probably Antipodes islands. Although the soft-plumaged petrel is common in the seas about Ile Amsterdam, Ile Saint-Paul, and Iles Kerguelen, there is still no evidence of breeding there. See map for pelagic range in the southern hemisphere. North Atlantic Ocean populations breed on Madeira Islands (extinct?), Ilhas Desertas, and Cape Verde Islands but do not cross the equator.

MOTTLED PETREL *Pterodroma inexpectata*

Nonbreeding Migrant

Identification. Plate 6. 13.5 in. (34 cm)/27 in. (69 cm). A medium-sized gray-backed gadfly petrel with a dark gray abdomen and a prominent black bar diagonally across the ulnar portion of the white underwing. The gray upperparts, which become brownish with wear, are darkest on the crown and leading edge of the wings, the latter forming an open M in flight. The forehead, throat, and undertail coverts are white, variably mottled with gray. In fresh plumage the back is lightly scaled with white, and white mottling is visible on the underparts. The bill is black; the feet are flesh with black toes. See the soft-plumaged petrel.

No characters are known for distinguishing first-year birds.

Flight and habits. Very rapid flight in great arcs. The wings are held slightly bent at the wrists and the tail appears shorter and more rounded when it is fanned than the tails of either the Kerguelen petrel or the soft-plumaged petrel. It is not given to following ships.

Voice and display. Silent at sea.

Food. Small squid and fish.

Molt. Adults probably molt in the northern hemisphere during the boreal summer, but prebreeders may molt in New Zealand waters in February.

Predation and mortality. Mortality of birds at sea is probably low, but man and introduced mammals have extirpated the mottled petrel from all but the most remote of its former breeding stations.

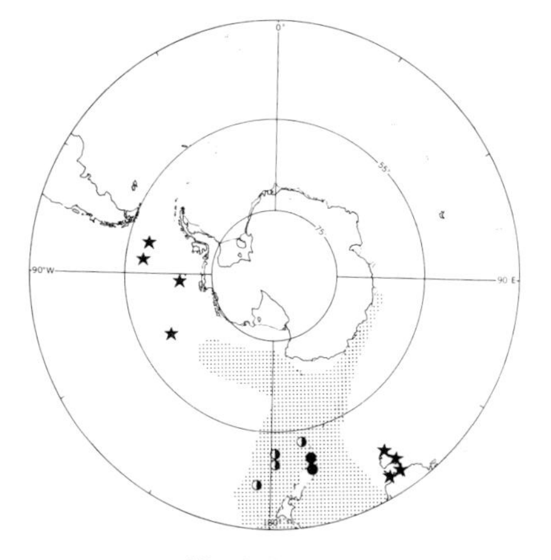

Mottled petrel

Habitat. Breeding birds probably frequent temperate and cold sub-Antarctic waters near the colonies, but in summers, presumed non-breeders or postbreeders are found in iceberg-strewn waters and pack ice near the Antarctic continent. Migrants occur in climatically similar areas in the northern hemisphere.

Distribution. At present the mottled petrel is only known to breed on islands in the Foveaux Strait and off Stewart Island and on the Snares Islands from November to April. Previously, it was more widespread in North Island and South Island, New Zealand, and on the Bounty, Chatham, Antipodes, and Auckland islands. Although at least part of the population migrates to the northern Pacific Ocean as far as Alaskan waters, significant numbers appear in the southernmost western Pacific and eastern Indian oceans after breeding, and they may spend the winter there.

GOULD'S PETREL *Pterodroma leucoptera*

Rejected Vagrant Record

When the U.S. Exploring Expedition 'ship *Peacock* was enveloped in a fog, latitude 68°S., longitude 95°W. of Greenwich' near Peter I Island on March 21, 1839, *Peale* [1848, pp. 294, 299] reported collecting two species of gadfly petrels 'amidst icebergs, buffeting the storms and fogs of the Antarctic Regions.' These included *Pterodroma inexpectata* (type of *Procellaria gularis* Peale, reported as *P. mollis* Gould by *Cassin* [1858, p. 410]) and *P. leucoptera brevipes* (type of *Pro-*

cellaria brevipes Peale, reported as *P. cookii* Gray by *Cassin* [1858, p. 414]). The far southern locality is not unreasonable for the first species, especially at a time when mottled petrels were far more abundant than they are at present, but it is undoubtedly an error for the Pacific Ocean subspecies of Gould's petrel.

There are no other antarctic records of this subtropical species, and the only eastern Pacific Ocean specimens are from near the Galapagos Islands. Peale's antarctic record can best be attributed to a mixup in labeling that occurred subsequently in Washington [*Peale*, 1848, p. 295; *Murphy*, 1936, p. 722].

COOK'S PETREL *Pterodroma cookii*

Rejected Vagrant Record

An old undated specimen of Cook's petrel (*Pterodroma cookii*) labeled only 'Macquarie Islands' is in the Rothschild collection in the American Museum of Natural History. The specimen has not been reported previously. The label is in H. H. Travers' handwriting but bears no date. Ornithologists visited Macquarie Island so infrequently in the past to collect specimens that it seems unreasonable to suppose that someone would have been unaware of the uniqueness of a Macquarie Island specimen of this otherwise subtropical New Zealand species. The southernmost breeding locality is at Codfish Island off Stewart Island. There are no more southerly records either on land or at sea. Traver's locality data from New Zealand sub-Antarctic Islands are not always reliable (R. A. Falla, personal communication, August 1974).

Juan Fernandez petrel

JUAN FERNANDEZ PETREL *Pterodroma externa*

Vagrant

A large (17 in., 43 cm/38 in., 97 cm) gray-backed gadfly petrel with white forehead and underparts and a small dark patch on the underwing at the wrist. The upperwing shows a diagonal dark bar across the ulnar portion. See the greater shearwater. The Juan Fernandez petrel is a tropical Pacific species that occurs as far south as 52°S (P. C. Harper, unpublished data, 1966) and has been recorded once on the Tristan da Cunha group ('*P.e. tristani*' [*Matthews,* 1931]).

SHEARWATERS

Shearwaters are long-billed medium to large petrels with long narrow wings and short rounded tails. The two genera of shearwaters occurring regularly in the antarctic area may be distinguished on the basis of bill characters and size. The beak is very slim and dark with a reduced nasal tube in the smaller *Puffinus* species and is heavier and pale with more prominent nostrils in the large *Procellaria* species. In flight the smaller shearwaters alternate short bursts of rapid flapping with long stiff-winged banking glides low over the surface. The larger species fly in a more relaxed deliberate manner similar to that of the large fulmarine petrels. Shearwaters feed while they are swimming on the surface or making shallow dives, taking small fish, squid, crustaceans, and occasionally floating offal. Six burrowing species breed on low-latitude antarctic and subantarctic islands; another, which breeds farther north, occurs sparingly in southern waters. Four species regularly migrate to the northern hemisphere, and the other three range in cool currents to the equator.

WHITE-CHINNED PETREL *Procellaria aequinoctialis*

Resident

Identification. Plate 6. 20 in. (51 cm)/54 in. (137 cm). A large black petrel with a conspicuous greenish white bill and variable amounts of white feathering on the chin and throat. The plumage becomes quite brown when it is worn. The feet are black. Birds breeding on Inaccessible Island of the Tristan da Cunha group (*P. a. conspicillata*) have extensive white markings on the face and head. See the giant fulmars, great-winged and Kerguelen petrels, and sooty and flesh-footed shearwaters.

No characters are known for differentiating first-year birds.

In the first down the chick is uniformly dark sooty brown or black with traces of white on the chin; the second down is grayer.

Flight and habits. The flight resembles that of a small albatross; wingbeats are slow, deliberate, and graceful, quite unlike the stiff-winged flight of the giant fulmars and smaller sooty shearwater. The white-chinned petrel keeps close to the surface instead of flying high like a gadfly petrel. Feeding birds gather in small flocks on the water.

Voice and display. The courtship calls have been described as a clacking or distinctive deep croaking 'gronk gronk' and a shrill trilling in the burrows accompanying mutual preening. Silent at sea.

Food. Primarily cephalopods and also some fish, euphausiids, and other crustaceans. The white-chinned petrel feeds both on the surface and in shallow dives.

Reproduction. Colonial. Breeding localities are usually on sheltered, grassy, or *Azorella*-covered slopes, where burrows 1-3 m long are excavated in soft damp soil. The nest is built of tussock and pieces of *Acaena* on an earth platform usually surrounded by water.

Arrival: Mid-October.

Eggs: Mid-November to early January. Laying may begin in late October on Inaccessible Island. Clutch, 1 white egg; averages 83.5 × 54.6 mm and 120-130 g.

Hatching: February. Both parents incubate and feed the young. Incubation period about 60 days; brood period 1-3 days.

Fledging and departure: April, about 95 days after hatching. Adults leave in April; young in May. Adults on Inaccessible Island are present in burrows in May after young have left.

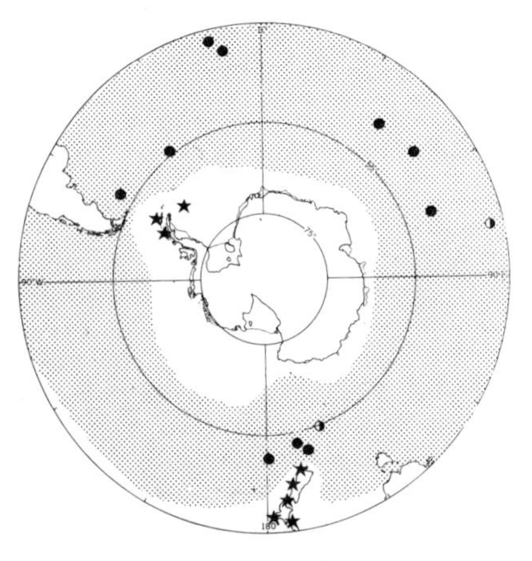

White-chinned petrel

Molt. January to May, more data needed.

Predation and mortality. Skuas attack adults on South Georgia, and rats kill smaller chicks on Iles Kerguelen. Inexperienced breeding adults tend to break or abandon eggs. Total mortality may range locally between 60 and 95% of eggs and chicks in active nests.

Ectoparasites. Feather lice, *Naubates fuliginosus* and *Trabeculus hexacon.*

Habitat. Widespread in pelagic and offshore waters of the sub-Antarctic, uncommon south of the Antarctic convergence.

Distribution. In the Antarctic the white-chinned petrel breeds on South Georgia, Prince Edward, Marion, Crozet, Kerguelen, and Macquarie (needs confirmation [*Warham,* 1969]) islands, and farther north it breeds on the Falkland, Tristan da Cunha group, Gough, Amsterdam (needs confirmation), Campbell, Antipodes, and Auckland islands. See map for pelagic range.

GRAY PETREL *Procellaria cinerea*

Resident and Breeding Migrant

Identification. Plate 5. 19 in. (48 cm)/50 in. (127 cm). A large gray shearwater with dark gray underwings and undertail coverts and white underparts. The upper surface is ash gray, much darker on the crown, tail, and wing tips. When the plumage is worn, it may appear quite brownish. The relatively long stout bill is greenish white on the sides and black on the nostrils, ridge, and tip. The legs are greenish flesh, black on the joints and outer toe and yellowish on the webs. See the greater shearwater and white-headed petrel.

No characters are known for distinguishing year-old birds.

The downy chick is medium gray, paler on the chin and breast.

Flight and habits. Similar to the flight and habits of the white-chinned petrel, but the wing flapping of the gray petrel appears ducklike. It frequently dives from the air and swims well submerged. It is usually seen in small flocks following ships for refuse or associating with whales.

Voice and display. A melodious 'aaargh-hooo-err-hoooer' and a breezy 'ped-i-unker.' Just after arrival at the breeding grounds, adults sit on tussock and bray continuously.

Food. Cephalopods and fish are captured by diving, and the refuse of ships and fish offal are eaten on the surface.

Reproduction. Colonial, a winter breeder nesting on sheltered

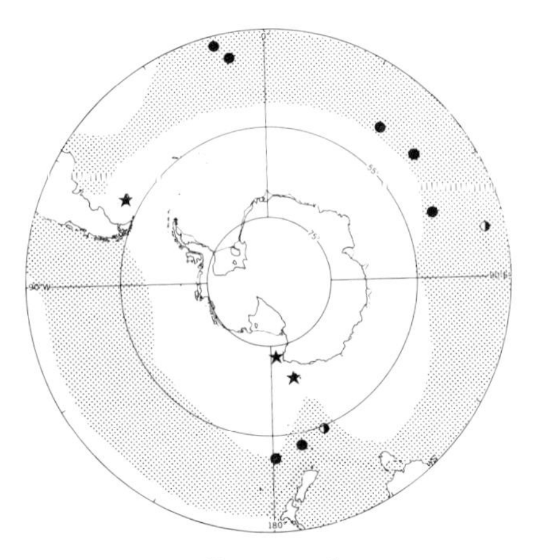

Gray petrel

Azorella- or *Acaena*-covered slopes or on the Tristan da Cunha group, between 300 and 600 m on the steepest seaward precipices. Burrows are variable in length, from 0.3 to 3.7 m, ending in a dry spacious nesting chamber that either is bare or has a tussock grass nest. Very little information is available on breeding schedule and habits owing to the winter schedule.

Arrival: Late February (needs confirmation).

Eggs: March to April. Fresh eggs have also been recorded in June and July. Clutch, 1 dull white egg; averages 83 × 56 mm and 110 g (only one weight).

Hatching: May. Incubation and brood periods unknown.

Fledging and departure: August to September, at an age of 80-90 days (Tristan Island) or about 100 days (Iles Crozet).

Molt. Primaries may be molted early in the breeding season.

Predation and mortality. Little information is available, the gray petrel being preyed upon by introduced cats and rats but apparently not by skuas.

Ectoparasites. Feather lice, *Trabeculus hexacon* and *Naubates fuliginosus;* flea, *Notiopsylla kerguelensis.*

Habitat. Pelagic, mostly in subantarctic waters.

Distribution. Breeds on Tristan da Cunha group, Gough, Marion, Crozet, Kerguelen, Campbell, and Antipodes islands and possibly on Ile Saint-Paul and Macquarie Island. See map for pelagic range.

Cory's shearwater

CORY'S SHEARWATER *Calonectris diomedea*

Questionable Vagrant

Very similar to the gray petrel in size (18 in., 46 cm/44 in., 112 cm) and general pattern, but the underwings are white, and the back is darker and decidedly brown. Some white feathers may be present at the base of the tail. The bill is straw colored. Cory's shearwater is a North Atlantic species that winters at sea off South Africa. Vagrants said to have been collected off Iles Kerguelen (types of *C. d. 'disputans'* [*Sharpe,* 1879]) may have actually been taken off South Africa [*Bourne,* 1955], but one has also been taken in New Zealand temperate waters [*Oliver,* 1934].

SOOTY SHEARWATER *Puffinus griseus*

Breeding Migrant

Identification. Plates 5 and 6. 18 in. (46 cm)/41 in. (104 cm). An all-dark shearwater with contrasting white underwing coverts and a slim black bill. In fresh plumage the upperparts are slaty black, and the underparts slaty gray, both becoming brownish with wear. The extent of white on the underwing is variable but is usually greatest in adults. The chin is grayish white in some birds. The feet are dark grayish brown, blue gray, or lilac. The more slender build and white flash in the underwing distinguish this species from the similar great-winged and Kerguelen petrels. See the short-tailed shearwater and dark-phase jaegers.

Immatures have a weaker bill and generally a darker underwing than adults.

In first down the chick is dark gray with paler underparts; the longer second down has a brownish cast.

Flight and habits. The distinctive stiff-winged rising and falling flight is both fast and graceful. The sooty shearwater flaps its wings rapidly to gather more height before banking steeply down toward

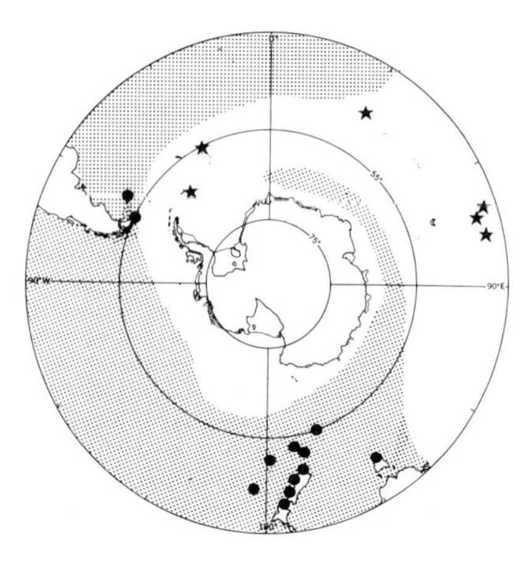

Sooty shearwater

the sea. It dives readily to procure food or escape predators. Gregarious, sometimes in huge flocks, especially on migration. It does not follow ships.

Voice and display. On the breeding grounds the birds either in flight or on the ground emit raucous screaming composed of inspirational 'oo's' and expirational 'ahh's' that gradually increase in tempo, the call usually ending in a howl or cough. Silent at sea.

Food. Squid, crustaceans, and small fish.

Reproduction. Colonial, breeding on offshore islands and exposed headlands. The burrows, which are usually hidden by vegetation in sloping terrain, are from 0.3 to 2 m long and terminate in an enlarged chamber containing a crude nest of leaves and small twigs.

Arrival: Late September.

Eggs: Late November. Clutch, 1 white egg; averages 77.4 × 48.3 mm.

Hatching: Early January to early February. Incubation period unknown; brood period 2-3 days.

Fledging and departure: Mid-March to late April, after about 97 days in the nest and shortly after the adults.

Molt. Complete molt from February to October or November, mostly in the northern hemisphere.

Predation and mortality. No information is available on Macquarie Island, elsewhere rats are predators.

Ectoparasites. Ticks, *Ixodes uriae, I. kerguelenensis,* and *I. auritulus;* feather mite, *Brephosceles puffini;* feather lice, *Trabeculus hexacon, Halipeurus diversus,* and *Austromenopon paululum;* flea, *Notiopsylla kerguelenis.*

Habitat. Generally colder temperate waters but dispersing into antarctic waters in summer.

Distribution. The southernmost breeding station is Macquarie Island, where the sooty shearwater is not common, but farther north, large numbers breed on the Falkland Islands and islands off southern South America, New Zealand (including Campbell, Auckland, Antipodes, Chatham, and Snares islands), and southeast Australia. Prebreeding birds range south into the pack ice, at least in the Australian sector, from January to March, but the whole population moves north in the Atlantic and Pacific oceans from May to October, the majority crossing the equator.

SHORT-TAILED SHEARWATER *Puffinus tenuirostris*

Nonbreeding Migrant

Identification. Plate 5. 15-16 in. (38-41 cm)/38 in. (97 cm). Closely resembles the sooty shearwater but is slightly smaller and disproportionately shorter billed and shorter tailed. The underwing linings are darker than those of most sooties, usually appearing as a smoky gray rather than a conspicuous silvery flash. The great variation in this character precludes its use in positively separating the two species at sea, but the general species composition of large flocks can be identified by the underwing character of the majority. The flight patterns of the two species are similar.

Distribution. Breeds on islands around Tasmania and southeastern Australia and migrates to the North Pacific Ocean. Specimens have been collected on Macquarie Island in November 1960 [*Warham,* 1969] and at sea south of the same island in February (P. C. Harper, unpublished data, 1966).

FLESH-FOOTED SHEARWATER *Puffinus carneipes*

Breeding Migrant

Identification. 19-20 in. (48-51 cm)/43 in. (109 cm). A heavily built dark brown shearwater with light-colored bill and feet. The relatively heavy bill is fleshy pink with a dark tip; the feet are pink. It resembles the sooty and short-tailed shearwaters but differs in having a richer browner color and light bill and feet. It differs from the sooty shearwater in having a dark underwing. It is smaller and browner than the white-chinned petrel with a pinkish, not greenish, bill. The pale edges on the feathers of the upperparts may become prominent with wear.

No character is known for distinguishing first-year birds.

The downy chick is medium gray above and light gray on breast, belly, and underwings. The bill is grayish blue; the feet are fleshy gray.

Flight and habits. Heavy flutter and glide on stiff wings low over the waves. The wingbeat is less rapid than that of the sooty. Moderately gregarious while it is feeding.

Voice and display. The flight call on the breeding grounds is a sharp mewing. The song, given in the air, on the ground, or in the burrow, is complex. It begins as a series of short 'gug gug gugs'; leads into a trisyllabic crooning 'ku koo ah' repeated 3-6 times, becoming a wild screaming sound with the first two notes exhaled and the last an inhaled sob; and finally dies away in a splutter almost as it began. The display of pairs is on the surface near the nest burrow, the birds sitting side by side while they mutually preen and nibble. The species is generally silent at sea except for high-pitched squabbling over food.

Food. Small crustaceans, squid, and sardinelike fishes obtained by diving.

Reproduction. On Ile Saint-Paul there is a single small colony of 20-30 pairs on the northern end of the island about 30 m above sea level. The burrows, which may have several entrances, are dug in loose volcanic soil interlaced with roots of dry tussock. They are 1.5-2

Flesh-footed shearwater

m long with a diameter of about 15 cm and open into a larger chamber, 15-60 cm below the surface, sparsely lined with grass roots and rush stems. Elsewhere the colonies of burrow are in bare flat ground or under vegetation including trees but with a bare takeoff and landing area nearby. The mouth of the burrow is often marked by white guano and may be clogged with grass while the birds, which are active exclusively by night, incubate. The Ile Saint-Paul colony has only been visited twice, in late January, during hatching, and in March. The following information is from Australian and New Zealand colonies and is consistent with observations on Ile Saint-Paul.

Arrival: September, occupying burrows in October.

Eggs: Late November to early December. Clutch, 1 white egg; 72-73 × 44.5-46 mm.

Hatching: Mid-January to early February. Incubation period unknown; brood period 2-3 days. Both adults incubate eggs and feed young.

Fledging and departure: In late April and early May, after 92 days.

Molt. Body molt begins in adults toward the end of the breeding cycle. Wing and tail molt are apparently completed rapidly in contranuptial quarters.

Predation and mortality. Rats probably prey heavily on eggs and

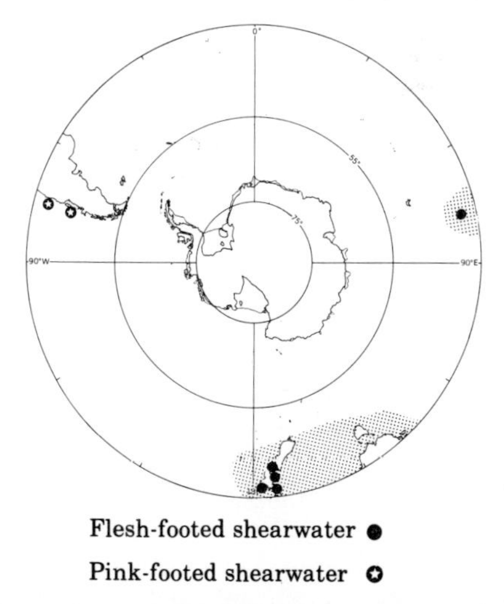

small chicks on Ile Saint-Paul, where breeding has only recently been confirmed. The total population there may be extremely small and possibly only recently established.

Ectoparasites. None recorded on Ile Saint-Paul birds.

Habitat. In most of its range the flesh-footed shearwater prefers offshore rather than pelagic waters near the Subtropical convergence, especially in areas of upwelling and high productivity.

Distribution. The flesh-footed shearwater breeds on Ile Saint-Paul [*Segonzac,* 1970], islands off southwest Australia (and formerly, it bred on the adjacent mainland), Lord Howe Island, and islands off northern New Zealand. From April to August, Australian birds migrate across the equator to the northern Indian Ocean off India, Pakistan, and Somalia (Somaliland); New Zealand birds presumably migrate to Japan, some possibly returning south by way of the North American and South American coasts. Migration of Ile Saint-Paul birds is unknown. The 'wedge-tailed shearwater,' *Puffinus pacificus,* collected at sea off Ile Amsterdam in January 1952 [*Paulian,* 1953, p. 193] was actually a flesh-footed shearwater. The pink-footed shearwater, *P. creatopus,* a bird with whitish underparts, usually finely barred with grayish brown, breeding on islands off western South America, is sometimes considered conspecific.

GREATER SHEARWATER *Puffinus gravis*

Breeding Migrant

Identification. Plate 6. 18-20 in. (46-51 cm)/43-46 in. (109-117 cm). A large brown-backed shearwater with a conspicuous dark cap, white underparts and underwings, and a white bar at the base of the tail. Immaculate white cheeks and an incomplete white collar contrast with the gray brown back. The rest of the underparts are mostly white with some variable dark smudging on the belly and wing coverts. The bill is slim and black; the feet are pinkish flesh with black markings on the outer toe and sides of the leg. See the gray petrel and Cory's shearwater.

The juvenile in fresh plumage has a marked gray bloom and pale feather edgings. The bill is mostly bluish gray.

The downy chick is bluish gray, paler on the underside of the neck and on the breast.

Flight and habits. The stiff-winged flight is less rapid and more deliberate than that of the sooty shearwater. Gregarious at sea when food is abundant.

Voice and display. The call at the breeding grounds is an excited

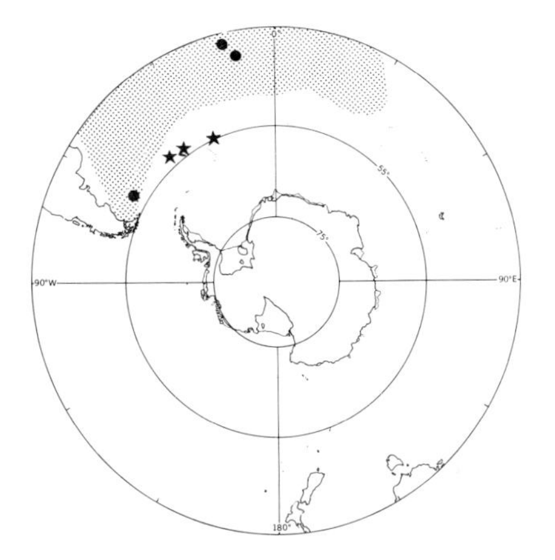

Greater shearwater

strident breathless 'ma-ma-ma-ma-ma-ma' or 'ha-ha-ha-ha-ha-ha.' The rhythm is variable: longer notes occur in the beginning, and the pace increases as the pitch rises. The greater shearwater also gives noisy harsh cries and screams at sea when it is squabbling over food.

Food. Fish, squid, and pteropod crustaceans. The greater shearwater is attracted to fishing vessels for offal and accompanies schools of whales and porpoises.

Reproduction. Colonial, burrowing in a variety of dry habitats including meadows, woods, and tussock slopes from the edges of low cliffs to the highest parts of the islands. Colonies are densest on Nightingale Island, where more than 2,000,000 pairs may breed and many birds, unable to find burrow sites, lay eggs on the surface and abandon them. The unlined burrow averages about 90 cm long and has a sharp turn just inside the entrance.

Arrival: August to September.

Eggs: Early November. Clutch, 1 white long oval egg; averages 80 × 51.5 mm.

Hatching: Early January. Incubation period 53-57 days; brood period unknown.

Fledging and departure: Late April to early May, at about 105 days. Adults leave in April when large chicks are still in down; juveniles leave during May.

Molt. Takes place rapidly in the northern hemisphere in July and August.

Predation and mortality. Skuas attack newly fledged young birds as they first emerge from the burrow.

Ectoparasite. Flea, *Parapsyllus longicornis.*

Habitat. Cool waters near the Subtropical convergence.

Distribution. The greater shearwater breeds on the Tristan da Cunha group, Gough, and the Falkland (one record [*Woods,* 1970]) islands, migrating to the North Atlantic Ocean from May to October. It has been recorded in January to March off Staten Island, off South Georgia, and east of the northernmost South Sandwich Islands [*Watson,* 1971]. Sightings in the South Pacific Ocean between Fiji and New Zealand [*Jenkins,* 1968] are highly dubious.

LITTLE SHEARWATER *Puffinus assimilis*

Resident

Identification. Plate 6. 12 in. (30 cm)/21 in. (53 cm). A very small black and white shearwater. The southernmost populations, *P. a. elegans,* have slaty gray upperparts with pale edgings to the feathers of the back in fresh plumage. The underparts, including the cheeks, throat, and underwings, are wholly white. The bill is dull gray blue; the feet are blue with pinkish webs.

Flight and habits. The little shearwater flies fairly fast, keeping low over the water with very rapid fluttering wingbeats. The only other southern seabirds that could be confused with this diminutive shearwater are the even smaller and chunkier diving petrels, which have a similar flight. The little shearwater swims and dives, often congregating in large flocks while it is feeding.

Voice and display. There are several flight calls on the breeding grounds (at Tristan da Cunha group) including a sharp whistling 'preep, preep,' a shrill scream, and a lower melodious dovelike coo. Adults' courting calls on the land are croaks and groans.

Food. Cephalopods and possibly also fish and crustaceans.

Reproduction. Breeds at most localities in winter. There is no information on breeding at Ile Saint-Paul, but in the Tristan da Cunha group and on Gough Island, burrows are dug in steep tussock grass banks in ravines near the sea. Adults arrive and clean out burrows in January and February and presumably lay a single white egg in June or July. Downy young are present until at least November.

Molt. In the Tristan da Cunha group and Gough Island, adults molt

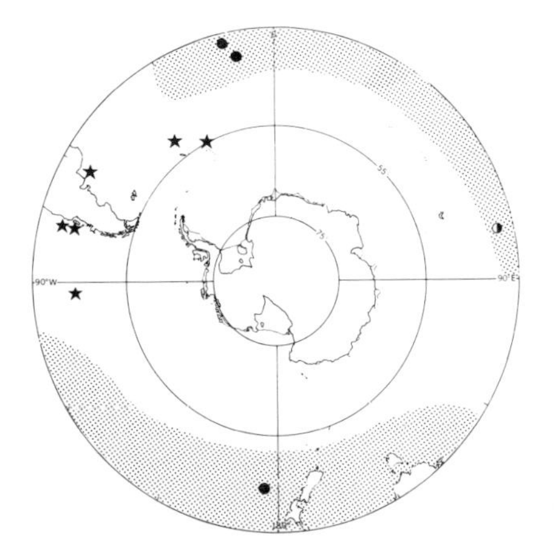

Little shearwater

body plumage in March to April and complete wing molt in November to December.

Predation and mortality. Skuas may take adults, but introduced rats are probably the most serious threat.

Ectoparasites. No information is available for southernmost populations.

Habitat. Offshore temperate sub-Antarctic waters, generally not far from land.

Distribution. Pelagic range in the southern hemisphere is nearly circumpolar between 30° and 45°S near the Subtropical convergence. The little shearwater breeds on Tristan Island, on Gough Island (*P. a. elegans*), possibly still on Ile Saint-Paul (a young bird was seen flying in late January, and the species, subspecies not determined, is common in nearby waters in April), on many islands off southern Australia and New Zealand (several other subspecies are not shown on the map) including Chatham Island (*P. a. 'kempi,'* i.e., *P. a. elegans*), on Rapa Island (*P. a. 'myrtae,'* i.e., *P. a. elegans*), and possibly on islands off the coast of Chile [*Jehl*, 1973]. Records at sea north of South Georgia and near the northernmost South Sandwich Islands (P. C. Harper, unpublished data, 1966; P. F. Vaughn, personal communication, 1967) probably represent vagrants from the population breeding in the Tristan da Cunha group. Other subspecies occur in climatically similar areas of the eastern North Atlantic Ocean.

STORM PETRELS: Oceanitidae

Storm petrels are very small pelagic petrels that are characteristically seen following in the wakes of ships. The wings are long and proportionately broader than those of all the Procellariidae, and the nostrils are united into a single tube with no visible median septum. Storm petrels fly rapidly and erratically low over the waves. In calms or light winds, flight is a butterflylike fluttering, but in moderate to heavy winds, flight is stronger, more direct, and swallowlike. When storm petrels are feeding, they pause to hover, their feet pattering the surface, appearing to 'tread water' while they pick at floating bits of food such as planktonic crustaceans, oil from debris, galley scraps, and small squid and fish. They rarely alight on the water but can swim buoyantly if it is necessary. On land, storm petrels walk upright only with difficulty and usually shuffle clumsily on their thin tarsi. The species inhabiting the Antarctic lay a single relatively large white egg in rocky crevices of talus slopes or, when soft soil is available, in short burrows. Adults are most active about the breeding colonies at night, when their nocturnal 'songs' and the piping of young are useful clues for locating nest sites. Egg and chick periods are prolonged, and both parents share incubation and feeding responsibilities. The downy chicks resemble balls of fluff and remain in the nest chamber until they are fully fledged. Three species of storm petrels occur in the Antarctic. They differ in dorsal color, presence or absence of a white rump, and color of the underparts. One species is abundant and widespread; the other two are less abundant and more restricted in range. Two other subtropical species breed in the Tristan da Cunha group and possibly on Ile Amsterdam, and a sixth northern hemisphere species has occurred as a vagrant.

For use of the family name Oceanitidae rather than the more generally used Hydrobatidae, see *Brodkorb* [1963, p. 246].

WILSON'S STORM PETREL *Oceanites oceanicus*

Breeding Migrant

Identification. Plate 7. 7 in. (18 cm)/16 in. (41 cm). A brownish black storm petrel with a conspicuous white rump and pale brown markings on the secondary coverts. The black legs are long and slender; the yellow-webbed toes (very difficult to see) extend beyond the end of the tail in flight. The bill is black. Birds breeding at northern localities (Cape Horn islands, Falkland Islands, South Georgia, and Iles Kerguelen, *O. o. oceanicus*) are about 10% smaller than those breeding at more southern sites (*O. o. exasperatus*). If only the back is seen, Wilson's storm petrel may be confused with the black-bellied storm petrel, which differs, however, in having mostly white underparts.

Juveniles are difficult to distinguish from adults but may have a light spot on the lores and white tips to the secondaries, scapulars, and belly feathers.

In first down the chick is dark gray with a slightly lighter V-shaped area (absent in second down) on the sides of the belly. The legs and feet are flesh colored, darkening with age, whereas the webs become increasingly yellow.

Flight and habits. Characteristically seen fluttering or gliding swallowlike in crisscross patterns through the wake of a ship. In feeding, Wilson's storm petrel patters, bounces, or skips low along the surface with the wings held vertically over the back, frequently pausing to hover over food. Gregarious both at sea and on the breeding grounds. It is attracted to lights at night.

Voice and display. On the breeding grounds a loud grating two- or three-note nasal cry 'aark-aark' given by one parent in the air elicits a similar response from the other bird in the nest. These calls may be repeated monotonously. Courtship in the burrow consists of mutual preening and chattering. Reports of a low nondirectional whistle from birds on the nest are probably due to confusion with the black-bellied storm petrel. When Wilson's storm petrel is feeding at sea, it utters a nearly inaudible rapid squeaking.

Food. Small crustaceans, mainly euphausiids, and cephalopods. On the Antarctic Peninsula, crustaceans predominate; at Géologie Archipelago, Adélie Coast, small squid are eaten. Wilson's storm petrel also skims animal oil and grease from about floating carcasses.

Reproduction. Loosely colonial, utilizing deep crevices in cliffs and rock screes along the coast and inland. Wilson's storm petrel also digs shallow burrows in soft soil or moss. Lichens, moss, small pebbles, egg shells, and mummified young of previous seasons accumulate in the nest chamber. The breeding schedule is highly variable, possibly being shortest on the continent and more prolonged elsewhere. Within colonies there is no synchrony in laying.

Arrival: Early November to mid-December.

Eggs: Early December to late January. Clutch, 1 dull white egg with a circle of reddish spots about the larger end; averages 34.8 × 24.7 mm and 11.0 g (Signy Island) or 32.2 × 23.2 mm and 10.8 g (Iles Kerguelen).

Hatching: Mid-January to late March. Incubation period 39-48 days but may be prolonged to 52 days by absence from nest; brood period 1-2 days.

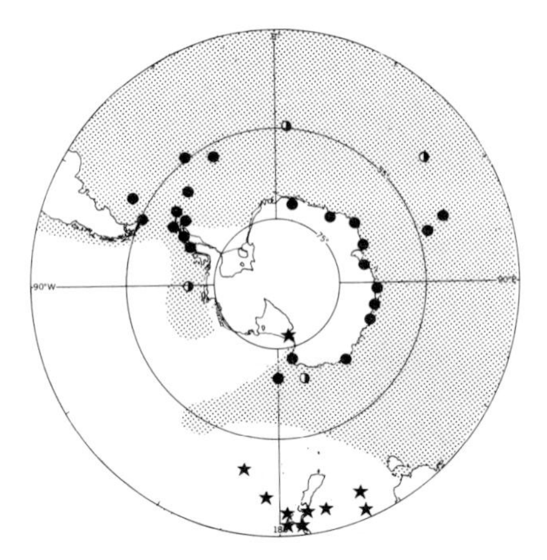

Wilson's storm petrel

Fledging and departure: Mid-March to early May, the nestling period averaging about 60 days.

Molt. A complete molt occurs in winter quarters in the North Atlantic Ocean from May to October, but it occurs up to 2 months later in the Indian Ocean. Wing molt lasts about 4 months, and the tail is the final area to finish molt. Some individuals, possibly unsuccessful breeders or prebreeders, may begin body molt in the south.

Predation and mortality. The major activity of Wilson's storm petrel on its breeding grounds takes place at twilight or at night, probably because of diurnal predation by skuas and gulls. Adult storm petrel remains are found in skua nests. In the far south, heavy snow may block burrow entrances and result in a mortality of up to 80%.

Ectoparasites. Feather mite, *Zachvatkinia hydrobatidii;* feather lice, *Halipeurus pelagicus, Philoceanus robertsi,* and *Saemundssonia marina.*

Habitat. Although this species is confined to colder pelagic and offshore antarctic and subantarctic waters during the breeding season, it migrates through the tropics to climatically similar northern hemisphere waters in the off-season.

Distribution. In the Antarctic, Wilson's storm petrel breeds on South Georgia, South Sandwich, South Orkney, South Shetland, Bouvet (needs confirmation), Crozet (probably Ile de l'Est only, needs confirmation), Kerguelen, Heard, Balleny (needs confirmation), Scott, and Peter I (very likely, but needs confirmation) islands and at numerous localities on the Antarctic Peninsula and Antarctic continent. Farther north it breeds on the Falkland Islands and other islands off southern South America. It migrates to the northern hemisphere from March to November and is abundant throughout the Atlantic and Indian ocean sectors but rare in the Pacific Ocean. Migrants have been noted in Marshall Islands in numbers [*Huber*, 1971], occasionally in Japanese waters, and very rarely in the central Pacific.

Leach's storm petrel

LEACH'S STORM PETREL *Oceanodroma leucorhoa*

Vagrant

A white-rumped black storm petrel (8 in., 20 cm/17-19 in., 43-48 cm) that closely resembles Wilson's storm petrel but has a forked tail and shorter all-dark feet. Its flight is stronger and involves more leaping and bounding than that of Wilson's storm petrel; it is not attracted to the wakes of ships. This northern hemisphere species migrates to the temperate zones of the southern hemisphere. A 'perfectly certain sight record' was made in the South Atlantic Ocean at 57°40'S, 5°00'E on January 3, 1947 [*Bierman and Voous*, 1950, p. 94].

BLACK-BELLIED STORM PETREL *Fregetta tropica*

Breeding Migrant

WHITE-BELLIED STORM PETREL *Fregetta grallaria*

Resident

Identification. Plate 7. 8 in. (20 cm)/19 in. (48 cm). A sooty black storm petrel with white rump, underwings, and belly, the last usually being bisected by an irregular black midventral line except on the birds on the Tristan da Cunha group, Gough Island, and possibly Ile Amsterdam and Ile Saint-Paul. A large triangular white patch on the underwing is conspicuous at a distance. The neck and breast are mostly black, but the throat may show some white feathers. The bill, legs, and feet are entirely black. Females tend to have longer wings than males (average 170.8 mm versus 162.3 mm for males on Signy Island). This species is a little larger and stockier than Wilson's storm petrel, which has dark underparts, and it is darker than the gray-backed storm petrel.

Juveniles and freshly molted adults have prominent pale edges on the dark feathers of the upperparts, imparting a scaled appearance to the back.

The downy chick is dark gray on the upperparts and paler below with a white midventral patch and black feet. The face, throat, and upper neck are bare. Some Wilson's storm petrel chicks also show a little white on the belly, but the legs are paler, and the feet have light-colored webs.

Flight and habits. The rapid, erratic, and butterflylike flight is more agile than that of Wilson's storm petrel. The black- and white-bellied storm petrels spring from side to side over the surface on outspread motionless wings, using both feet as 'springboards.' Wilson's storm petrel flutters and patters, the wings being held high rather than just above the horizontal. Black-bellied storm petrels are usually seen flying ahead of or around a ship but rarely follow in the wake. They are usually solitary or with a group of Wilson's storm petrels, but they are sometimes gregarious in small parties at sea.

Voice and display. The most common nocturnal call is a thin high-pitched nondirectional piping or whistling 'hüüüüüü' lasting about 4 seconds. It is given by birds that either are inside the nest or are standing at the entrance and carries up to 300 m on calm nights. If the call is imitated with the lips or on a piccolo fife in the key of G or F, it may attract flying adults. Calling begins at dusk soon after arrival in November and continues until April, although it has a reduced intensity after mid-February. Another call heard less frequently and presumably related to courtship is given intermittently when both adults are present in the nest. It is similar in quality to the common nocturnal call but is softer and carries only a few meters. It consists of a sequence of single notes 'PEE-EEP-pip-pip-pip-pip-PEE-EEP-pip-pip,' each sequence separated by a pause of about 4 seconds.

The first two notes of this call are occasionally uttered in high courtship flights, in which a pair glides down in unison, one bird remaining just above and behind the other.

Food. Small cephalopods, crustaceans, and fish picked from the surface. Small pebbles or pumice are also frequently found in the stomach.

Reproduction. Loosely colonial. Nests are usually well hidden in crevices in stable small boulder scree or in rock and lava slopes near the sea, but sometimes short burrows are dug in bare soil. No nest material is used, except in the Tristan da Cunha group and on Gough Island, but egg shells, mummified young, and other debris accumulate in the nest as they do for Wilson's storm petrel. On the Tristan da Cunha group and Gough Island the nest, which is in a short burrow, consists of leaves and stems of decaying tussock grass or other vegetation. Both sexes incubate, the male taking the first shift. Changeovers occur about every 3 days, during which time weight loss is 12%.

Arrival: Mid-November to early December on Signy Island. The female departs to feed at sea 8-15 days prior to laying, but the male remains at the nest and keeps the entrance free of snow.

Eggs: Early December to late January. Clutch, 1 egg similar to that of Wilson's storm petrel but slightly larger; averages 37 × 27 mm and 15 g (Signy Island).

Hatching: Late January to February. Incubation period 38-44 days. The chick is not brooded. Adults visit the nest only to feed the chick after hatching and only at night.

Fledging and departure: By mid-April, after 65-71 days in the nest.

Molt. Probably in the tropics after the breeding season (needs confirmation).

Predation and mortality. Remains of adults have been found in skua nests.

Ectoparasites. Feather mite, *Zachvatkinia hydrobatidii;* feather lice, *Halipeurus pelagicus* and *Philoceanus fasciatus.*

Habitat. Antarctic and subantarctic waters in the breeding season, probably wintering in the central tropics. Highly pelagic.

Distribution. Breeds on South Georgia, South Orkney, South Shetland (particularly abundant on Elephant Island [*Furse and Bruce,*

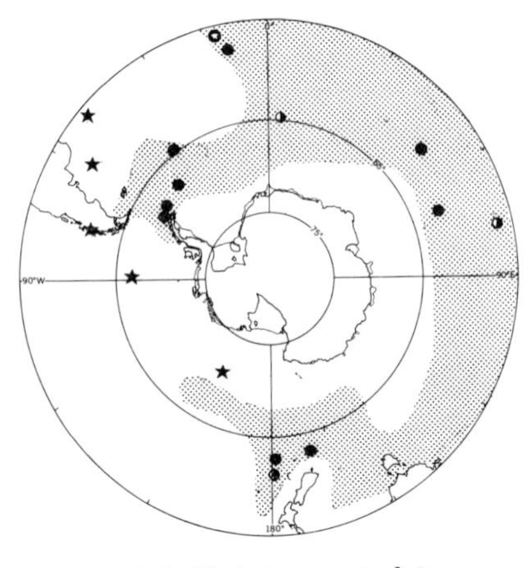

Black-bellied storm petrel ●

White-bellied storm petrel ✪

1971]), Bouvet (needs confirmation), Crozet (Ile de l'Est only [*Despin et al.*, 1972]), Kerguelen, Auckland, Bounty (needs confirmation), and Antipodes islands. White-bellied birds breeding on Gough Island and others that occur around Ile Amsterdam and Ile Saint-Paul (may breed) possibly also belong to this species, but the white-bellied storm petrel, which breeds in the Tristan da Cunha group, belongs to the closely related widespread subtropical species *F. grallaria.* This species is far less common than Wilson's storm petrel both on its breeding grounds, where they occur together, and at sea. On Signy Island, South Orkney Islands, population estimates are 100-200 breeding pairs of black-bellied storm petrels and 200,000 pairs of Wilson's storm petrels [*Beck and Brown*, 1971]. See map for pelagic range of the black-bellied storm petrel.

GRAY-BACKED STORM PETREL *Garrodia nereis*

Resident

Identification. Plate 7. 6.5 in. (17 cm)/13 in. (33 cm). A small ashy gray storm petrel with white belly and underwings. The entire head and breast, the primaries, the leading edge of the underwing, and the terminal band on the tail are sooty black. The rump and base of the tail are light gray. The bill and feet are black. This species is smaller and paler than the conspicuously white-rumped Wilson's and black-bellied storm petrels.

PLATE LEGENDS

Plate 1. Penguins.

Plate 2. Penguins and shags.

Plate 3. Great albatrosses, sooty albatrosses, and giant fulmars.

Plate 4. Mollymauk albatrosses.

Plate 5. Fulmarine petrels, gadfly petrels, and shearwaters.

Plate 6. Fulmarine petrels, gadfly petrels, and shearwaters.

Plate 7. Blue petrel, prions, storm petrels, and diving petrels.

Plate 8. Ducks, blue-eyed shag, weka, sheathbills, and Macquarie Island and South Georgia land birds.

Plate 9. Skuas, pomarine jaeger, gull, and terns.

Plate 10. Parasitic jaeger, gull, and terns.

Plate 11. Tristan da Cunha group and Gough Island rails and land birds.

Adult
Juvenile
KING
PENGUIN
Adult
EMPEROR
PENGUIN
Juvenile
Juvenile
GENTOO PENGUIN
Adult
Adult
ADÉLIE PENGUIN
Juvenile
Adult
CHINSTRAP
PENGUIN
Juvenile
Adult
Juvenile
MAGELLANIC
PENGUIN
EMPEROR
Chick
KING
GENTOO
CHINSTRAP
Chick
ADÉLIE

Plate 1

After young hatch
BLUE-
EYED
SHAG
Adults
With courtship
crest
Juvenile
Adult
KING SHAG
Juvenile
Adult, southern race
Juvenile
Adult
MACARONI
PENGUIN
Juvenile
Adult,
northern
race
ROCKHOPPER
PENGUIN
Adults
Erect-crested race
Snares
Island
race
CRESTED
PENGUIN
Light
phase
Dark phase
'ROYAL
PENGUIN'
MACARONI
Chick
'ROYAL'
ROCKHOPPER
KING
SHAG
Kerguelen
race
BLUE-EYED
SHAG

Plate 2

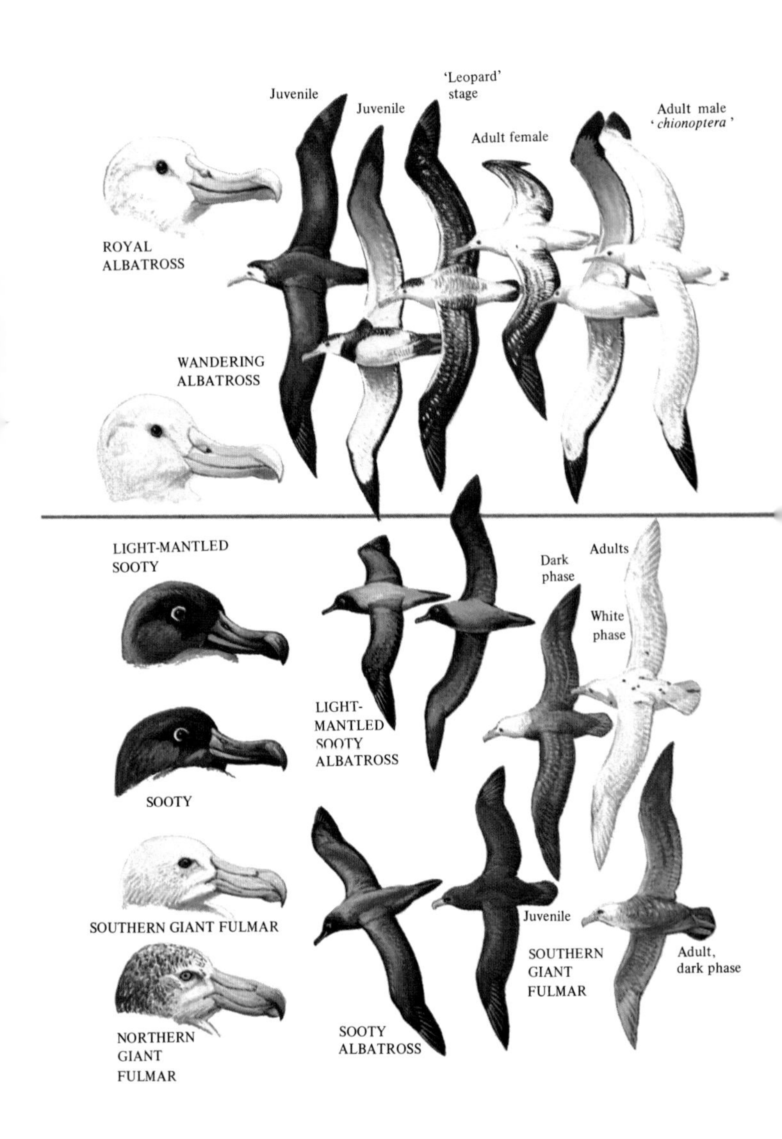

Juvenile
Juvenile
'Leopard' stage
Adult female
Adult male 'chionoptera'
ROYAL ALBATROSS
WANDERING ALBATROSS
LIGHT-MANTLED SOOTY
Dark phase
Adults
White phase
LIGHT-MANTLED SOOTY ALBATROSS
SOOTY
SOUTHERN GIANT FULMAR
Juvenile
SOUTHERN GIANT FULMAR
Adult, dark phase
NORTHERN GIANT FULMAR
SOOTY ALBATROSS

Plate 3

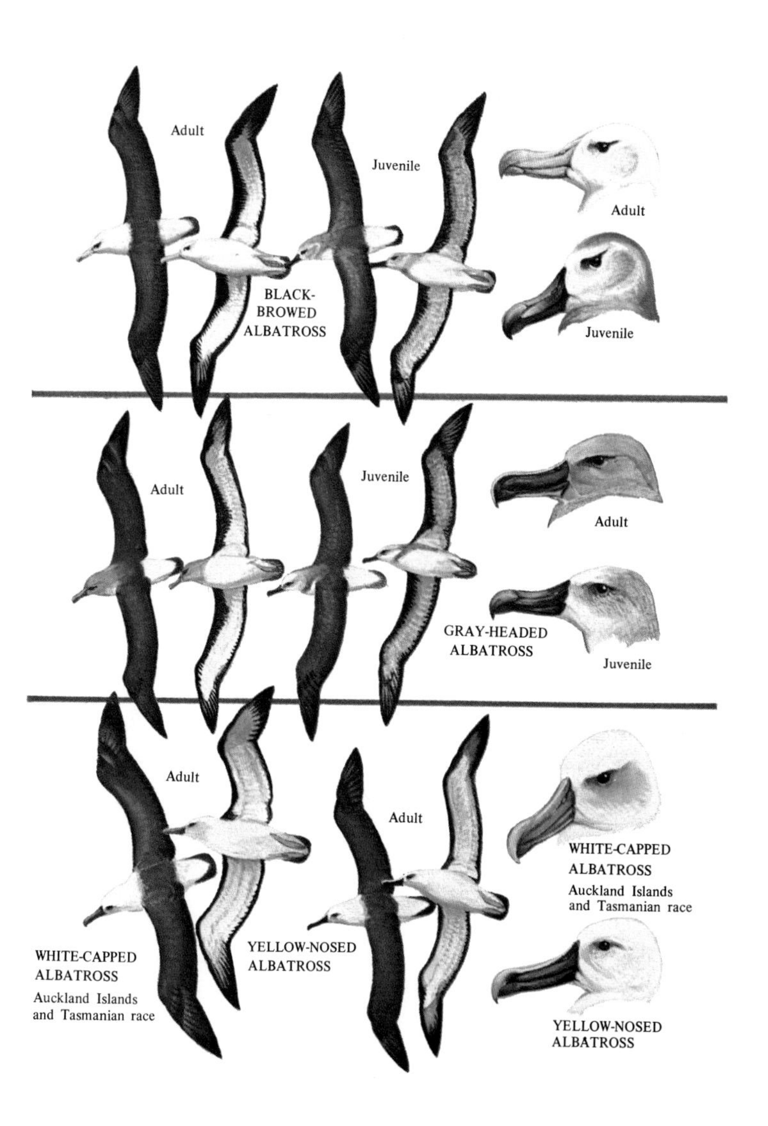
Adult
Juvenile
Adult
Juvenile
BLACK-
BROWED
ALBATROSS
Adult
Juvenile
Adult
GRAY-HEADED
ALBATROSS
Juvenile
Adult
Adult
WHITE-CAPPED
ALBATROSS
Auckland Islands
and Tasmanian race
WHITE-CAPPED
ALBATROSS
Auckland Islands
and Tasmanian race
YELLOW-NOSED
ALBATROSS
YELLOW-NOSED
ALBATROSS

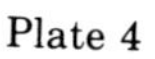
Plate 4

SOUTHERN
FULMAR
ANTARCTIC
PETREL
CAPE
PIGEON
GRAY
PETREL
WHITE-
HEADED
PETREL
ATLANTIC
PETREL
SOOTY
SHEARWATER
SHORT-TAILED
SHEARWATER
GREAT-WINGED
PETREL

Plate 5

SNOW PETREL
LITTLE SHEARWATER
GREATER SHEARWATER
MOTTLED PETREL
SNOW PETREL
KERGUELEN PETREL
Light phase
WHITE-HEADED PETREL
SOFT-PLUMAGED PETREL
Dark phase
ANTARCTIC PETREL
KERGUELEN PETREL
SOUTHERN FULMAR
SOOTY SHEARWATER
aequinoctialis race
WHITE-CHINNED PETREL
WHITE-CHINNED PETREL
conspicillata race

Plate 6

NARROW-
BILLED
PRION
ANTARCTIC
PRION
BLUE
PETREL
BLUE
PETREL
NARROW-BILLED
PRION
FULMAR
PRION
ANTARCTIC
PRION
salvini
salvini
vittata
vittata
BROAD-BILLED
PRION
WILSON'S
STORM
PETREL
BLACK-BELLIED
STORM PETREL
FAIRY PRION
GRAY-BACKED
STORM PETREL
WHITE-
FACED
STORM PETREL
FULMAR PRION
SOUTH GEORGIA
DIVING PETREL
SOUTH GEORGIA
DIVING PETREL
KERGUELEN
DIVING PETREL
KERGUELEN
DIVING
PETREL
Southern
race
Northern race

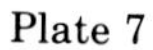

Plate 7

BLUE-EYED SHAG
Before breeding
GRAY DUCK
KERGUELEN PINTAIL
YELLOW-BILLED
PINTAIL
Extreme form
GRAY
DUCK
Juvenile
Adult
males
KERGUELEN
PINTAIL
Adult female
Adult
YELLOW-
BILLED
PINTAIL
Juvenile
WEKA
Dark
phase
Light phase
REDPOLL
SOUTH
GEORGIA
PIPIT
LESSER
SHEATHBILL
Juvenile
Chick
Adult
AMERICAN SHEATHBILL
STARLING

Plate 8

BROWN SKUA
Adults
Adult
Juvenile
SOUTH POLAR SKUA
Light phase
Dark phase
POMARINE JAEGER
Breeding adult
Nonbreeding adult
Juvenile
First year
Breeding adult
ANTARCTIC TERN
BROWN SKUA
Adult
SOUTH POLAR SKUA
Juvenile
SOUTHERN BLACK-BACKED GULL

Plate 9

Light phase
Adults
Second year
Third year
Dark phase
SOUTHERN
BLACK-BACKED
GULL
PARASITIC JAEGER
Juvenile
KERGUELEN
TERN
Nonbreeding
Breeding
Breeding
Nonbreeding
BROWN NODDY
ARCTIC TERN
Juvenile
Breeding adult
Breeding
BROWN NODDY
Juvenile
Juvenile
ANTARCTIC TERN
KERGUELEN TERN
ARCTIC
TERN
Nonbreeding

Plate 10

WILKINS' BUNTING
GOUGH BUNTING
Adult
TRISTAN
BUNTING
Female
Juvenile
Male
First year
Immature
Second
year
Adult female
Adult male
TRISTAN
THRUSH
Juvenile
Nightingale
Island race
Tristan
Island race
Immature
GOUGH
MOORHEN
Adult
Immature
Adult
INACCESSIBLE ISLAND
FLIGHTLESS RAIL

Plate 11

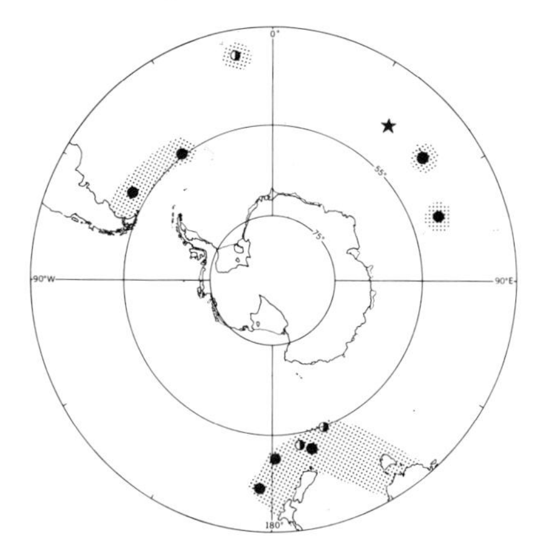

Gray-backed storm petrel

Freshly molted adults and possibly juveniles have prominent pale edges to the dark feathers of the upperparts, imparting a scaled appearance to the back and wings.

The downy chick is pale gray, somewhat lighter on the belly, and has black feet.

Flight and habits. The gray-backed storm petrel skips and bounces from side to side with wings extended very low over the water while it is feeding. It hovers buoyantly over food.

Voice and display. A high-pitched twittering in the air at the breeding grounds and a 'crakelike' scratching call.

Food. Small cephalopods and other mollusks (more data required).

Reproduction. Loosely colonial. The birds excavate either small scrapes or short burrows in the bases of tussock and *Acaena* clumps or other low vegetation near the sea, occasionally using rock crevices. Small amounts of vegetation may be found in the nest.

Arrival: Late October to early November.

Eggs: Mid-November to mid-December. Clutch, 1 white egg with reddish dots mostly about the larger end; 31.5-37.0 × 22.7-27 mm. The largest examples are from the New Zealand region.

Hatching: Unknown.

Fledging and departure: April.

Remarks: The breeding biology of this species is still virtually unstudied.

Molt. One adult in full molt in February.

Predation and mortality. No data are available, but its crepuscular habits on the breeding grounds suggest that adults may fall prey to skuas.

Ectoparasites. Feather lice, *Saemundssonia nereis* and *Philoceanus garrodiae.*

Habitat. Highly pelagic, largely in the sub-Antarctic zone.

Distribution. Breeds on Falkland, South Georgia (no records since 1903), Crozet (Ile de l'Est only [*Despin et al.,* 1972]), Kerguelen, possibly Macquarie [*Keith and Hines,* 1958], Auckland, Antipodes, and Chatham islands. The gray-backed storm petrel's status on Campbell and Gough islands is unclear. Pelagic range, especially in the Indian Ocean, is poorly known.

WHITE-FACED STORM PETREL *Pelagodroma marina*

Breeding Migrant

Identification. Plate 7. 8 in. (20 cm)/17 in. (43 cm). A gray storm petrel that differs from the smaller gray-backed storm petrel in having wholly white underparts and underwings and a narrow black subocular eye stripe contrasting with the white forehead, superciliary, and cheek. The tail and wings are nearly black, markedly darker than the light gray rump. The bill and legs are black; the webs are yellow.

No characters are known for distinguishing juvenile birds.

Flight and habits. The white-faced storm petrel dangles its long legs in flight as it 'dances' from right to left in pendulum motion and jumps over the waves, hitting the water simultaneously with both feet. Its flight is stronger and more prionlike than that of Wilson's storm petrel. It does not follow ships closely.

Voice and display. The call at the breeding grounds is a mournful 'wooo' repeated about once a second or expanded into a sirenlike moaning 'ooooaaaoooo.'

Food. Copepods and other small planktonic crustaceans and possibly small squid. The white-faced storm petrel is usually seen feeding in or outside kelp beds around islands.

Reproduction. Live and dead birds have been found in burrows on Nightingale Island in the Tristan da Cunha group in September and October, and a half-grown chick was found on Gough Island in December, but there is no other evidence of current breeding in the

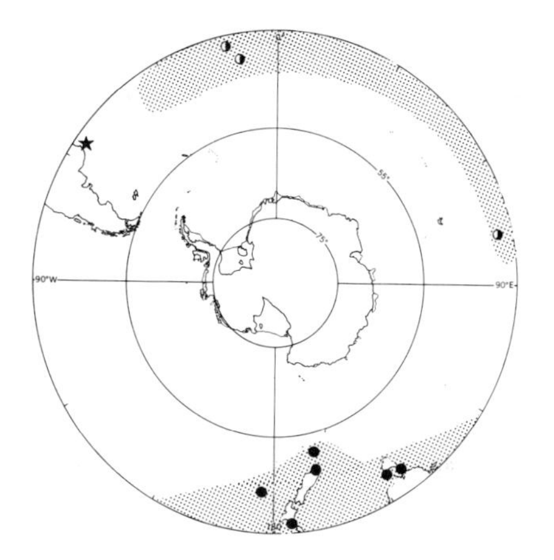

White-faced storm petrel

area of this manual. Elsewhere the species is colonial, breeding during the summer in burrows lined with vegetation. There is a single egg similar in size and markings to eggs of the black-bellied storm petrel.

Molt. Body molt may begin near the end of the breeding cycle, but the wings and tail are molted later at sea.

Ectoparasites. No information is available for subantarctic populations.

Predation and mortality. No information is available, but presumably the species was extinguished by rats on Ile Amsterdam.

Habitat. In the breeding season the white-faced storm petrel frequents the area of the Subtropical convergence.

Distribution. Breeds on temperate and subtropical islands in the North and South Atlantic oceans, probably including the Tristan da Cunha group and Gough Island, and on islands off Australia and New Zealand, including Auckland and Chatham islands. The white-faced storm petrel was also recorded at sea in February in the southern Indian Ocean [*Mörzer Bruyns and Voous,* 1964], where at least formerly, it bred on Ile Amsterdam [*Jouanin and Paulian,* 1960].

DIVING PETRELS: Pelecanoididae

Diving petrels are diminutive black and white chunky-bodied petrels closely resembling the small alcids of the northern hemisphere in both appearance and behavior. The paired nasal tubes that open upward rather than forward like those of other petrels are probably an adaptation for diving. Flight consists of labored flurries of rapid wingbeats and short glides in a straight line just above, or even through, the waves. Usually, after flying a few hundred meters, diving petrels drop abruptly either to swim on the surface or to plunge underwater, but some flights may be surprisingly long and direct. They feed on crustaceans, cephalopods, and possibly small fish. The wings molt abruptly, so that diving petrels may be flightless while the primaries are in growth. The birds breed on islands, coming ashore only at night. They dig long tortuous burrows in which a single white egg is laid. Diving petrels are frequent victims of skuas, and the presence of their dismembered carcasses about predators' nests is often the only visible sign that these secretive petrels are present. They do not follow ships but may come aboard at night when they are attracted to the lights. Two species breed on islands near the Antarctic convergence. Identification of diving petrels at sea is difficult, since the diagnostic characters, scapular bars, bill shape, distribution of markings on the throat and flanks, and underwing pattern, are of use only at close range.

SOUTH GEORGIA DIVING PETREL *Pelecanoides georgicus*

Resident

Identification. Plate 7. 8 in. (20 cm)/13 in. (33 cm). The upper surface is black except for grayish white scapular feathering, which forms a broken V across the shoulders. The underparts are white, the underwing coverts being a nearly immaculate white and the inner webs of the primaries being a pale gray. The cheeks and sides of the lower neck are variably mottled with gray, occasionally covering the throat but not forming a pronounced breastband. When the short black bill is viewed from the underside, it is broad at the base, the narrow rami converging gradually to a pointed 'Gothic' arch toward the tip. The nostrils are shorter in relation to bill length than those of the Kerguelen diving petrel. The feet are blue.

The juvenile can only be distinguished from the adult if both can be compared in the hand so that the fainter mottling on the throat, the whiter scapulars, and the narrower weaker bill of the young bird become apparent.

Nestlings are covered with light gray down, paler on the belly and sparsely covering the throat and sides of the head. As the chick develops, the face becomes feathered, and the body down becomes thicker. Juvenile feathers first appear on the head, wings, and tail, but down remains on the belly until the fledgling leaves the nest.

Flight and habits. See the family introduction for flight and habits.

Voice and display. 'A nasal squeaking like a rusty door' given on the ground and in the air at the breeding grounds.

Food. Copepods, amphipods, and small euphausiids, some probably being taken at night (needs confirmation). At Iles Kerguelen, where both diving petrels occur, this species feeds at a greater distance from the islands.

Reproduction. Colonial, burrowing in bare soft soil, sand, or volcanic ash, generally at higher altitudes than the Kerguelen diving petrel and usually well inland. On Iles Kerguelen and Heard Island, burrows are dug in flat ground, whereas on Marion Island and South Georgia they are located in the slopes of terminal moraines and hills of volcanic debris. The twisting burrows are usually less than 1.2 m long and 5-8 cm wide, the entrance generally being hidden by a stone. The enlarged nest chamber is usually bare but may contain a few pebbles, some down, or lichens. The earliest dates are from Iles Kerguelen, where the entire schedule is about 1 month earlier than it is elsewhere. See the Kerguelen diving petrel.

Arrival: Mid-September to November, but on Heard Island, South Georgia diving petrels are present all year, and some may come ashore to roost in burrows in any month.

Eggs: Early November to early January. Clutch, 1 white egg; averages 38.8 × 32.5 mm and 19 g.

Hatching: Late December to February. Incubation period about 55 days.

Fledging and departure: Little information, probably in March.

Molt. No information is available.

Predation and mortality. Skuas and rats kill both adults and chicks at the breeding grounds. Diving petrels are attracted to bright lights, and many crash into lighthouses or come aboard ships at sea.

Ectoparasites. Tick, *Ixodes kerguelenensis;* feather louse, *Pelmatocerandra enderleini.*

Habitat. Coastal and offshore waters around islands near the Antarctic convergence. The South Georgia diving petrel possibly feeds south of the convergence and farther out at sea than *P. (u.) exsul.*

Distribution. Breeds on South Georgia, Marion, Crozet, Kerguelen, Heard, and Auckland islands (very rare) and possibly on Macquarie Island, where it has seldom been recorded. Diving petrels can be seen at sea between 40° and 60°S and occasionally even farther south in the Ross Sea. Most of these records pertain to one of the two antarctic

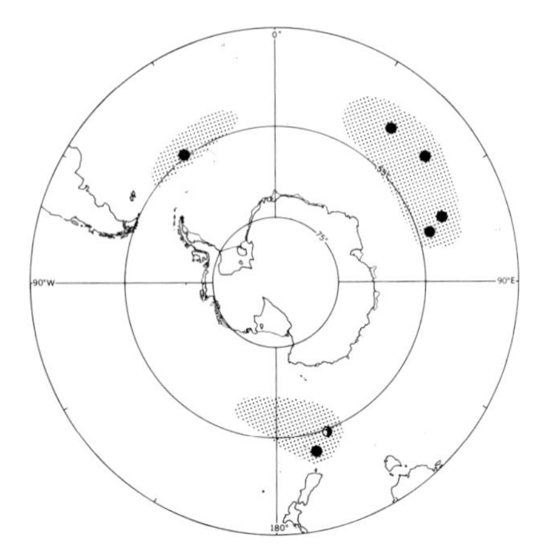

South Georgia diving petrel

species, although off southern South America, Gough Island, and New Zealand, other temperate zone forms may be involved. Specimen records at sea are needed to ascertain the pelagic range of the two antarctic species.

KERGUELEN DIVING PETREL *Pelecanoides (urinatrix) exsul*
(subantarctic diving petrel)

Resident

Identification. Plate 7. 8 in. (20 cm)/13 in. (33 cm). Differs from the South Georgia diving petrel in lacking light scapular bars and having darker underwings, a distinct mottled breastband, and a more elongated and heavier bill. The underwing coverts are gray with black shafts, and the inner webs of the primaries are dark. Gray mottling on the lower throat is heavy, forming a complete breastband and continuing along the sides of the breast to the flanks. When the heavy rami of the lower mandible are viewed from below, they are nearly parallel and converge in a rounded 'Romanesque' arch at the tip. The nostrils are more elongated than those of the South Georgia diving petrel. The feet are blue. Birds from the Tristan da Cunha group, Gough Island, and Ile Amsterdam (*P. u. dacunhae*) are smaller than birds from the antarctic area (*P. (u.) exsul*) and have relatively weaker bills than they do.

The juvenile has a more slender bill than the adult.

The downy nestling is probably indistinguishable from the nestling of the South Georgia diving petrel until it is old enough to show bill structure differences and the all-dark scapular feathering.

Flight and habits. See the family introduction for flight. The Kerguelen diving petrel cannot rise from level ground. It swims underwater with its wings partly open.

Voice and display. A gentle crooning with a marked rising inflection somewhat resembling the low mew of a cat or the cooing of a dove. The Kerguelen petrel calls in flight over burrows. On the Tristan da Cunha group the call is a resonant loud nasal bleating or braying 'kerrraaa-ek' or 'bee-aw-ak' followed by a repeated 'ek, ek, ek.' Silent at sea.

Food. Amphipods, copepods, and other small crustaceans. The Kerguelen diving petrel may feed so voraciously that it cannot rise from the water.

Reproduction. Colonial, breeding on low-lying coastal slopes with a heavy vegetation cover. On Marion and Heard islands the burrows are in tussock- and moss-covered slopes of the coastal plain, on Iles Kerguelen in abandoned rabbit burrows, and on Nightingale Island in the Tristan da Cunha group in nearly vertical rock faces with a few centimeters of turf cover. The burrow is dry and shallow, and the nest is usually lined with grass and feathers. The schedule of this species is about 1 month ahead of that of the South Georgia diving petrel except on Iles Kerguelen, where it lays 1 month later. More information is needed on the reproductive cycle.

Arrival: August to October.

Eggs: Late November to early January and early September on Nightingale Island. Clutch, 1 dull white egg; averages 37.9 × 31 mm.

Hatching: January to February. Incubation period about 55 days; brood period about 10 days.

Fledging and departure: February to March, at an age of about 54 days.

Molt. January to February.

Predation and mortality. Like the predation and mortality of the South Georgia diving petrel.

Ectoparasites. Feather lice (on *P. u. urinatrix* only), *Pelmatocerandra setosa, Halipeurus pacificus,* and *Austromenopon elliotti;* flea, *Notiopsylla kerguelensis.*

Habitat. Coastal and offshore waters, *P. (u.) exsul* possibly feeding north of the Antarctic convergence.

Distribution. *P. (u.) exsul* breeds on South Georgia, Marion, Crozet (Ile de la Possession only), Kerguelen, Heard, Macquarie (needs confirmation), Auckland, Campbell (needs confirmation), and Antipodes

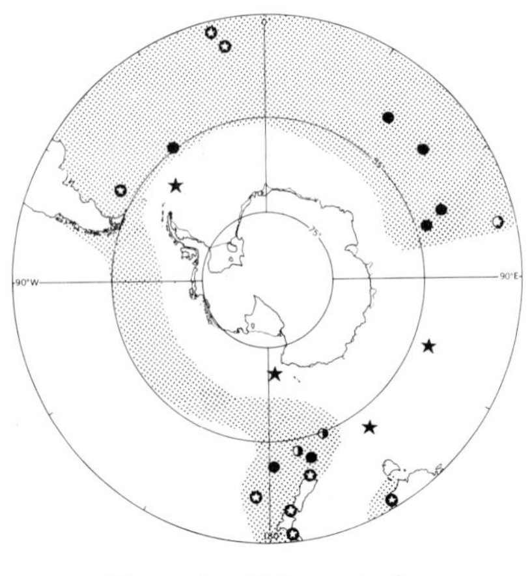

Kerguelen diving petrel

P. (u.) exsul ●

P. u. subspecies ✪

islands. This heavy-billed form is thought by some authorities to be conspecific with *P. urinatrix,* a species that is widespread on subantarctic and temperate islands in the southern Atlantic Ocean (Falkland Islands, Tristan da Cunha group, and Gough Island), off southeast Australia, and on New Zealand sub-Antarctic Islands (including Chatham, Stewart, and Snares islands) and was recorded once on Signy Island, South Orkney Islands, in February 1967 [*Beck,* 1968*a*]. Specimens taken at sea near Ile Amsterdam, where there is still no evidence of breeding, are similar to Tristan and Gough island birds (*P. u. dacunhae*). See map for presumed pelagic range of diving petrels of these forms based largely on sight records. Specimens are needed to confirm identity.

CORMORANTS: Phalacrocoracidae

Cormorants are long-necked coastal birds that have thin hooked bills and long stiff tails. In the antarctic species the cobalt blue eyelids and yellowish fleshy nasal caruncles of the adults are strikingly colorful. On land the glossy blue black upperparts, white underparts, and upright stance suggest penguins. In the water, however, cormorants swim with their longer slimmer necks held erect, not hunched into the shoulders as is typical of penguins. Like ducks or geese they fly in long lines or V formations low over the water, alternating short bursts of rapid wingbeating with brief glides. Cormorants feed in shallow coastal waters, where they catch fish and possibly some crustaceans. They dive from the surface by lunging forward, pursuing their prey using both feet and wings. After emerging from the water they perch upright, holding their broad rounded wings outspread to dry the flight feathers. When cormorants become overheated, especially while they are incubating, they pant by fluttering the throat pouch. On the Antarctic Peninsula they breed in small colonies on steep coastal cliffs or offshore stacks. The nest, which is repaired and reused annually, is a truncated cone of seaweed and land vegetation that becomes cemented together with mud and guano. The 2-5 pale greenish elongate eggs are covered with a chalky white deposit, which soon becomes soiled and stained. Newly hatched chicks are naked and black and must be closely brooded by the adults until the dark sooty brown down appears.

Two closely related species, blue-eyed and king shags, breed together in southern South America, but on the Antarctic Peninsula and southern islands where cormorants occur, either species may breed alone. Useful identification characters include the distribution of white on the cheeks and neck and the presence or absence of a wing bar and dorsal patch. Another all-dark temperate zone species occurs as a vagrant on Macquarie Island.

Vagrants of two other pelecaniform families, the boobies Sulidae and the frigatebirds Fregatidae, have occurred on Ile Amsterdam and the Tristan da Cunha group.

BLUE-EYED SHAG *Phalacrocorax atriceps*

Resident

Identification. Plates 2 and 8. 24 in. (61 cm)/49 in. (124 cm). A white-breasted black-backed cormorant with largely white cheeks and neck. The border of black and white on the face lies above an imaginary line drawn from the gape across the cheeks. A prominent white middorsal patch and white wing bars are present in adults while they have chicks. The Heard Island population (*P. a. nivalis*) acquires only scattered white feathers on the shoulders. The bill is dark brown, lighter on the tip of the lower mandible; the iris is brown, sometimes mixed with gray; the nasal caruncles are pale yellow to orange, varying with season and geography; and the feet are pink. Elongated head plumes are present during most of the winter and during courtship but are lost by molt after the young hatch. The length of the head plumes, size of nasal caruncles, and amount of white on the face vary geographically. In southern South America, the only area where this species and the king shag breed together, their differences are marked. The blue-eyed shag has shorter plumes and much more white on its face. Its nasal caruncles are small and bright yellow or orange rather than large and dull to greenish yellow like those of the king shag.

The juvenile is similar to the adult in overall pattern, but the back is dark brown, and the wing bar and dorsal patch are pale brown. Whether the light wing bar and dorsal patch are retained all year is unknown. Year-old birds are intermediate with some new glossy black feathers on the back but with the wings remaining brown. The eyelids and nasal caruncles are purplish brown, and the feet are dark gray or pinkish edged with gray.

At hatching the black shiny skin of the chick is devoid of down, but during the first week a dark brownish black down develops over the entire body. Tufts of white down appear on the head, throat, and belly at about the same time as the wing and tail quills begin to show.

Flight and habits. See the family introduction for flight. Gregarious in small flocks. The blue-eyed shag never occurs far from land or the ice edge.

Voice and display. During courtship a soft croaking and low-pitched droning while the head and neck are swayed from side to side. When the blue-eyed shag is disturbed, it hisses loudly. Fledglings give a low mellow whistle.

Food. Mainly fish. Crustaceans, occasionally found in stomach contents, may be derived from the fish eaten. The blue-eyed shag can dive to depths of at least 25 m [*Conroy and Twelves*, 1972].

Reproduction. Gregarious in small colonies, usually between 20 and 40 pairs, in the Antarctic. The colonies are often associated with chinstrap and gentoo penguin rookeries. The seaweed, mud, and guano nests are built on coastal cliffs and steep slopes, mainly on offshore islets and stacks. The breeding season commences nearly a month earlier on low-latitude islands and is more prolonged than it is in the Scotia ridge and Antarctic Peninsula.

Arrival: Adults are present at many breeding stations throughout the year. Courtship activities begin in late August to early October.

Eggs: October to early January. Clutch, 2-3 elongate white eggs with chalky surfaces in the far south and up to 5 farther north; 51-70 × 38-41 mm and 57-63 g.

Hatching: November to February. Incubation period not recorded.

Fledging and departure: Fledging occurs in January to March. Departure from colonies takes place in April.

Molt. Twice per cycle. A complete molt begins during breeding and

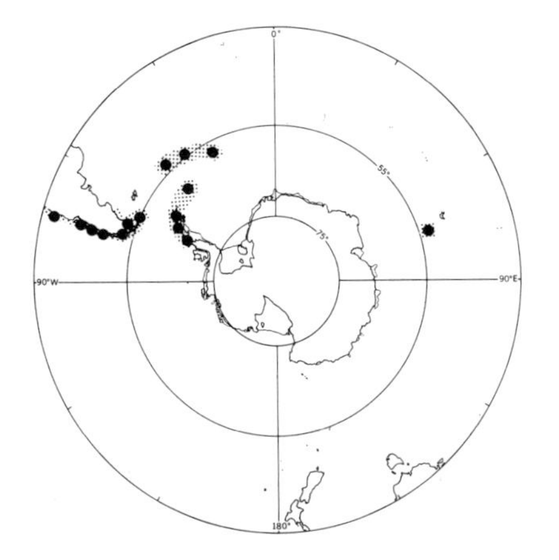

Blue-eyed shag

lasts into June. This produces the elongated crest plumes and leaves the back and wings entirely black. Another partial molt takes place in the spring just after courtship and involves loss of the crest plumes and rapid growth of the white middorsal patch and wing bars. The timing and extent of these two molts need further study.

Predation and mortality. On land, sheathbills take eggs. The leopard seal attacks cormorants at sea, and skuas pirate fish from feeding adults.

Ectoparasites. Feather mite, *Scutomegninia phalacrocoracis;* feather lice, *Piagetiella caputincisa* and *Pectinopygus turbinatus.*

Habitat. Low-latitude antarctic and subantarctic coastal waters. The blue-eyed shag requires open water for feeding, and flocks of birds may fly considerable distances in search of ice-free areas.

Distribution. Resident on the west coast of southern South America; on Shag Rocks, South Georgia, South Sandwich, South Orkney, and South Shetland islands; and on the Antarctic Peninsula south to 68°S. An isolated Indian Ocean population occurs on Heard Island.

KING SHAG *Phalacrocorax albiventer*
(includes Kerguelen shag) (*P. verrucosus*)

Resident

Identification. Plate 2. 28 in. (71 cm)/49 in. (124 cm). Similar to the blue-eyed shag but having much more black on the head and neck and never developing a white middorsal patch. The border of black and white on the face lies below an imaginary line drawn from the gape across the cheek. The prominent nasal caruncles are yellowish olive, dull yellow, or bright orange depending on the season and varying geographically. The iris is brown to hazel gray; the feet are pinkish white. Adults of the smaller (25 in., 64 cm/43 in., 109 cm) Iles Kerguelen population, *P. a. verrucosus,* sometimes considered a distinct species, normally lack the white wing bar and have yellow caruncles and more brown on their feet (see the section on distribution). In the intermediate populations on Marion Island and Iles Crozet (*P. a. melanogenis*) the wing bars are somewhat reduced.

The juvenile king resembles the first-year blue-eyed shag, but the underparts may have dark streaking, there is never a pale dorsal patch, and the pattern of dark and light on the face is similar to that of the adult king shag. On Iles Kerguelen all juveniles have brown streaking below, and some have entirely dark underparts.

Nestlings are probably indistinguishable from blue-eyed shags of the same age.

Flight and habits. Because of the greater extent of black on the neck and body this species may appear heavier than the blue-eyed shag.

Voice and display. Hoarse croaking and hissing sounds probably similar to those of the blue-eyed shag.

Food. Mainly fish. Sea urchins have been recorded from Macquarie Island birds, and euphausiids from Marion Island birds.

Reproduction. Colonial, in relatively small groups on steep coastal cliffs, frequently on sheer ledges overgrown with *Cotula.* The nest itself is a bulkier structure than that of the blue-eyed shag, incorporating quantities of vegetation up to 40 cm high. The schedule is prolonged: eggs and well-developed chicks may be present simultaneously in the same colony.

Arrival: Adults are sedentary around most breeding islands and begin courtship from August to January.

Eggs: October to early January. Clutch, 1-4 (usually 2 on Marion Island) elongate chalky greenish white eggs; average 64 × 40 mm and 51 g. The eggs are slightly smaller on Iles Kerguelen (average 60 × 38 mm).

Hatching: Early December to mid-February. Incubation period unknown. Chicks are brooded at least until they are covered with down.

Fledging and departure: February to March, about 7 weeks after hatching. Immatures disperse around the islands from February to April.

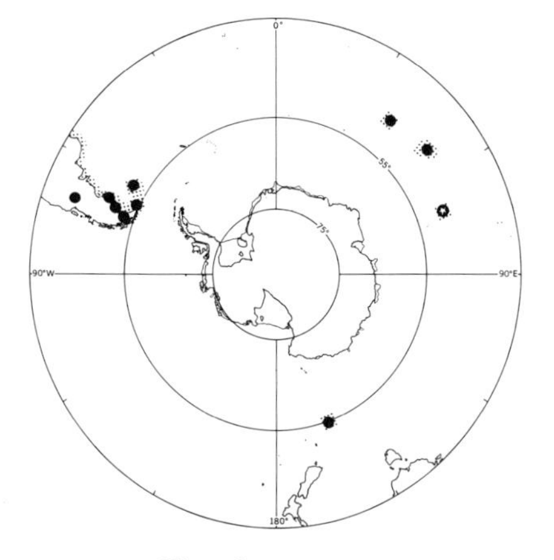

King shag
P. a. subspecies ●
P. a. verrucosus ✪

Molt. Twice per breeding cycle. A prolonged complete molt begins during breeding and produces the short head plumes and the white wing bar. Wings do not complete growth until midwinter. Another partial molt takes place just before breeding. The long head plumes and dark wing coverts are acquired in this molt. More data are needed on the timing and extent of molt.

Predation and mortality. Skuas, gulls, and sheathbills take eggs, and sheathbills also kill small chicks. No predation has been recorded on Marion Island. Leopard seals catch adults at sea.

Ectoparasites. Tick, *Ixodes uriae;* feather lice, *Eidmanniella pellucida, Pectinopygus turbinatus,* and *Piagetiella caputincisa;* flea, *Parapsyllus magellanicus.*

Habitat. Probably confined to coastal waters around subantarctic breeding islands.

Distribution. Resident on the eastern coast and inland lakes of southern South America and on Falkland, Marion (but not Prince Edward), Crozet, Kerguelen, and Macquarie islands.

Observations on Iles Kerguelen in 1971 showed that a small number of shags showing *P. albiventer* characters occur in breeding colonies of *P. a. verrucosus* and that the two forms interbreed successfully [*Derenne et al.,* 1974]. This interbreeding, whether it is of long standing owing to the natural appearance of *P. albiventer* mutants on Iles Kerguelen or of recent occurrence owing to accidental transfer of king shags from Iles Crozet to Iles Kerguelen, argues against *Voisin*'s [1970] suggestion that the Iles Kerguelen population should be considered specifically distinct. The situation on Iles Kerguelen should be studied further.

GREAT CORMORANT *Phalacrocorax carbo*

Vagrant

A large (37 in., 94 cm/54 in., 137 cm) black cormorant with dark underparts. A white area surrounds the base of the bill, and a well-defined crest extends onto the nape. In the breeding season, from May to October, small white flecks appear on the head and neck, and a large white patch appears on each flank. The sexes are alike in color, but the female is smaller. First-year birds are dark brown above with some white on the underparts but are never pure white below like adult and immature king and blue-eyed shags. This cosmopolitan species, which breeds on New Zealand, has been recorded on at least three occasions on Macquarie Island [*Keith and Hines*, 1958].

GANNET *Morus bassanus*

Vagrant

A mollymauk-sized seabird (36 in., 91 cm/66 in., 168 cm) with long pointed wings and tail and a heavy pointed bill. Gannets characteristically dive from the air to capture their fish prey underwater. Adults are mostly white, the quills of the wings and some or all of the tail being black. The head is washed with yellowish buff. The bill is bluish white or gray, the bare facial skin is black, and the feet are black or dark brown marked with green. Immatures are dark grayish

Great cormorant Gannet

brown above, liberally streaked and speckled with white, and whitish below, streaked with grayish brown. A gannet was observed briefly off the coast of Ile Amsterdam in early March 1970 [*Segonzac*, 1972]. It presumably was from the South African population (*M. b. capensis*), which has an all-black tail as an adult, rather than from the Australian-New Zealand population (*M. b. serrator*), which has only the central tail feathers black.

FRIGATEBIRD *Fregata* sp.

Vagrant

A large (29-43 in., 74-109 cm/72-90 in., 183-229 cm) long-winged predatory seabird with a deeply forked tail and long hooked bill that shows extraordinary powers of flight in chasing smaller seabirds to rob them of their prey. Males are generally all black, females have white breasts, and immatures have white heads and underparts sometimes tinged with rufous. Frigatebirds are found in tropical oceans, occurring south of 20°S only as vagrants. Frigatebirds of unknown species were observed in March 1951 on Inaccessible Island and December 1951 on Tristan Island [*Elliott*, 1957].

The frigatebird recorded from Iles Kerguelen [*Cabanis and Reichenow*, 1876] was actually from Ascension Island [*Studer*, 1889, p. 88].

HERONS: Ardeidae

Herons are long-legged slim-necked aquatic birds that frequent shallow coastal waters. The wings are surprisingly long and broad for such slim-bodied birds. At least five migratory species of this widespread family have occurred as vagrants on the Tristan da Cunha group, South Georgia, Marion Island, Ile Amsterdam, and Macquarie Island.

Frigatebird White-faced heron

WHITE-FACED HERON *Ardea novaehollandiae*

Vagrant

A medium-sized (26 in., 66 cm/46 in., 117 cm) bluish gray heron with a conspicuous white face and chin and long plumes on the back. The crown is blackish gray; the breast is streaked with chestnut. The black wing tips contrast with the rest of the plumage in flight. The bill is mostly black; the feet are greenish or reddish yellow. This Australian species, which has established itself naturally in New Zealand during the past 30 years, has been recorded once on Macquarie Island, in March 1957 [*Keith and Hines*, 1958].

GREAT EGRET *Egretta alba*

Vagrant

A large (35 in., 89 cm/63 in., 160 cm) all-white heron with blackish brown legs and a yellow or black bill. This cosmopolitan species straggled once or twice from South America to South Georgia in March 1964 and possibly in 1960 [*Tickell*, 1965; *Jeffries*, 1965] and twice from Australia or New Zealand to Macquarie Island in April 1957 and May 1951 [*Keith and Hines*, 1958; *Serventy*, 1952*a*] and possibly to Nightingale Island in April 1950 [*Elliott*, 1957].

YELLOW-BILLED EGRET *Egretta intermedia*

Vagrant

A medium-sized (26 in., 66 cm/49 in., 124 cm) all-white heron with a yellow bill and black legs. At some times of the year the exposed tibial portion of the leg may be yellowish. The yellow-billed egret is probably not distinguishable in the field from the great egret unless a direct size comparison can be made. This African and South Asian species has been recorded on Marion Island in mid-April 1951 [*La Grange*, 1962]. The same species may also have been responsible for

Great egret

Yellow-billed egret

an earlier record of 'eight small storklike birds, white with long necks and their legs sticking out behind, . . . seen flying around in a flock several times' on Marion Island in late March 1948 [*Crawford*, 1952]. The observer, however, suggested that they may have been South African 'tickbirds' (i.e., the cattle egret, *Bubulcus ibis*), a notorious old-world straggler across open oceans. That species is slightly smaller (20 in., 51 cm/37 in., 94 cm) than the yellow-billed egret and has a yellow or red bill, black or magenta feet, and buff plumes on its back in the breeding season.

Another unidentified white 'ardeid' also suggested to be a cattle egret was observed twice on Ile Amsterdam, in 1950 and in May 1951 [*Paulian*, 1953, p. 140; *Jouanin and Paulian*, 1954], and small flocks of white herons feeding among cattle have been reported on Tristan Island [*Elliott*, 1953].

SNOWY EGRET *Egretta thula*

Vagrant

A trim and delicate medium-sized (22 in., 56 cm/36 in., 91 cm) all-white heron with a black bill and legs and conspicuous yellow toes. In the breeding season, filamentous egret plumes adorn the back and breast (European little egret, *E. garzetta,* has plumes only on its back). This widespread American species, which breeds as far south as central Argentina, has been collected once on Tristan Island in early May [*Elliott*, 1953].

BLACK-CROWNED NIGHT HERON *Nycticorax nycticorax*

Vagrant

A stocky (24 in., 61 cm/43 in., 109 cm) short-legged heron with a heavy bill. The adult is black on the crown and back, gray on the

Snowy egret

Black-crowned night heron

wings and tail, and white on the underparts. It has long white crown filaments. The immature is brownish gray above with white spotting and white below with brown streaking. A heron observed on Ile Amsterdam in October 1969 was referred to this cosmopolitan species that migrates from Eurasia to Africa and India [*Prévost and Mougin*, 1970, p. 129].

WATERFOWL: Anatidae

Waterfowl are a diverse group of web-footed aquatic birds with flattened lamellate bills. The four resident antarctic ducks are mostly drab mottled brown, but drakes of some of the vagrants may be more brightly colored. Head and wing patterns, especially colors of the underwing and iridescent speculum on the secondaries, are useful for identification. The flight is fast and direct, with rapid wingbeats in ducks and slower more powerful beats in geese and swans. Waterfowl waddle awkwardly on land but are adept in the water. Ducks occur around inland ponds and glacial streams and, especially in the winter, frequent open coastal water. Their food consists of a variety of plant and animal matter, some taken on the land but most obtained by upending in shallow water. During the breeding season, waterfowl are territorial, but at other times they are gregarious. The distinctive and elaborate displays of dabbling ducks of the genus *Anas* have been studied in detail and show remarkable similarity from species to species. Their comparative study has contributed much to behavioral theory and duck classification. Courtship displays of the four species that breed in the Antarctic and sub-Antarctic are discussed at length by *Johnsgard* [1965], and therefore they will not be described in the species accounts. The nest, located on the ground, usually near water, is built of grass and lined with down from the female's breast. Clutches vary from 3 to 9 greenish or cream-colored eggs. The downy young, which are able to swim soon after hatching, are attended by both parents in the four resident

Black-necked swan　　　　Upland goose

species. Flight feathers are shed simultaneously, so that the birds are flightless for a brief period during molt. In addition to the four ducks that occur naturally on islands near the Antarctic convergence, a South American sheldgoose (now extirpated) was introduced on South Georgia, and six other temperate zone waterfowl species have occurred as vagrants on Macquarie Island and in the American sector.

BLACK-NECKED SWAN *Cygnus melanocoryphus*

Vagrant

A large (49 in., 124 cm/75 in., 190 cm) heavy-bodied white swan with black neck and head and a fleshy red knob on the forehead. The sexes are alike. The immature has gray flecks on its neck, a rusty gray tinge on its body, and no knob. An emaciated individual of this southern South American swan was captured in 'Charlotte Channel, South Shetland Islands' (i.e., Charlotte Bay, Antarctic Peninsula) in 1916-1917 [*Bennett,* 1922].

UPLAND GOOSE *Chloephaga picta*

Extirpated Introduction

Identification. 27 in. (69 cm)/64 in. (163 cm). A large white or reddish goose with a prominent white patch and metallic green speculum on the gray wing and black barring on the body. The sexes differ strikingly. The adult male is largely white with black barring on the upper back and flanks. The female is smaller and mostly deep reddish cinnamon with black barring on the upper back and underparts. The bill is black in both sexes; the feet are black in the males and yellow in the females. Young birds are duller and less clearly barred.

Flight and habits. Typical gooselike flight with powerful heavy wingbeats. The white wing patch is the most conspicuous flight character. The upland goose generally avoids water except when it is protecting its young during breeding and when it is flightless during molt. Gregarious outside of the breeding season.

Voice and display. Male has a high soft whistle; the female a deep harsh cackle.

Food. Grass, especially in close-cropped areas.

Distribution. This southern South American species was introduced from the Falkland Islands onto South Georgia to provide fresh meat for whale and seal hunters in 1910 or 1911. It inhabited tussock grass hillsides from sea level to the upper limits of vegetation but never became abundant. No information is available on its breeding biology on South Georgia. It was extirpated by 1950; another introduction in 1959 was extirpated shortly thereafter (W. L. N. Tickell, personal communication, 1968).

GRAY DUCK *Anas superciliosa*

Resident

Identification. Plate 8. 23 in. (58 cm)/35 in. (89 cm). A somber brown duck with white underwings and a striking pattern of black and buff stripes on the face. The back, flanks, and breast are blackish brown, each dark feather having a light margin, imparting a scaled appearance to the entire bird. The chin, throat, and remainder of the underparts are pale buff. The speculum is glossy green and black with white borders. The bill is lead gray with a black nail; the feet are yellowish brown with gray webs. The female is similar in color to but smaller than the male.

The juvenile is lighter brown with longitudinally streaked, rather than scaled, underparts.

The downy chick is brown above and buffy yellow on the face, underparts, and spots on the back and wings. A dark line runs from the bill through the eyes.

Flight and habits. A typical large dabbling duck similar in flight and habits to the northern hemisphere mallard, *Anas platyrhynchos*, or the North American black duck, *A. rubripes*. The gray duck tends to be crepuscular and timid on Macquarie Island.

Voice and display. A quacking, soft and subdued in the male and loud and harsh in the female. The courtship and other displays are identical in the female to and only slightly different in the male from those of the mallard [*Johnsgard,* 1965, pp. 174-176].

Food. The only food recorded from Macquarie Island is kelp fly larvae, but elsewhere the gray duck feeds on a variety of aquatic plants and animals.

Reproduction. Little information is available from Macquarie Island, where a 7-egg clutch was recorded in December and a 9-egg clutch in January and young were seen in late November, late January, and early February. The grass and down nest is presumably hidden in a tussock clump. The following information is from New Zealand and Kermadec Islands.

Eggs: The average clutch consists of 7 pale greenish or creamy white eggs; 50-60.5 × 41-42.5 mm.

Hatching: Incubation period 28 days.

Predation and mortality. No information is available. Although the gray duck was formerly hunted on Macquarie Island, it is now protected by law and consequently is fairly plentiful.

Ectoparasites. Tick, *Ixodes anatis;* nasal mite, *Rhinonyssus rhinolethrum.*

Habitat. Generally found on the coast, where flocks occur from March until May. The gray duck also frequents ponds, marshes, streams, and brackish lagoons.

Distribution. Breeds on Australia, Tasmania, and New Zealand and on Chatham, Snares, Auckland, Campbell, and Macquarie islands. Winter records on Macquarie Island suggest that the birds are resident.

MALLARD DUCK *Anas platyrhynchos*

Introduction, Status Uncertain

A large (24-26 in., 61-66 cm/33-44 in., 84-112 cm) duck with metallic green head, maroon breast, vermiculated brownish gray upperparts, and black tail. The speculum is blue and white. The female all year and the drake in eclipse are similar to the gray duck but lack a striped face pattern, are much lighter in color, and have a blue rather than green speculum. The mallard duck, a holarctic species that has been introduced on New Zealand, was first recorded on Macquarie Island in August 1949 [*Gwynn,* 1953*b*] and has been seen there several times subsequently [*Merilees,* 1971*a*]. The mallard may

Mallard duck

now be an established breeding bird on Macquarie Island. Four pairs were introduced on Iles Kerguelen in January 1959 but disappeared within 5 years. Others have been introduced subsequently, but their current status is unknown [*Prévost and Mougin,* 1970].

YELLOW-BILLED PINTAIL *Anas georgica*

Resident

Identification. Plate 8. 15 in. (38 cm)/28 in. (71 cm). A small mottled brown duck with reddish crown, light brown cheeks and throat, dark gray underwings, and pointed tail. The brown back, flanks, and breast are scalloped with buff; the underparts are buffy white, heavily mottled with brown. The bill is yellow with a blue and black line on the culmen and tip; the feet are greenish gray. In the male the upperwing coverts are uniformly brownish gray, and the speculum is glossy greenish black bordered with buff. The female is smaller and has mottled upperwing coverts and a dull brown speculum. The South American subspecies *A. g. spinicauda,* sometimes considered a separate species, is larger (21 in., 53 cm/32 in., 81 cm) and paler and has 14 tail feathers, whereas the South Georgia bird has 16.

The juvenile resembles the adult but presumably has underparts broadly streaked with brown (needs confirmation).

The downy chick is dark brown above, the throat, underparts, and markings on the face, wings, and back being yellowish buff.

Flight and habits. Flies infrequently and only for short distances in a typical ducklike fashion. The yellow-billed pintail is quite tame in summer but more wary in winter, when it hides in tussock grass.

Voice and display. The drake's call is a short shrill repeated whistle, frequently given in flight. The female has a soft quack and a gurgling note 'suggesting a bursting bubble.' The calls and displays are similar to the calls and displays of the northern hemisphere pintail, *Anas acuta,* with only minor differences in emphasis [*Johnsgard,* 1965, pp. 185-188].

Food. Amphipods, other small crustaceans, and filamentous marine algae in inland freshwater ponds and glacial streams and, when low tide exposes rocks and kelp, on coastal flats. Fjords are also favored feeding grounds. The bill is used for dabbling and straining food.

Reproduction. Solitary pairs. The nest is hidden in tussock, often some distance from the shore or water, and is composed of grass and down.

Pairing: Mid-November.

Eggs: Early December and probably also some in November. Eggs in February and March may be replacement clutches necessary because of predation. Clutch, 5 yellowish cream eggs; measurements unrecorded.

Hatching: Most in late December. Incubation period not recorded. Young leave the nest and can swim soon after hatching.

Fledging: Throughout January.

Molt. Little information is available; body molt occurs in February.

Predation and mortality. The main enemy is the brown skua, which preys on both eggs and young. The duck may also fall victim to introduced rats (needs confirmation). Males outnumber females, a high mortality of breeding hens thus being suggested (needs further study). When the female is disturbed, it feigns wing injury, allowing the young to scatter and hide. The yellow-billed pintail was previously the only native game bird on South Georgia. It was hunted for food and consequently became rare, but the population is not in danger at present.

Ectoparasites. Nasal mite, *Rhinonyssus rhinolethrum.*

Habitat. Inland valleys, coastal flats, and offshore islands, where it occurs on both freshwater and salt water. Pintails prefer ponds surrounded by tussock and other grasses. They gather in flocks of hundreds along the fjord coasts in winter but are more widespread in small family groups during the summer.

Distribution. Resident all year on South Georgia but not recorded from the exposed south coast. The South American form occurred on Deception Island, South Shetland Islands, in 1916 and 1917 [*Bennett,* 1922] and possibly on Signy Island, South Orkney Islands, in late October 1965 [*Burton,* 1967]. An individual seen on Breaker Island off Anvers Island in mid-January 1975 (D. F. Parmelee, personal communication, 1975) may have been from either the South Georgia or the South American populations.

KERGUELEN PINTAIL *Anas (acuta) eatoni*

Resident

Identification. Plate 8. 15 in. (38 cm)/28 in. (71 cm). A small reddish brown duck with mottled underwings and bluish gray bill. At all times both sexes have the head, foreneck, and breast reddish brown spotted with black, the remainder of the underparts are light brown, and the feet are olive brown. The wing coverts of the male are uniformly gray, and the speculum is glossy green bordered with buff and white, whereas the female's wing coverts are mottled with buff, and the brown speculum has no gloss. Males between November and April, and females all year, are blackish brown above, each feather being edged with buff, imparting an overall mottled appearance to the plumage. In courtship plumage, from May to October, the male is somewhat more brightly colored: the flanks, wing coverts, and back are barred and mottled with black; the black central tail feathers and scapulars are elongated. Rarely, a drake may develop a more extreme plumage suggesting the pattern of the northern hemisphere pintail, *A. acuta*. In this plumage the head is dark brown with whitish bands on the cheeks and sides of the neck. The upper breast is rusty, and the lower breast and sometimes all the underparts are grayish white. The buffy white upperparts and flanks are coarsely mottled and vermiculated with black.

The juvenile resembles the adult female, but the underparts are streaked longitudinally.

The downy chick is brown above and grayish white below with a narrow dark line through the eyes. The eyebrows, cheeks, and bars on the back and wings are rufous.

Flight and habits. A strong flyer that rises vertically from land or water and tends to fly high. The Kerguelen pintail is agile on the ground, like a grouse rather than a duck. It is tame and easy to approach. In the water it dives without using its wings like a true diving duck (tribe Aythini) rather than with open wings like other dabbling ducks (tribe Anatini).

Voice and display. Males whistle, and females quack or 'wheeze' in flight. The calls are like those of a green-winged teal, *Anas crecca*, and are a little higher in pitch than those of the northern hemisphere pintail. Although the distinctive male courtship plumage has been lost by the drake Kerguelen pintail, all of the numerous elaborate displays of its northern hemisphere relative are present [*Johnsgard*, 1965, pp. 182-185].

Food. Both plant and animal matter including seeds of Kerguelen cabbage, roots of *Azorella*, grass seeds, earthworms, insect larvae, crustaceans (isopods and amphipods), and small fish. The Kerguelen pintail feeds on tidal flats at low tide.

Reproduction. Solitary pairs. Nest sites are variable: coastal cliffs, rocky clefts in inland precipices, or the ground under tussock grass or *Acaena* clumps. The nest is a carefully concealed cup of grass lined with moss or down and measures 13-18 cm in diameter.

Pairing: Early November.

Eggs: Mid-December. Eggs recorded in February are probably replacement clutches. Clutch, 3-6 pale olive green eggs; average 51 × 36 mm.

Hatching: No information. Incubation period unknown. Young leave the nest and are able to swim soon after hatching.

Fledging: At latest by the end of March.

Molt. Twice per breeding cycle. Complete molt ends in April; body molt in October (both need confirmation). Abrupt wing molt leaves adults flightless in April.

Predation and mortality. Eggs and young are taken by skuas and possibly also by introduced rats. The Kerguelen pintail was formerly hunted by man but is now abundant on Iles Kerguelen. It is apparently uncommon on Iles Crozet (more data needed).

Ectoparasites. No information is available.

Habitat. Wet areas from sea level to snow line but rarely above 450 m. The Kerguelen pintail also occurs at sea around the breeding islands.

Distribution. Resident on Iles Kerguelen and Iles Crozet and introduced on Ile Amsterdam and Ile Saint-Paul. The species *Anas acuta* is holarctic and somewhat migratory. Presumably, the Indian Ocean populations are relicts of migratory vagrants that reached the islands during the Pleistocene.

SPECKLED TEAL *Anas flavirostris*

Newly Arrived Resident

Identification. 16 in. (41 cm)/30 in. (76 cm). A small brown duck with strongly spotted grayish white underparts and a finely barred and speckled gray head. The flanks are grayish brown. The gray wing has a velvety black speculum with a metallic green inner margin and narrow buffy white edging on the secondaries. The bill is yellow on the sides and black on the ridge and nail of the culmen. The iris is brown; the feet are dark gray. The sexes are similar, but the female is duller and slightly smaller than the male and has a less bright yellow

Speckled teal

bill. This species is grayer and less reddish than the yellow-billed pintail on South Georgia and has more pronounced spotting on the breast.

The juvenile is duller still than the female and less heavily spotted on the underparts.

The chick is generally dark brown above and yellowish below with strong light and dark horizontal streaks on the face and flanks.

Flight and habits. A tame and confiding species but capable of rapid and agile flight. During breeding, pairs and family parties are seen together, but flocks gather at other times.

Voice and display. Similar to the voice and display of the closely related northern green-winged teal, *Anas crecca* [*Johnsgard,* 1965, pp. 152-155], but the drake whistles with no head movement and in the greeting ceremony stretches his head and neck horizontally while he utters a rapid multisyllabic twittering. Elongated feathers at the back of the neck are partly erected during courtship. The female has a harsh scraping quack.

Food. In the summer the speckled teal eats mainly plant material from the margins of small freshwater ponds but also small crustaceans (*Daphnia* and *Cladocera*), other invertebrates, and aquatic insects obtained from the surface by straining through the bill and by upending in shallow water. Its winter food habits are unknown.

Reproduction. On mainland South America, pairs adopt a variety of nesting sites hidden in vegetation on the ground or in holes in banks. They breed from August to October with a possible second brood later in the summer. The clutch is small for a duck, 5-6 pale cream eggs (average 52 × 38 mm) that hatch in 26 days. The male assists in caring for the young. A brood of three ducklings was seen on South Georgia in early December.

Molt. No information is available.

Predation and mortality. Little information is available on South Georgia, where the total population was estimated at 40-50 birds in November and December 1971. Males outnumbered females 2.5:1.

Ectoparasites. Feather lice, *Anaticola crassicornis, Anatoecus* sp., *Trinoton querquedulae,* and *Holomenopon clypeilargum.*

Habitat. In the breeding season the teal frequent deep permanent glacial ponds that are rich in invertebrate life and the tussock grass surrounding them. Possibly, the teal use unfrozen streams or coastal waters in winter, or they may even migrate northward.

Distribution. The species is widespread in southern South America, the Andes, and the Falkland Islands. It was discovered breeding at Cumberland Bay, South Georgia, in 1971 [*Weller and Howard,* 1972]. Whether this is a natural colonization or an artificial introduction of a species that does well in captivity is unknown. It is resident in the Falkland Islands but is a migrant in southern South America.

GRAY TEAL *Anas gibberifrons*

Vagrant

A small (17 in., 43 cm/28 in., 71 cm) dark brown duck. The gray teal is similar to the gray duck but lacks the prominent black and buff head striping of that species. The head is dark brown, the cheek and the throat being a grayish white. The sexes are alike. This fresh-water Australian and New Zealand species has been reported several times from Macquarie Island [*Keith and Hines,* 1958].

CHILOE WIGEON *Anas sibilatrix*

Vagrant

A moderately large (24 in., 61 cm/34 in., 86 cm) black and white duck with a white face and white wing patches. The male has a

Gray teal Chiloe wigeon

greenish black head and neck and a black back and breast streaked and barred with white. The flanks are rusty; the rump and underparts are white. The female is a little duller than the male. An emaciated male of this South American species was found dead on Signy Island, South Orkney Islands, in mid-October 1966 [*Beck,* 1968*a*].

UNIDENTIFIED TEAL *Anas* sp.

Vagrant

A flock of small wild ducks, several of which were shot, was reported in freshwater ponds on Tristan Island 'some years' before 1950 [*Elliott,* 1953].

A few bones of a small duck, similar in size to a garganey teal, *Anas querquedula,* were found on Ile Amsterdam [*Jouanin and Paulian,* 1954].

ARGENTINE RUDDY DUCK *Oxyura vittata*

Vagrant

A small (16 in., 41 cm/24 in., 61 cm) chunky duck that usually carries its stiff tail erect while it is swimming. The breeding male has a

Argentine ruddy duck

black head and neck, a reddish chestnut body, and usually a whitish abdomen. The bill and feet are blue. The female and the drake in eclipse plumage are brown with indistinct black vermiculation. A horizontal stripe on the cheek, the chin, and the underparts are silvery white. This southern South American species was recorded on Deception Island in 1916-1917, when numbers arrived following a severe drought in Argentina [*Bennett,* 1920].

BIRDS OF PREY: Falconiformes

The birds of prey are a diverse group of terrestrial predators including hawks, eagles, harriers, falcons, and vultures. They are characterized by hooked bills and strong legs and feet with curved talons. They hunt visually on the wing by day, pounce on their prey, and carry it off in their talons. Some species eat carrion. Two vagrant birds of prey have occurred on Macquarie Island.

MARSH HARRIER *Circus aeruginosus*

Vagrant

A medium-sized (23 in., 58 cm/46 in., 117 cm) brown bird of prey with long wings and tail. Adults are dark brown above with a white nape and rump. The underparts are buff with brown streaks, and the underwings are pale rufous buff. Harriers are highly variable in color, becoming lighter with increasing age. Some are nearly white below and on the head and pale gray on the wings and tail. The immature is dark chocolate brown with a rufous rump and a buff patch on the nape. The wings show a pronounced dihedral as the bird hunts in a crisscross pattern low over marshes and wet meadows. This Australian harrier, *C. a. approximans,* here considered conspecific

Marsh harrier

Banded rail

with a widespread old-world species, has been recorded 3 times as a vagrant on Macquarie Island, in late October 1949 and June-July and September 1960 [*Gwynn,* 1953*b*; *Warham,* 1969].

HAWK

Vagrant

Another 'hawk, distinctly smaller than the swamp harrier' (i.e., marsh harrier) was seen on Macquarie Island during September and October 1957 [*Keith and Hines,* 1958].

RAILS AND COOTS: Rallidae

Rails belong to a cosmopolitan family of somber-colored semi-aquatic birds with strong legs and feet. A flightless small rail and a flightless gallinule occur in the Tristan da Cunha group and on Gough Island, the flightless New Zealand weka has been introduced on Macquarie Island, coots have occurred there and at Tristan Island as vagrants, and another gallinule has occurred on South Georgia and the Tristan da Cunha group. A population of the widespread western Pacific banded rail (*Rallus philippensis macquariensis*) formerly occurred on Macquarie Island but was extirpated by 1894 [*Falla,* 1937].

INACCESSIBLE ISLAND FLIGHTLESS RAIL *Atlantisia rogersi*

Resident

Identification. Plate 11. 6.5 in. (17 cm)/7 in. (18 cm). A diminutive all-dark rail with indistinct buff or white transverse barring on the wings, flanks, thighs, belly, and undertail and a fiery red eye. The furlike plumage of the back is reddish brown; the face, throat, and breast are slate gray. The bill is black with some reddish on the sides of the lower mandible; the feet are blackish brown. Birds of intermediate age are brown above and slate gray below but lack any barring. Adult and intermediate males have blackish cheeks and ear coverts, whereas in females these areas are gray. The juvenile is almost wholly sooty black, sometimes washed on the back with brown or on the belly with cinnamon. The eyes are brown. The downy chick is black.

Flight and habits. Flightless. The movements of the Inaccessible Island flightless rail are deliberate, but it runs with great speed and takes cover in tussock. It stays in family groups of three to five birds. It is active by day and at dusk.

Voice and display. The warning note is a high metallic 'pseep,' 'tick,' or 'tjupp'; the contact and feeding call is a mournful repeated chicklike 'tee ap' or a wooden 'chunk, chunk'; the meeting call is a high whistled 'dabchicklike' trill.

Food. Seeds, berries, and insects. Grit is also ingested.

Reproduction. Territorial in pairs. The nest, hidden in a tussock or under a patch of flattened water grass, is a leaf-lined chamber about 8 cm in diameter in a large ball of tussock approached through a 30-cm-long tunnel in the undergrowth.

Pairing: No information.

Eggs: October to November, late (replacement?) layings occurring in December and January. Clutch, 2-3 grayish milk white eggs tinged with buff and spotted all over, most strongly at the apex, with rufous chocolate and lavender; 35 × 25 mm.

Hatching: Little direct information. Chicks seen into late February. Incubation period unknown. Chicks probably leave the nest soon after hatching.

Fledging: No information.

Molt. No sign of molt from November to March.

Predation and mortality. The Inaccessible Island flightless rail is presumably preyed on by skuas and is known to be harried by thrushes. The total population is estimated at 1200-10,000.

Ectoparasites. No information is available.

Habitat. Dense tussock, rushes and ferns in bogs, and steep slopes and the highest ridges up to at least 550 m. Paths in heavy undergrowth function as burrows.

Distribution. Resident on Inaccessible Island, Tristan da Cunha group.

WEKA *Gallirallus australis*

Introduced Resident

Identification. Plate 8. 21 in. (53 cm)/23 in. (58 cm). A large dark flightless rail with a strong bill, webless feet, and reduced wings. The weka occurs in two color phases, both of which are mainly brown with black barring, streaking, and spotting. In the light phase the head, back, and breast are mostly chestnut; the remainder of the underparts are grayish brown; and the chin, throat, cheeks, and eyebrows are gray. In the dark phase the body feathers are brownish black narrowly fringed with chestnut, the facial and breast patterns are indistinct, and the chin and belly are dark gray. The bill and feet are reddish brown; the iris is red. In both phases, males are markedly larger than females.

The juveniles of both phases resemble the adults, but the plumage has a softer texture, the iris is dark brown, and the legs are gray. In the light-phase juvenile the color is generally darker, and the pattern is less defined than that of the adult.

The light-phase downy chick is uniformly grayish brown with no pattern, whereas the dark-phase chick is said to be all black (needs confirmation).

Flight and habits. Flightless. The weka is inquisitive in disposition but secretive and crepuscular in habits. It uses tussock grass as shelter and cover but may occasionally feed in the open. It frequently flicks its tail while it is walking and can run rapidly. It swims well.

Voice and display. A shrill loud repeated whistle 'coo-eet.' Several birds may call simultaneously.

Food. On Macquarie Island the weka occupies the same scavenging and predatory niche as sheathbills do elsewhere. It takes kelp fly larvae, small fish, and marine invertebrates on beaches; farther inland it feeds on earthworms, rodents, and birds' eggs and young and eats refuse around the weather station. The weka turns over debris with its bill and uses its powerful legs for digging out petrel burrows. It is also a capable ratcatcher.

Reproduction. Solitary pairs. The nest is a loosely constructed cup of dead grass and leaves usually well hidden under vegetation. The breeding season is long and irregular; chicks have been recorded on Macquarie Island from October to May (probably representing multiple broods). Wekas mature early and may breed at the age of 9 months.

Eggs: Clutch size and egg data are unrecorded on Macquarie Island, but clutches of 2 or 3 eggs have been reported on Stewart Island, and up to 6 have been reported farther north.

Molt. No information is available.

Predation and mortality. No predation has been recorded. The weka was originally introduced on Macquarie Island to provide food for whalers and is presently abundant there.

Ectoparasites. Feather louse, *Rallicola harrisoni.*

Habitat. Open sea coast and dense vegetation of moist banks, hillsides, and marshes. The weka occurs on tussock slopes from the beach to the upper limits of growth in the highlands.

Distribution. Resident on New Zealand and introduced from

Stewart Island onto Macquarie Island, where it is now numerous, in 1872.

GOUGH MOORHEN

Gallinula nesiotes
(includes *G. comeri*)

Resident

Identification. Plate 11. 10.5 in. (27 cm)/18 in. (46 cm). A nearly black heavyset chickenlike bird with a bright red bill and heavy yellow legs and feet irregularly marked with red. The head is black; the body plumage is sooty gray washed with brown on the back and wings. Undertail coverts and the extreme tips of some flank feathers are white. The conspicuously swollen frontal plate and base of the bill are coral red, but the tip is yellow.

The immature is mostly brown, lighter on the sides and whitish on the belly. The bill, small frontal shield, legs, and feet are olive greenish.

The downy chick is entirely black including its legs and feet. The bill is horny yellow, banded and tipped with black.

Flight and habits. The wings are so reduced that they are useless for flight but are used for balance in running swiftly. The Gough moorhen bathes in freshwater but avoids salt water. It generally stays hidden in vegetation, especially when it is disturbed, but appears in the open when there is no danger.

Voice and display. The call is a repeated harsh 'tcherk aaa kak.' The alarm note is a shrill metallic whistle or staccato 'chack-chack.' The white undertail coverts are displayed in courtship and alarm.

Food. Follows the high-tide line in search of stranded littoral organisms and feeds on grass heads with a scythelike motion of the beak. The Gough moorhen also scratches in ground litter.

Reproduction. Little information is available. The nest, which is well hidden in heavy cover, is a shallow scrape on the ground lined with dead grass. Presumably, breeding takes place early in the spring, and some pairs are double brooded. A downy chick was found in mid-December, and juveniles were found in February. Eggs have not been discovered.

Molt. In progress in February.

Predation and mortality. The Gough moorhen is generally common, but its absence from open areas suggests potential predation by skuas.

Ectoparasites. No information is available.

Habitat. Confined to the forest and tree fern zone of the island, avoiding open ground on the upper mountains. The Gough moorhen is especially common along the shore, in boggy areas, and in grass and fern undergrowth along streams.

Distribution. Resident on Gough Island (*G. n. comeri*) but extirpated from Tristan Island (*G. n. nesiotes*) in the late nineteenth century [*Beintema,* 1972].

PURPLE GALLINULE *Porphyrula martinica*

Vagrant

A stocky (10.5 in., 27 cm/21 in., 53 cm) green-backed water bird with purple head and underparts, white undertail coverts, and long yellow legs. The bill is red, and the frontal shield is bluish white. The immature has a brown head, tan underparts, and duller bill and feet than the adult has. An immature of this tropical American species, which ranges from the southern United States south to northern Argentina, was recorded on South Georgia in mid-June 1943 (type of *Porphyrula georgica* [*Pereyra,* 1944]), whereas numerous individuals, mostly immatures, have been recorded on Tristan Island in March to July and once in November.

EURASIAN COOT *Fulica atra*
UNIDENTIFIED COOT *Fulica* sp.

Vagrant

A chunky sooty black aquatic rail (15 in., 38 cm/30 in., 76 cm) with a small white patch on the trailing edge of the wing and a bluish white bill and frontal shield (*F. atra*). The grayish green toes have fleshy lobes. The immature is grayish brown above and grayish white on the throat and upper breast. The coot swims with a head-bobbing motion and dives readily. In taking off it patters its feet along the

Purple gallinule

Eurasian coot

surface before becoming airborne. The legs trail in flight. The cosmopolitan Eurasian coot breeds in Australia, Tasmania, and New Zealand. Several stragglers arrived on Macquarie Island in May, and some survived on the island until October 1957 [*Keith and Hines,* 1958].

A 'quite young' coot was collected on Tristan Island before 1909 [*Sclater,* 1911]. The specimen was studied by A. J. Beintema (personal communication, 1973), who reports it to be one of the South American species, possibly the red-gartered coot, *F. armillata* (the adult has no white on the trailing edge of the wing, the bill is yellow, the frontal shield is red, bordered with yellow, the uppermost portion of the leg is red, and the toes are olive green).

UNIDENTIFIED RAIL

Vagrant

A poorly preserved mummy of a small rail similar in size to a corn crake, *Crex crex,* was found on Ile Amsterdam but disintegrated during collection before it could be identified precisely [*Jouanin and Paulian,* 1954].

PLOVERS: Charadriidae

Plovers are small to medium-sized shorebirds that frequent beaches and mud flats. They have large heads, short pigeonlike bills, and short thick necks. During breeding they have brighter markings than they do at other times. Two or possibly three vagrant species have occurred in the Antarctic.

BLACK-BELLIED PLOVER *Pluvialis squatarola*

Vagrant

A stocky plover (11 in., 28 cm/24 in., 61 cm) with a pale silvery gray dappled back, white rump and tail, and black axillary feathers. The face and underparts are white from October to February but become black from March to August. The black-bellied plover is a rare holarctic migrant in Australia and New Zealand that has been recorded once on Macquarie Island [*Simpson,* 1965].

CHILEAN DOTTEREL *Zonibyx modestus*

Vagrant

A large-headed (9 in., 23 cm/19 in., 48 cm) brown-backed plover with an orange chestnut breast separated from the white abdomen and underwings by a black band. The face is gray; the forehead and

Black-bellied plover Chilean dotterel

wide eyebrows are white; the bill is black. Nonbreeding adults and immatures are grayish brown above, light brown on the breast, and white on the rest of the underparts. A vagrant of this southern South American and Falkland Island species was collected on Tristan Island in mid-May 1952 [*Elliott,* 1953].

UNIDENTIFIED SHOREBIRD

Vagrant

Two large shorebirds with a conspicuous dark and white pattern on the wings were seen on Ile Amsterdam in October 1969 [*Prévost and Mougin,* 1970, p. 132]. They may possibly have been stray white-tailed plovers, *Chettusia leucura,* from central Asia, which winter in Egypt and India, or, less likely, American willets, *Catoptrophorus semipalmatus,* which may occur as vagrants in Europe.

SANDPIPERS: Scolopacidae

Sandpipers are small to medium-sized wading birds that occur on beaches, mud bars, and shallow ponds. They have longer bills, smaller heads, and longer legs than plovers. Most are cryptically colored, brown barred or spotted with black, in the breeding season and grayer at other times. The 13 species that have occurred in the Antarctic are well-traveled migrants that breed in the Arctic and winter in the southern hemisphere temperate zone.

UPLAND SANDPIPER *Bartramia longicauda*

Vagrant

A medium-sized (11 in., 28 cm/23 in., 58 cm) brown and buff streaked shorebird with a relatively small head, short bill, long tail, and long yellow legs. There is no wing pattern, and the rump is dark.

The upland sandpiper tends to fly with stiff wingbeats like common or spotted sandpipers (*Actitis*). This North American migrant straggled to Deception Island in February 1923 [*Dabbene,* 1923], to Tristan Island in mid-October 1952 [*Elliott,* 1953], and possibly to Signy Island, South Orkney Islands, in December 1962 and January 1963 (tentative secondhand identification by *Holdgate* [1965*a*]).

COMMON CURLEW *Numenius arquata*

Vagrant

A very large (22 in., 56 cm/49 in., 124 cm) heavy-bodied wader with a long (4-in., 10-cm, to 6-in., 15-cm) decurved bill, white rump, and barred tail. The back is streaked and mottled brown and buff. The head, neck, and upper breast are distinctly streaked, whereas the rest of the underparts are creamy white. The common curlew breeds in Eurasia and winters south to southern Africa and southeast Asia. It was recorded by an unreliable observer on Ile Saint-Paul in November 1874 [*Vélain,* 1877]. A Far Eastern species, the eastern long-billed curlew, *N. madagascariensis,* which migrates to New Zealand and might turn up on Macquarie Island, lacks the white rump.

The whimbrel, *N. phaeopus,* a somewhat smaller species (17 in., 43 cm/40 in., 101 cm) with two distinct dark brown bars down the center of the crown, may have been sighted on Ile Saint-Paul in early February 1970 [*Segonzac,* 1972].

BAR-TAILED GODWIT *Limosa lapponica*

Vagrant

A large (15 in., 38 cm/32 in., 81 cm) wader with a long (3.5-in., 9-cm) lightly upturned bill and long legs and neck. The lower back and

Upland sandpiper Common curlew

Bar-tailed godwit Greenshank

rump are white; the tail is barred brown and white. The upperparts are mottled brown and gray; the underparts are grayish white. In January or February, males assume the brick red underparts of breeding dress. Females are larger and show only traces of buff on underparts. The bar-tailed godwit breeds in eastern Siberia and northwestern North America and migrates to New Zealand, where it is abundant in the austral summer. Stragglers were recorded on Macquarie Island in October 1912 [*Falla,* 1937] and in December 1951 [*Lindholm,* 1952].

GREENSHANK *Tringa nebularia*

Vagrant

A large (12 in., 30 cm/25 in., 64 cm) wader with a white lower back and rump and a faintly barred white tail. The upperparts are light gray flecked with black and white; the wings appear black in flight. The underparts are pure white from September to March, but the breast has small black spots from April to August. The slightly upturned 2-in.-long (5-cm-long) bill is shorter than that of a godwit. A regular Palaearctic migrant to Africa and Indian Ocean islands that is rare in both Australia and New Zealand, the greenshank has occurred several times on Ile Amsterdam from August to January [*Paulian,* 1956; *Segonzac,* 1972], once on Iles Kerguelen in May 1952 [*Jouanin and Paulian,* 1954], once on Macquarie Island in 1962 [*Simpson,* 1965], and on Iles Crozet [*Prévost and Mougin,* 1970, p. 124].

COMMON SANDPIPER *Actitis hypoleucos*

Vagrant

A small (8 in., 20 cm/14 in., 36 cm) sandpiper with chocolate brown back and rump and white underparts. When it is standing, it shows a dark smudge at the side of the breast and nervously bobs its tail. In its distinctive low flight, shallow stiff wingbeats alternate with brief glides on downcurved wings. A Palaearctic species that migrates to South Africa, Indian Ocean islands, southern Asia, and Australia, the common sandpiper was recorded on Ile Amsterdam in mid-December 1953 and October 1956 [*Paulian,* 1956; *Segonzac,* 1972]. Possibly, this was also the small charadriiform species seen on Ile Amsterdam in 1952 [*Jouanin and Paulian,* 1954].

SPOTTED SANDPIPER *Actitis macularia*

Vagrant

In nonbreeding plumage the spotted sandpiper is indistinguishable from the common sandpiper, but in breeding plumage it has strong black spotting on the underparts. This North American species that migrates to Brazil, northern Chile, and occasionally Argentina has been collected once on Tristan Island in early February 1952 [*Elliott,* 1953].

RUDDY TURNSTONE *Arenaria interpres*

Vagrant

A short-legged ploverlike shorebird (7 in., 18 cm/12 in., 30 cm) with a chestnut, black, and white harlequin pattern on the back. An irregular black breastband crosses the white underparts and extends onto the cheeks. The nonbreeding adult and immature have a less striking brown and white marbled pattern on the back and a black or brown breastband. Two shorebirds observed on Ile Amsterdam in October 1969, three on Ile Saint-Paul in March 1971, and three on Ile de l'Est, Iles Crozet, in February 1971 were identified as this abundant

Common sandpiper

Ruddy turnstone

holarctic species that migrates to tropical islands [*Prévost and Mougin,* 1970; *Segonzac,* 1972; *Despin et al.,* 1972].

JAPANESE SNIPE *Gallinago hardwicki*

Vagrant

A short-legged long-billed (3-in., 8-cm) wader (13 in., 33 cm/21 in., 53 cm) with strongly streaked brown, black, and tan upperparts, buffy underparts, and barred flanks. The Japanese snipe takes off suddenly, almost from underfoot, calling harshly in a twisting and rapid flight. *Gwynn* [1953*b*] saw a snipe 'presumably' of this species on Macquarie Island in November 1949, and *Warham* [1969] made one 'probable' sighting in November 1960. The Japanese snipe breeds in Japan, migrates to Australia, and is uncommon on New Zealand.

KNOT *Calidris canutus*

Vagrant

A stocky short-legged wader (10 in., 25 cm/22 in., 56 cm) with a short (1-in., 3-cm, to 1.5-in., 4-cm) straight bill, barred rump, and indistinct whitish wing bar. In nonbreeding dress the upperparts are gray lightly speckled with black. The underparts are pale grayish white. From February to April the head, neck, and breast molt to rufous, and the black and silver speckling of the mantle becomes more intense. The knot is a regular migrant to New Zealand from eastern Siberia that was recorded on Macquarie Island in November 1913 [*Falla,* 1937].

WHITE-RUMPED SANDPIPER *Calidris fuscicollis*

Vagrant

A very small (7 in., 18 cm/15.5 in., 39 cm) wader with a wholly white rump, indistinct wing stripe, and short bill. The breeding

Japanese snipe

Knot

White-rumped sandpiper Sharp-tailed sandpiper

plumage is rusty above; the nonbreeding plumage is gray above; in both the underparts are white. The white-rumped sandpiper is a North American migrant that was photographed on South Georgia in November 1958 [*Tickell*, 1960].

Tickell [1960] also alludes to an 'unidentified wader' seen on Signy Island, South Orkney Islands, in November 1954.

SHARP-TAILED SANDPIPER *Calidris acuminata*

Vagrant

A long-necked (9 in., 23 cm/18 in., 46 cm) sandpiper with a pointed tail and lightly spotted and streaked buff and white underparts. The back, which is brown in breeding season and gray otherwise, shows snipelike buff streaks. In flight the sharp-tailed sandpiper lacks wing bars, but white patches at the sides of the rump contrast with very dark central tail feathers. A long-distance vagrant of this northeast Siberian species that migrates to Australia, New Zealand, and southwest Pacific islands was collected on Tristan Island in mid-June 1950 [*Elliott*, 1953].

CURLEW SANDPIPER *Calidris ferruginea*

Vagrant

A small (7 in., 18 cm/12 in., 30 cm) long-legged and long-necked shorebird with a slender curved bill and white rump. The breeding plumage is rusty red with grayish wings; the nonbreeding plumage is gray brown above and white below. A group of five sandpipers seen at Ile Amsterdam on October 7-10, 1969, was referred to this Palaearctic species that migrates to South Africa, India, and tropical Indian Ocean islands [*Prévost and Mougin*, 1970].

Curlew sandpiper

Sanderling

SANDERLING *Calidris alba*

Vagrant

A small (6.5 in., 16 cm/11 in., 28 cm) active shorebird with a conspicuous white wing stripe and contrasting dark 'wrists' and wing tips. The back is reddish brown in breeding plumage and very pale gray in nonbreeding plumage. A specimen of this holarctic migrant that commonly reaches tropical oceanic islands was collected on Tristan Island by Keytel before 1910 [*Winterbottom*, 1958].

PHALAROPES: Phalaropodidae

Phalaropes are dainty sandpiperlike northern hemisphere shorebirds with dense plumage and peculiar lobed toes that permit them to swim buoyantly while they feed on small surface organisms. They have a distinctive spinning motion while they are swimming. Females are larger than males and, in their distinctive breeding plumage, brighter. Two species have occurred as vagrants in the Antarctic.

RED PHALAROPE *Phalaropus fulicarius*

Vagrant

A pelagic (8 in., 20 cm/15 in., 38 cm) phalarope with a white wing bar and dark rump with white sides. From about July to March the back is pale bluish gray; the face and underparts are white with a conspicuous dark line through the eyes. The bill is black, occasionally with a little yellow at the base. From April to July it is mostly reddish chestnut with white cheeks and a strongly scaled buff pattern on the back. The bill is bright yellow with a dark tip. The red phalarope breeds in the holarctic tundra and winters far out at sea, regularly south to both coasts of southern South America and western South Africa. It has occurred as a vagrant in New Zealand. An adult male in chestnut breeding plumage with bright yellow bill was collected on

Red phalarope

Wilson's phalarope

Anvers Island, Antarctic Peninsula, in mid-January 1970 (specimen in the National Museum of Natural History, Smithsonian Institution).

WILSON'S PHALAROPE *Steganopus tricolor*

Vagrant

A slim (9 in., 23 cm/15 in., 38 cm) phalarope with white underparts, dark gray wings, gray back and tail, and conspicuous white rump. The face and neck are mostly white in nonbreeding plumage, but the neck shows a broad chestnut stripe extending from the eye to the back during the breeding season. An extreme vagrant of this North American species that winters inland in South America was found dead at Fossil Bluff, Alexander Island, in mid-October 1968 [*Conroy*, 1971*a*].

SHEATHBILLS: Chionididae

All-white land birds that in appearance and behavior strongly suggest either diminutive domestic fowl or pigeons. The face is adorned with fleshy wattles at the base of the bill and below the eyes, and a large horny sheath covers the nasal openings on the compressed conical bill. The sheathbills' pigeonlike flight is strong and capable of carrying them considerable distances, but in traveling short distances they fly low, their feet dangling awkwardly. Their legs and feet are robust with only rudimentary webs between the front toes. Generally, sheathbills prefer to use their legs rather than their wings to keep them out of trouble. Unafraid and highly inquisitive, these birds will pause to investigate and peck at anything left on the ground. They are quarrelsome and use the short carpal spurs on their wings in fighting among themselves. During the summer months, sheathbills usually inhabit penguin and cormorant colonies, where they impudently distract adult birds from feeding their young, causing them to drop the food, which the sheathbills quickly retrieve. Working in pairs and using both patience and cunning, sheathbills harvest considerable numbers of eggs from attending penguins during the laying and incubation period. They also consume littoral marine life, bird carcasses, and placentas from calving seals, and especially during the winter they are frequent visitors to camp refuse dumps. Between breeding seasons they occur in small flocks and feed communally. Although sheathbills can swim well, they usually avoid water except when they are bathing. The nest is a slovenly heap of tussock grass, sticks, feathers, bones, and decayed food hidden under a large rock or concealed in a deep crevice. The conspicuous adults standing on an exposed rock near the nest site betray its presence. Petrel or rabbit burrows may be utilized where they are available. The 2-5 creamy eggs are heavily blotched with brown and purple markings. The brown- and gray-speckled chicks are fed at the nest by both adults until they are ready to fly. Two species of sheathbills, which differ in facial wattlings, bill shape and color, and leg color, are found in the Antarctic Peninsula and Scotia Sea region and islands in the southern Indian Ocean. The American population is partly migratory.

AMERICAN SHEATHBILL *Chionis alba*

Resident and Partial Migrant

Identification. Plate 8. 16 in. (41 cm)/31 in. (79 cm). A plump all-white bird with white facial papillae, variegated bill, and bluish gray feet. The bill is mostly brown with subdued yellow, red, and black markings. The nasal openings are covered with a hard sheath that adheres closely to the bill contours. Females tend to be smaller than males and have reduced facial papillae and smaller bills.

The juvenile may be identified immediately after fledging by the traces of down still adhering to the tips of the white body feathers, causing the plumage to appear dirty. Its bill is weaker than that of the adult, and the bare facial areas are less wattled. How long yearling sheathbills retain these characters of immaturity is not known. Another aging character, only of use in the hand, is the shape of the tips of the primaries: those of the immature are pointed rather than somewhat rounded like those of the adult. Because of wear, however, this character is usually only good for inner and unworn primaries.

The downy chick at hatching is brown, a little lighter on the head and tufted with white on the chin and below the eyes. This feather down is subsequently pushed out on the tips of the developing white juvenile feathers. When the chick is about 12 days old, a dark bluish gray body down commences to grow in the bare areas between the feather tracts, gradually producing an overall gray appearance just before the white feathers begin to show.

Flight and habits. Wingbeats are strong and rapid, and the tail is spread in flight. On the ground the American sheathbill walks with a pigeonlike bobbing of the head. It is very tame, on occasion even permitting capture by hand or short-handled nets.

Voice and display. A harsh and throaty crowlike call and low muttered sounds accompanied by mutual deep bowing are conspicuous in courtship.

Food. Omnivorous, taking any seasonally available food. Euphausiid shrimp obtained by direct interference with penguins feeding chicks seem to be the primary food during the summer, but penguin and cormorant eggs, excrement, and, to a lesser extent, young chicks are also taken. Seal placentas and excrement, algae, and limpets are important food items during the nonbreeding season.

Reproduction. Solitary pairs. The number of sheathbills nesting in an area is probably correlated with the size of nearby penguin or cormorant colonies. The loosely constructed untidy nest, which is built of any readily available debris, most commonly penguin tail feathers and egg shells and membranes, is hidden in a deep crevice under a large boulder. Incubation begins with the first egg, and the chicks of a family are of different ages.

Arrival: Early October to early November.

Eggs: Early December to mid-January. Clutch, 1-4, mostly 2-3, off-white eggs speckled with gray and brown markings denser on the broad end; 52.5-64.5 × 36.0-40.5 mm and 40-52 g.

Hatching: Early January to early February. Incubation period 28-32 days.

Fledging and departure: Late February to late March, when the chicks are 50-60 days old. Departure takes place from March to late May, but some birds remain near the breeding site if food is available throughout the winter.

Molt. A complete molt begins in adults in January before the chicks have fledged and lasts into April or May in the South Shetland Islands and into June on South Georgia.

Predation and mortality. Practically no predation occurs, although skuas may occasionally take wandering chicks. The main cause of nest mortality is egg stealing by other, nonbreeding sheathbills.

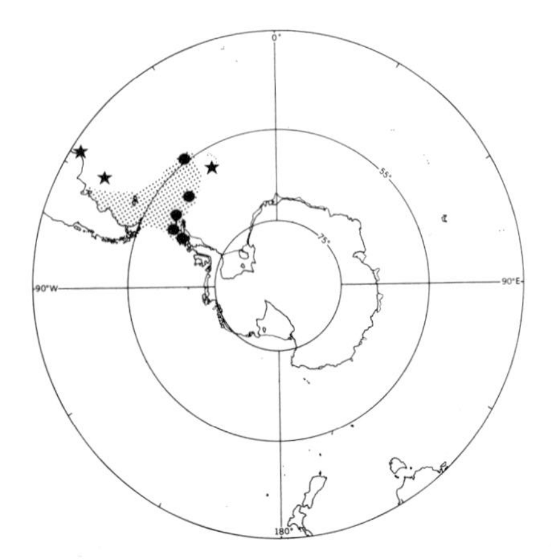

American sheathbill

Ectoparasites. Feather mites, *Alloptes aschizurus* and *A. chionis;* feather louse, *Quadraceps antarcticus.*

Habitat. During the breeding season the American sheathbill frequents penguin or cormorant colonies on rocky coasts; in the winter it may remain in the far south only if a steady supply of food is available at a station refuse heap. In southern South America it is confined to the sea coast.

Distribution. Breeds on South Georgia, South Orkney, and South Shetland islands and on the Antarctic Peninsula south to about 65°S. Some birds migrate to South America and the Falkland Islands, but at least part of the population remains resident at the breeding grounds. An individual recorded on Saint Helena Island in May 1968 probably arrived aboard a ship from the Antarctic [*Loveridge*, 1969].

LESSER SHEATHBILL *Chionis minor*

Resident

Identification. Plate 8. 15-16 in. (38-41 cm)/29-31 in. (74-79 cm). Similar to the American sheathbill but with a wholly black bill. The uplifted saddle-shaped sheath is conspicuously concave in Iles Kerguelen (*C. m. minor*) and Heard Island (*C. m. nasicornis*) populations, the structure superficially resembling the nasal tube of a petrel. The feet are brown (Iles Kerguelen) or dull pinkish white (Heard Island). In the Prince Edward and Marion island (*C. m. marionensis*) and Iles Crozet (*C. m. crozettensis*) populations the sheath is less pronounced, the facial skin tends to be mauve rather than fleshy pink like that of the Iles Kerguelen and Heard Island birds, and the legs are black (Prince Edward and Marion islands) or grayish brown (Iles Crozet).

The juvenile lacks the well-developed sheath and caruncles of the adult and has traces of down on leaving the nest. Presumably, it may also be identified by the pointed tips of its inner primaries.

The downy chick is similar in appearance to the young of the American sheathbill except that the underparts have white patches. The feather and body down sequence is also similar.

Flight and habits. Like the flight and habits of the American sheathbill.

Voice and display. 'Short rattling croaks' and a 'cluck' like that of a domestic fowl about the nest. In flight the lesser sheathbill has 'a peculiar note strongly suggestive of the "chat" of the common [American?] black bird'; other flight calls have been likened to the piping flight note of plovers.

Food. Omnivorous, taking advantage of the same sort of varied food sources in penguin and cormorant colonies as the American sheathbill but also including insects, mussels, isopods, stranded plankton, excrement of penguins and seals, and carrion. The uplifted bill sheath serves as a hook for carrying snatched eggs. Because of its sedentary nature the lesser sheathbill forages on inland meadows for insects in the fall and scours beaches for food during the winter.

Reproduction. Solitary pairs, but several nests may be in the same penguin rookeries. The nest is similar to that of the American sheathbill but also incorporates quantities of vegetation. Nests may occasionally be located in crevices of steep cliffs some distance from the nearest penguin and cormorant rookeries or in burrows made by rabbits and larger petrels.

Pairing: Sedentary on the breeding islands. Courtship and occupation of the nesting site occur in early October to November.

Eggs: Mid-December to late January. Clutch, 1-3 creamy white eggs with brownish irregular blotches more numerous on the larger end; 53-60 × 37.0-40.5 mm and 33-40 g.

Hatching: Mid-January to early February. Incubation period about 29 days.

Fledging and departure: Early March.

Molt. Presumably after breeding, like the American sheathbill, but little information is available.

Predation and mortality. Occasionally skuas may take isolated

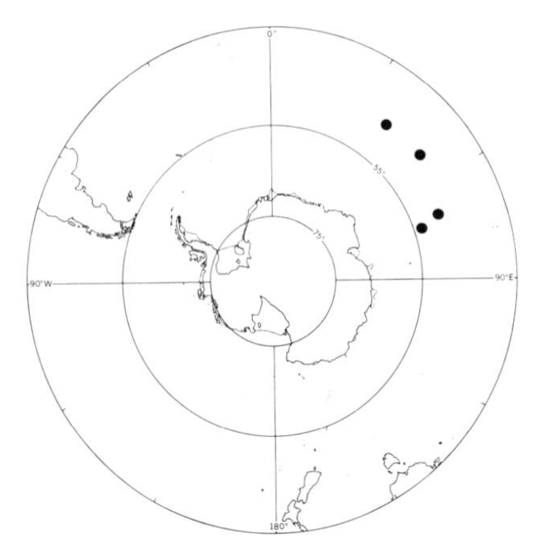

Lesser sheathbill

chicks or weakened adults. Addled eggs are common, and only one chick may fledge from a clutch for reasons not yet understood.

Ectoparasites. Feather mites, *Alloptes aschizurus* and *A. chionis;* feather lice, *Actornithophilus pauliani, Saemundssonia australis,* and *Quadraceps vaginalis.*

Habitat. Like that of the American sheathbill, but the lesser sheathbill also occurs in damp inland meadows.

Distribution. Resident on Prince Edward, Marion, Crozet, Kerguelen, and Heard islands.

SKUAS AND JAEGERS: Stercorariidae

Skuas and jaegers are dark gull-like predatory seabirds with conspicuous white flashes in the primaries that show in flight. The dark bill is strongly hooked and is covered at the base with a flat horny sheath, and the black feet are strongly clawed. Females are on the average larger than males. During the breeding season, adult skuas and jaegers have a conspicuous light collar formed by the golden pointed tips to the feathers of the cheeks and nape. In nonbreeding adults and immatures these feathers are rounded at the tips, and their light areas are either less conspicuous or entirely absent. More precise information on aging and seasonal plumage changes must await detailed studies of birds of known age through banding.

The all-dark broad-winged skuas are well known for their aggressive rapacious habits. Remarkably maneuverable for such heavy-bodied birds, they have a rapid, sustained, and powerful flight, enabling them to overtake and rob many birds. The victims are relentlessly pursued and forced to disgorge their catch, which the skuas adeptly seize before it hits the water. Skuas also prey on chicks and eggs, particularly those of penguins, and take a heavy toll of small petrels. In addition to their plundering, skuas feed on carrion and capture fish, crustaceans, and cephalopods in short shallow dives into the water. Skuas characteristically proclaim jurisdiction over a territory by boldly challenging intruders with open wings raised over the back, the broad white flashes thus being displayed to best advantage. This challenge is accompanied by a harsh screaming 'charr charr charr.' They are particularly aggressive in defense of young chicks and engage in a series of swooping dives, striking trespassers with their wings and feet. Skuas usually nest near breeding colonies of their penguin or petrel prey, laying 1-3 olive brown splotched eggs in a shallow nest. Two species of skuas are found in the Antarctic; one nests on the continent (*Catharacta maccormicki*), and the other farther north on subantarctic islands (*C. lonnbergi*). The two breed together on the Antarctic Peninsula and adjacent islands. Other skuas, which nest in South America (*chilensis*), the Falkland Islands (*antarctica*), the Tristan da Cunha group (*hamiltoni*), and islands in the North Atlantic Ocean (*skua*), are probably conspecific with one or the other of the antarctic skuas.

Jaegers resemble skuas in habits but are smaller and slimmer. They have narrower more pointed wings with less conspicuous white wing flashes, and their central tail feathers are elongated. Two color phases occur, a rarer all-dark one and a light phase with white underparts. Jaegers victimize small seabirds and are usually associated with pelagic flocks of feeding terns. Two species have been recorded in the Antarctic as rare vagrants from the northern hemisphere, but because jaegers are common elsewhere in the southern hemisphere, they may prove to follow Arctic terns regularly into the far south.

SOUTH POLAR SKUA *Catharacta maccormicki*

Resident and Breeding Migrant

Identification. Plate 9. 21 in. (53 cm)/50 in. (127 cm). The smaller (weight usually less than 1800 g) and shorter legged (tarsus less than 70 mm) of the two antarctic skuas. The south polar skua occurs in two color phases, one with pale brownish gray underparts and prominent golden hackles and the other with darker and browner underparts and less conspicuous hackles. The back in both phases is dark brown with no rufous feathers (see the brown skua). The plumage of the light phase tends to wear rapidly and more extremely than that of the dark phase, so that by late summer the head, neck, underparts, and some of the dorsal feathers appear buffy or whitish gray. Dark-phase birds remain mostly dark until they begin molting in March. On the continent the light phase predominates, whereas the dark phase is most common on the Antarctic Peninsula. The bill and feet are black, the latter occasionally being marked with white. The south polar skua is stockier and broader winged than either species of jaeger and much darker and more robust than the fledgling southern black-backed gull.

For subadult and seasonal characters see the family introduction.

The juvenile is uniformly sooty gray with no pale feathers about the head or neck. The white area is present on the wing but is less prominent than it is in the adult until the primaries are fully grown. The bill is noticeably weaker, and the legs and feet retain transitory traces of gray, especially at the joints.

Much uncertainty exists about the color of skua chicks, which may vary individually or reflect color phase differences. In the far south, newly hatched chicks are pale sandy gray with a bluish cast. On the Antarctic Peninsula the dorsal color of chicks varies from sandy to dark brownish gray with a tendency toward a bluish tone; the underparts are paler than the back. Both extremes may be found in the same brood. Even the palest chicks turn browner as they mature (and also when they are preserved as study specimens), becoming nearly indistinguishable from brown skua chicks. They tend, however, to be somewhat smaller at all ages. The hatching weight of south polar chicks is usually 70 g or less. The feet are distinctly bluish gray.

Flight and habits. See the family introduction for general comments and the brown skua for comparative notes. The south polar skua is the more sociable of the two skuas, and small groups gather together to bathe in freshwater ponds.

Voice and display. In addition to the challenge or 'long call' described in the family introduction, a low-pitched plaintive quacking alarm call is given by uneasy incubating adults. Although skuas are generally silent at sea, they quarrel noisily when they are feeding communally.

Food. During the summer the south polar skua preys heavily on eggs and young of breeding penguins near the coast and of Antarctic and snow petrels in inland mountain colonies. Some skuas are independent of bird colonies and feed largely on fish and krill secured at sea. Indeed, on the Antarctic Peninsula, competition between the two skuas may result in this species' depending to a large extent on non-avian prey (needs further study). Gulls, cormorants, and many other birds are victimized and forced to disgorge their food both in the Antarctic during the summer and at sea at other times of the year. Skuas feed on galley refuse and follow ships at sea.

Reproduction. Although the south polar skua is not strictly colonial, groups of breeding pairs nest in close association with their prey. Sites are variable but are usually in sheltered locations on rocky outcrops, moss-covered cliffsides, or valley floors both near the coast and near inland petrel colonies. The nest, which is a shallow depression on the ground, may be moss lined on the Antarctic Peninsula. Distances between adjacent nests vary from less than 10 to more than 50 m depending on local conditions. Inasmuch as the south polar skua generally breeds farther south and under more extreme conditions than the brown skua, its schedule is about 1 month later, but even on the Antarctic Peninsula, where the two occur together, the south polar skua breeds 2 or 3 weeks later. Most subadult birds return to the breeding grounds at the age of 4-5 years and begin to breed at the age of 5-6 years.

Arrival: Late October to mid-December.

Eggs: Mid-November to late December. Clutch, 2, rarely 1, eggs of variable shades of tan or light green with brown splotches; 47-52 × 65-79 mm and average 78.8 g.

Hatching: Late December to late January. Incubation period 24-34 days; brood period 12 hours to 3 days. The second chick to hatch seldom survives and is harried by the older chick.

Fledging and departure: Late February to mid-March, 49-59 days after hatching. Departure takes place from March to early April.

Molt. Twice annually. A complete molt begins with the body feathers in February or March (date of completion unknown) and a partial molt, involving at least the head and neck, is completed by October (date of beginning unknown).

Predation and mortality. Breeding success ranges from 20 to 50% of the eggs laid. Giant fulmars and other skuas are infrequent predators on unattended nests and wandering chicks. Egg loss through chilling is much lower than chick mortality from starvation and exposure.

Ectoparasites. Feather mite, *Alloptes stercorarii;* feather lice, *Harrisoniella grandis* and *Saemundssonia stresemanni.*

Habitat. Adults stay near breeding colonies in the Antarctic zone during the summer but during the off-season may range north to subtropical waters. Subadults may be found far out at sea all year. During the summer, vagrants have been seen far inland over the continent.

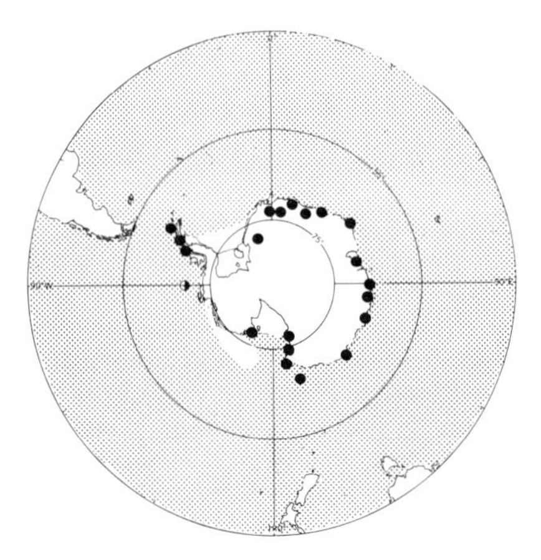

South polar skua

Distribution. Breeds in the South Shetland Islands (probably rare) and on the Antarctic Peninsula and continent (including inland mountain ranges) and adjacent islands including the Balleny and probably Peter I islands. The south polar skua disperses northward after breeding to spend the winter at sea as far as the subtropics and occasionally into the northern hemisphere (India, Japan, and California). At-sea skua records are not usually identifiable as to species, but this form is probably widespread, although it is not abundant in all parts of the southern ocean outside the breeding season. The map shows the presumed combined summer and winter range; there are very few at-sea specimen records and practically no sight records in the Pacific Ocean sector.

BROWN SKUA *Catharacta lonnbergi*

Resident and Breeding Migrant

Identification. Plate 9. 25 in. (64 cm)/58 in. (147 cm). Similar to the dark-phase south polar skua but considerably larger (usually more than 1800 g) with a more massive bill and longer heavier tarsi (more than 70 mm). Scattered rufous feathers occur on the back and underparts, and in some populations (especially on Iles Kerguelen) the underparts are reddish chestnut. The overall body color appears browner than that of the south polar skua, and in breeding adults the golden portions of the pointed hackles are less pronounced. By February the plumage becomes abraded: the breast lightens, and the feathers of the back become pale buff, imparting a hoary appearance to the dorsal surface. The bill and feet are black.

For intermediate age and seasonal characters see the family introduction.

Other southern hemisphere populations, *antarctica* in the Falkland Islands, *hamiltoni* in the Tristan da Cunha group and Gough Island (both small and dark), and *chilensis* of southern South America (small and rusty red below), are usually considered conspecific with the brown skua and with the North Atlantic skua, *Catharacta skua.*

The juvenile is dark brown with some chestnut flecking on the neck and underparts. The white wing area is more restricted than that of the adult, and the bill is weaker. The black feet have some gray mottling.

The downy chick is grayish brown like the dark-phase south polar skua chick, but it is probably heavier, and the gray feet may lack the bluish tinge (needs confirmation).

Flight and habits. The brown skua is more solitary, bolder, and more aggressive than the south polar skua, even attacking the greater albatrosses at sea.

Voice and display. Like the voice of the south polar skua but possibly higher pitched (comparison needed).

Food. Generally similar to that of the south polar skua, but prions and diving petrels form a large part of the summer diet in some areas. On Macquarie Island, brown skuas capture young and weakened rabbits and have even been seen taking milk from lactating elephant seals [*Johnston*, 1973].

Reproduction. Solitary pairs select and stoutly defend an extensive nesting territory usually located near a penguin or petrel colony. Nests are located on either grass tussocks, clumps of *Azorella* or moss with little or no nest material, or high rocky areas overlooking a bird colony and are constructed of tussock, dried kelp, and other vegetation and thickly lined with moss and lichens. The age of first breeding may be as much as 7 years, but more data on birds of known age are needed.

Arrival: Early September to mid-October.

Eggs: Late October to late November. Clutch, 2 or 3 olive brown eggs with blotches and spots of grayish brown; average 76.6 × 52.6 mm and 100 g. The eggs are smaller on the Tristan da Cunha group.

Hatching: Mid-November to mid-January. Incubation period 29-32 days. The chick can leave the nest within 24 hours.

Fledging and departure: Late December to early February, at 55-60 days, but young remain dependent on parents for up to another month. Departure takes place from March to late April.

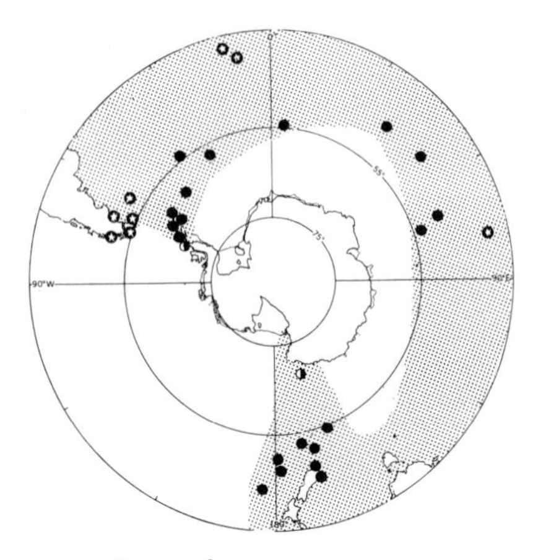

Brown skua
C. lonnbergi ●
Related skuas (see the section on identification) ✪

Molt. Twice annually, presumably like the south polar skua but a little earlier. Information is needed. Tristan da Cunha group skuas molt from December to February.

Predation and mortality. Like the predation and mortality of the south polar skua.

Ectoparasites. Tick, *Ixodes uriae;* feather mite, *Alloptes stercorarii;* feather louse, *Saemundssonia stresemanni;* flea, *Parapsyllus magellanicus.*

Habitat. Sub-Antarctic and low-latitude Antarctic zones. During the austral summer the brown skua is essentially coastal, but at other times it is highly pelagic.

Distribution. In the Antarctic the brown skua breeds on South Georgia, South Sandwich, South Orkney, and South Shetland islands; on the Antarctic Peninsula south to at least 65°S and perhaps to 68°S; on Bouvet, Prince Edward, Marion, Crozet, Kerguelen, Heard, and Macquarie islands; and possibly also in the Balleny Islands. Farther north this skua or related skuas breed in southern South America and on the Falkland Islands, the Tristan da Cunha group, Gough Island, Ile Amsterdam, Ile Saint-Paul, and many islands in the New Zealand area (Campbell, Auckland, Snares, Stewart, Antipodes, Bounty, and Chatham islands). Postbreeding dispersal of the brown skua is not well documented owing to the presence of south polar and other skuas, especially in the Atlantic and Pacific oceans, but a brown skua banded on Deception Island was recovered on Guadeloupe, West Indies [*Hudson*, 1968*a*], and there are several specimens from the tropical Indian Ocean. The map shows combined summer and winter ranges based on the specimens collected. The species may be expected throughout its pelagic range but nowhere abundantly except near the breeding grounds.

POMARINE JAEGER *Stercorarius pomarinus*

Vagrant

Identification. Plate 9. 22 in. (56 cm)/48 in. (122 cm). This larger and proportionately stouter billed of the two jaegers in the Antarctic is characterized by its long, blunt-ended, and curiously twisted central tail feathers. It occurs in two color phases, a rarer (about one of every seven) all-dark sooty gray phase and a light phase in which the face and underparts are white and the dorsal surface, breast-band, and underwings are dark brown. In both phases the crown is black, the bill dark yellowish brown, and the legs black. Because jaegers molt on migration, those seen in the far south are usually in off-season dress and lack the golden cheeks and nape of the breeding plumage. In the light phase the white areas are barred with sooty brown, and the dark areas with grayish white. The dark phase shows no noticeable seasonal change. See the gadfly petrels, sooty shearwater, and juvenile southern black-backed gull.

In subadult stages of both phases the upperwing and underwing coverts, axillaries, and rump are barred with white instead of being uniformly sooty brown like those of adults. In addition, the central tail feathers are shorter and pointed, but since they may be broken off in adults, the tail shape is not a completely reliable age or species character (see the parasitic jaeger).

The short-tailed juvenile is mostly dark brown barred with buff. The neck, face, and underparts are paler than the back.

Flight and habits. Resembles a small slender skua in the air but flies more adeptly with the wings bent at the wrist like those of the gadfly petrels. The pomarine jaeger is almost exclusively pelagic except when it is breeding.

Food. Largely fish, which it pirates from other birds.

Distribution. Breeds on the arctic tundra generally a little north of the parasitic jaeger and migrates south, mostly to the tropics. In the eastern Atlantic Ocean the pomarine jaeger reaches South Africa, and in the western Pacific Ocean it occurs as far south as Tasmania and New Zealand. Vagrants, presumably of this species, were recorded on the Antarctic Peninsula south of 65°S in February of both 1937 and 1953 [*Sladen*, 1954] (second bird photographed) and in December 1946 and January 1947 [*Beck*, 1968*b*] (the same records were cited as 'unidentified "visitors" ' by *Tickell* [1960]).

PARASITIC JAEGER *Stercorarius parasiticus*

Vagrant

Identification. Plate 10. 18 in. (46 cm)/36 in. (91 cm). Smaller and slimmer, but in overall plumage pattern, including light and dark phases and seasonal changes, this species is very similar to the pomarine jaeger. It differs, however, in having less prominent white flashes in the primaries and a paler broader breastband. The bill is both shorter and much slimmer, and the tips of the central tail feathers in the adult are pointed and not twisted. The proportions of light- and dark-phase birds are about equal. See the gadfly petrels, sooty shearwater, and skuas.

The mottled juvenile is not distinguishable from the juvenile pomarine jaeger unless direct size and bill comparison can be made.

Flight and habits. More agile and aggressive than the pomarine jaeger.

Food. Food habits are similar to those of the pomarine jaeger.

Distribution. Breeds on arctic tundra and migrates south to tropical and temperate regions of the Atlantic and Pacific oceans as far as Tierra del Fuego and New Zealand. Vagrants were collected on Signy Island, South Orkney Islands, in January 1951 [*Sladen*, 1952] and were seen there in January 1966 [*Beck*, 1968*a*] and at about 61°S, 92°W in early January 1973 by D. F. Parmelee (personal communication, 1973).

GULLS AND TERNS: Laridae

Gulls and terns are cosmopolitan long-winged coastal birds usually gray or black above and white below. The bill and webbed feet are usually brightly colored. The nostrils open as slits on the bill and are not enclosed in tubes like those of the albatrosses and petrels. The family contains two well-defined subfamilies: the gulls (Larinae) and the terns (Sterninae).

GULLS: Larinae

Gulls are scavengers with relatively long broad wings and short mostly square tails. The only species that breeds in the Antarctic is a widely distributed southern representative of the northern hemisphere black-backed/herring gull complex. Like skuas, gulls fly with deliberate wingbeats, but they are more graceful in the air and are notable for their soaring and gliding abilities. They alight freely on the water and float buoyantly when they are swimming but seldom dive deeply from either the air or the water. Unlike terns they have sturdy legs and walk about freely on land. Gulls are extremely vocal; their ringing cries are a dominant feature of any seacoast. They are omnivorous, combing beaches for stranded plankton and carrion and characteristically following ships within sight of land. Although they are egg and chick predators, gulls are not as a rule as rapacious as skuas. In the Antarctic, small groups breed at secluded coastal sites, building nests of shells, moss, and other vegetation. The 2 or 3 eggs are cryptically patterned. Young gulls, which are cared for by both parents, leave the nest and can swim well soon after hatching. The definitive plumage is not attained for several years, immatures passing through a bewildering series of dark mottled stages. Two other southern hemisphere temperate zone species have been recorded on South Georgia and Marion Island, and a northern hemisphere migrant has been recorded on Gough Island.

SOUTHERN BLACK-BACKED GULL *Larus dominicanus*

Resident

Identification. Plates 9 and 10. 22 in. (56 cm)/52 in. (132 cm). A large mostly white gull with slate black mantle and wings. The flight feathers are tipped with white, broadly so on the secondaries and inner primaries, forming a pronounced bar when the wings are folded. The outer one or two primaries have a white subterminal spot. The lemon yellow bill has a red spot near the tip of the lower mandible. The iris is yellowish gray, and the eyelid is deep orange. Foot color varies from yellow to olive green or bluish gray. Females are slightly smaller and more slender than males and have relatively weaker bills. During molt, when most of the wing coverts may be lost simultaneously, the white bases of the secondaries are exposed, a second light wing bar or spot thus being produced.

The downy chick is pale grayish brown with irregular dark brown and black markings on the buffy head and throat and a creamy white belly. Its stout bill is black with a flesh-colored tip, and the feet are grayish pink.

The juvenile is grayish brown with a regular pattern of buff mottling on the back and buff streaking on the head and underparts, both produced by the pale edgings of individual feathers. Wing and tail quills are dark brown with the outer rectrices barred with buff.

The underwings are mostly dark gray brown. The rump is white conspicuously barred with brown. A dark area about the eyes imparts a sullen look to the young bird. The bill is black, the iris is brown, and the legs and feet are a grayish brown. See the skuas.

A complicated series of semiannual partial and full molts begins in March of the year of hatching and continues during the next 3 or 4 years to produce a bewildering array of plumage changes until the definitive plumage is reached in the fourth or fifth year.

The first-year molts result in little change of appearance from that of the juvenile except that the back is darker because the new feathers lack the pale buff edges. The neck and underparts may develop light areas as abrasion exposes the white feather bases. The tip of the bill becomes flesh colored, but the iris and feet remain like the iris and feet of the juvenile.

The second-year molts produce variable results. The back can be either brown or black. The wings are mostly brown with light tips on some of the inner primaries and most of the secondaries, whereas the tail has variable amounts of white but always shows a marked brown subterminal bar. The bill gradually becomes flesh colored with dark markings near the tip of the lower mandible, and later in the year it turns yellow with a pale orange spot on the lower mandible. The iris and feet are gray.

The third-year gull is similar to the adult except for variable brown mottling on the head, neck, and upper breast and the dark subterminal bar on the tail. The bill is slightly paler than that of the adult with an orange, rather than red, spot on the lower mandible. The iris, legs, and feet remain gray as in the late second year.

Most fourth-year birds are indistinguishable from adults, but some retain a little dark mottling on the head and have a weakly colored bill.

Flight and habits. Flies in a slow leisurely manner with considerable gliding. Gregarious. The southern black-backed gull is less bold near its nest than most other antarctic species.

Voice and display. Various vocalizations, the most common being the 'long call,' a harsh high-pitched repeated laugh 'ha-ha-haro,' and a loud mewing. Young begging for food give a piping trilled whistle.

Food. Primarily limpets (method of securing needs study), which are usually swallowed whole. The shells are regurgitated after their contents are digested. Gulls also scavenge for a variety of marine and terrestrial animals including crustaceans, mollusks, fish, insects, and birds' eggs and young. Hard-shelled organisms too large to be swallowed whole are broken by being dropped from the air onto a hard surface. Carrion and refuse are also important items in the diet.

Gulls are partly nocturnal in feeding habits. They hunt for small petrels and may feed on carcasses at night, perhaps to avoid competition from more aggressive skuas. The number of gulls present throughout the winter depends on weather and the availability of food and open water.

Reproduction. Loosely colonial. In the far south, as few as three or four pairs may nest together, and occasionally individual pairs are found. Breeding sites are in secluded areas near the sea, on beaches or low rocky or vegetation-covered headlands but rarely on cliff ledges. The nest varies from a sparse lichen- and seaweed-lined scrape to a bulky structure of tussock grass, moss, seaweed, bones, and feathers. Accumulations of limpet shells invariably litter the nearby ground.

Arrival: Migrants return to colonies and pairing occurs in mid-September to late October.

Eggs: Early November to early December. Clutch, 2 or 3 eggs various shades of greenish or brown heavily mottled with dark brown or black. Eggs in the same clutch may show considerable range in color; average 71 × 49 mm and 80 g. Eggs are laid, and chicks hatch, several days apart. Occasional late layings are probably replacement clutches.

Hatching: Early December to early January. Incubation period about 27 days. The young are capable of leaving the nest soon after hatching and hide in vegetation or rock crevices. When they are pursued, they also take to the water, where adults gather to protect them from skuas.

Fledging and departure: Early January to February. They fly at 5-6 weeks and are independent of adults at 7-8 weeks.

Molt. Twice per breeding cycle. A complete molt begins late in the breeding period and lasts about 2 months; a partial molt of the head and neck before breeding does not result in any change in plumage color (dates for antarctic birds needed).

Predation and mortality. No predation has been recorded, but early nest failure may be great judging from the high proportion of nonbreeding adults observed about the breeding grounds during the chick stage. Egg and chick cannibalism by other gulls, skuas, and sheathbills results in mortality as high as 50% in some island colonies, whereas weather may produce early nest failure on the Antarctic Peninsula.

Ectoparasites. Feather mite, *Alloptes obtusolobus;* feather lice, *Actornithophilus piceus, Austromenopon transversum, Saemundssonia lari, Quadraceps ornatus fuscolaminulatus,* and *Q. punctatus sublingulatus;* flea, *Notiopsylla kerguelensis.*

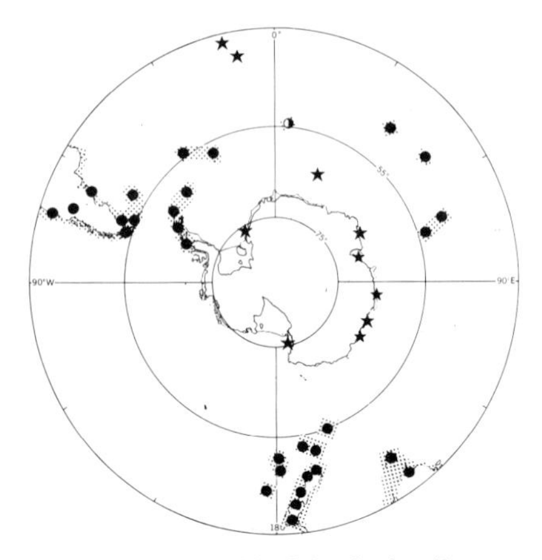

Southern black-backed gull

Habitat. The gull frequents a variety of coastal environments, concentrating its feeding in the intertidal zone of beaches or just beyond the breaking waves. The southern black-backed gull is seldom seen out of sight of land. It displays a wide latitude of zonal tolerance from the Antarctic to the subtropics.

Distribution. In the Antarctic the southern black-backed gull breeds on South Georgia, South Sandwich, South Orkney, and South Shetland islands; on the Antarctic Peninsula south to about 68°S; and on Bouvet (needs confirmation), Prince Edward, Marion, Crozet, Kerguelen, Heard, and Macquarie islands. Several recent records on the continent possibly indicate a future expansion of breeding range [*Johnstone and Murray,* 1972]. Farther north it also breeds in South America (north to 7°S), the Falkland Islands, South Africa, southern Australia (recent colonization and spreading [*Thomas,* 1967]), and New Zealand and its sub-Antarctic Islands (Campbell, Auckland, Snares, Antipodes, Bounty, and Chatham islands). A few birds winter south to 65°S on the Antarctic Peninsula, where refuse provides a reliable source of food, but most birds in the southern American sector move north after the breeding season. Island breeding populations elsewhere are largely sedentary.

BAND-TAILED GULL *Larus belcheri*

Vagrant

A medium large (22 in., 56 cm/49 in., 124 cm) black-backed gull with a broad black subterminal band on the tail and no white spots

Band-tailed gull

on the wing tips. The band-tailed gull looks like a small subadult southern black-backed gull, but the yellow bill has red and black on the tips of both mandibles, the iris is brown, and the feet are yellow. The smaller Pacific Ocean subspecies (*L. b. belcheri*) differs from the Atlantic Ocean subspecies (*L. b. atlanticus*) in having a pale gray wash on the head and neck in summer and a dull brownish black hood in winter; the underwing is gray, not white; and the back and upperwing are brownish black, not slaty black. The immature of both forms is mottled brown like that of the southern black-backed gull, but the head is markedly darker than the back, the underwing is gray, not mottled, and the bill is pale greenish yellow with black on the tip. Intermediate-aged birds of both forms have a dark hood and gray underwings (with some white feathers in Atlantic Ocean individuals). The species occurs in Peru, northern Chile, Uruguay, and northern Argentina and has been recorded in Tierra del Fuego (subspecies unknown). An immature specimen of the Atlantic Ocean subspecies, collected in January 1949 on South Georgia, is in the Buenos Aires Museum (C. C. Olrog, personal communication, August 1968).

FRANKLIN'S GULL *Larus pipixcan*

Vagrant

A small (14 in., 36 cm/35 in., 89 cm) gray-backed dark-headed gull with a broad white trailing edge to the wing and a white 'window' between the white-spotted black wing tips and the gray mantle. In the breeding season the entire head is black, but in other seasons the forehead is white with a gray half hood across the hind crown. The bill is dark red; the legs are brown. The immature is gray brown above with no wing pattern and white below with a black subterminal bar on the white tail. See the silver gull. Franklin's gull is a North American species that migrates to South America, possibly as far as the Strait of Magellan [*Peterson and Watson,* 1971], and was recorded once on Gough Island, in February 1956 [*Swales and Murphy,* 1965].

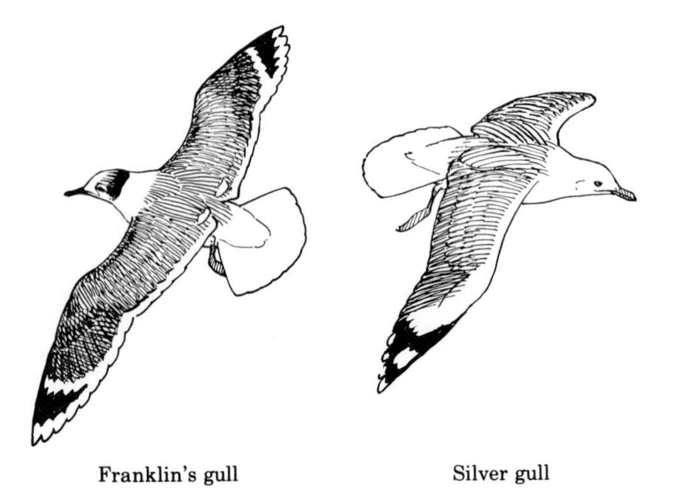

Franklin's gull Silver gull

SILVER GULL *Larus novaehollandiae*

Vagrant

A small (15 in., 38 cm/36 in., 91 cm) white-bodied and white-headed gull with a light gray mantle and considerable white in the wings. The bill and feet are red. The leading edge of the wing has a wedge of white, and the black primaries have white spots at or near the tips. The immature shows the same general wing pattern but has a brownish bar on the ulnar portion and less white in the primaries. The head and body are washed with brown. The silver gull is a South African, Australian, and New Zealand species that has been recorded once on Marion Island, in late June 1962 [*Rand,* 1963].

TERNS: Sterninae

Terns are generally smaller and slimmer than gulls and have narrow pointed wings, very short legs, and moderately webbed feet. The deeply forked tail and aerial maneuverability have resulted in their popular name 'sea swallow.' Terns fly gracefully and buoyantly with the sharp bill characteristically pointed downward. They frequently hover over suspected prey in the water and plunge abruptly from the air with the wings partially folded over the back to capture small fish and crustaceans on or near the surface. Although terns may submerge entirely, they do not remain under the water and seldom swim. The calls of the adults are harsh and grating. Flocks band together in noisy communal aerial attacks to drive off skuas and other intruders from nesting areas. Terns form small loose colonies and lay 1-3 heavily spotted eggs in a scanty nest. The young are fed by both parents and remain in or near the nest until they are fledged. The parents continue to feed them after fledging. Three very similar black-capped terns are found regularly in the Antarctic: one widespread species is circumpolar, a second is restricted to islands in the south Indian Ocean, and a third northern hemisphere migrant is present only during the austral summer. A fourth species, the all-dark subtropical brown noddy, reaches the southern limit of its breeding range in the Tristan da Cunha group and on Gough Island. Another subtropical species has been recorded as a vagrant in the Drake Passage.

BLACK-CAPPED TERNS: *Sterna* spp.

Three medium-sized black-capped terns frequent the Antarctic and are so similar in structure and color that they are difficult to distinguish in flight. In breeding dress all have gray backs and underparts with contrasting white moustachial streak and cheeks, black caps, and bright red bills and feet. After the breeding season, molt alters the plumage considerably. The crown becomes streaked with white, and the forehead and lores are almost entirely white. The underparts whiten (except in the Kerguelen tern), the bill darkens to black or reddish black, and the feet darken to dull red. The colors of tail and underwing are useful species characters for identifying adults throughout the year. First- and second-year birds retain white foreheads all year and have a dark cubital bar on the ulnar portion of the upperwing surface. Juveniles, which have light foreheads similar to those of the nonbreeding adults, are strongly barred with black on the wings and back in the two resident species and tinged with light tan in the northern migrant.

ANTARCTIC TERN *Sterna vittata*

Resident

Identification. Plates 9 and 10. 15 in. (38 cm)/31 in. (79 cm). This widespread resident tern in the Antarctic has the outermost tail streamers white or only lightly washed with pale gray on the outer web. The underwing coverts and most of the primaries are white, only one third of the inner web of the outer primary being dark gray (Figure 9). The intensity of gray on the back and underparts varies, being darkest in Iles Kerguelen birds and palest in those breeding on subantarctic islands. All Antarctic tern populations, however, are

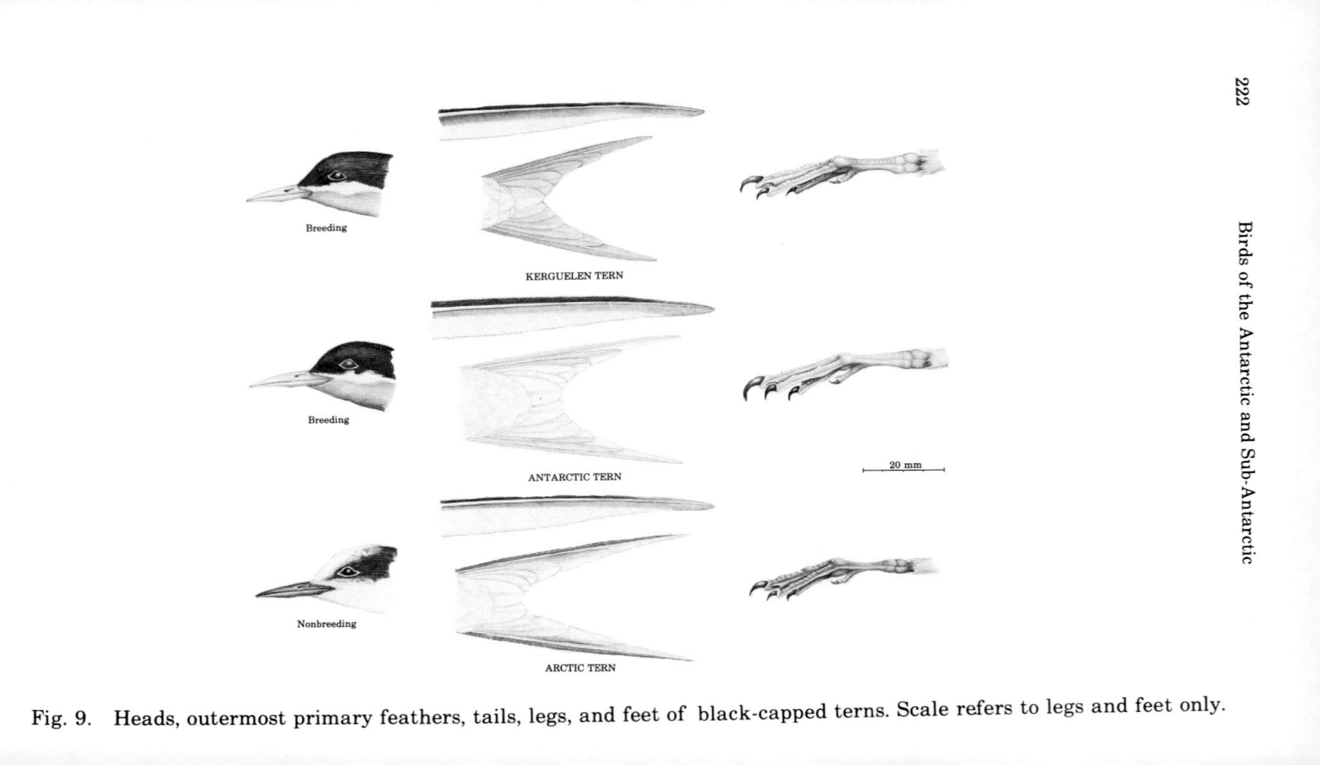

Fig. 9. Heads, outermost primary feathers, tails, legs, and feet of black-capped terns. Scale refers to legs and feet only.

lighter than the Kerguelen tern and darker than the Arctic tern. The tail is moderately forked and relatively short, extending only slightly beyond the tips of the folded wings when the bird is perched, except in Ile Amsterdam and Ile Saint-Paul birds, which have longer more deeply forked tails. The Antarctic tern has a more robust body and heavier bill than the Arctic tern, and the tarsus is generally longer (16.5-20 mm), except in the South Georgia population (16-17.5 mm) (Figure **9**). The molt schedule may also be useful in distinguishing the two species (see the sections on molt for both the Arctic and Antarctic terns and the section on identification for the Arctic tern).

The juvenile is gray on the back and wings, strongly barred with black and washed with buff when it is in fresh plumage. The forehead is white with some black or brown streaking, but the back of the crown and the nape are mostly black washed with buff. The white underparts and rump are finely barred with black or brown, strongly so on the throat and breast. These dark tips abrade rapidly, so that the underparts and rump become almost pure white. Tail feathers have darker gray outer webs than those of the adults and black and buff bars at the tips. The bill and feet are black.

A partial molt of the juvenile plumage begins in March, new gray first-winter feathers appearing to a variable degree on the back. Some barred juvenile secondary wing coverts and wing and tail quills are retained until a complete molt takes place just before the next breeding season. The immature, then in its second year, looks like a nonbreeding adult with white forehead and underparts and black bill and feet, but the tail is shorter and more strongly marked with gray, and a dark gray cubital bar is present.

The newly hatched downy chick is cryptically patterned with gray and black. The chin and throat are dark gray, the rest of the undersurface being light gray or white. As the chick grows, the underparts become lightly peppered with brownish gray. The bill is black; the feet are pale flesh.

Flight and habits. The Antarctic tern flies in a graceful undulating fashion, frequently hovering over the water. Gregarious, generally to be seen fishing in small parties just beyond the surf line. In winter, flocks rest on icebergs and ice floes near the ice edge.

Voice and display. A shrill and high-pitched 'trr-trr-kriah.' 'Another common chattering note is like the rattle of pebbles or the gritting of teeth.' Comparisons between the calls of this species and those of the other two terns in the Antarctic are needed. The display has not been studied.

Food. Small fish (principally nototheniids) and various crustaceans. Antarctic terns also scavenge in the intertidal zone for stranded littoral organisms and may infrequently follow ships for scraps.

Reproduction. Colonial, but rarely more than forty widely separated nests in any one locality. Nest sites are often associated with those of gulls and are usually on moraines or scree slopes and less frequently on rock-strewn beaches or inaccessible cliffs (Ile Amsterdam). The eggs are laid in a shallow pebble- or shell-lined scrape on the ground. The nests are very difficult to locate owing to the cryptically patterned eggs and chicks.

Arrival: September to October. Adults are generally sedentary around many insular breeding stations, moving only to the nearest open water in winter.

Eggs: Late October to early January. Clutch, 1-3 elongate slightly glossy light blue to buffy olive eggs with brown and black spots or blotches; average 48 × 33 mm and 26.3 g.

Hatching: Late November to late January (into February on South Georgia). Incubation and brood periods unknown.

Fledging and departure: January to May. Dispersal is delayed until June on South Georgia. Parents attend young for several weeks after fledging.

Remarks: The entire cycle is at least 1 month later on Iles Crozet, Iles Kerguelen, and Heard Island, where laying begins in late December. The clutch size, which is 2-3 in the American sector, is almost invariably 1 on Iles Kerguelen and Heard Island, possibly because of the late breeding date.

Molt. Twice per breeding cycle. Complete molt begins in February or March (data on timing are needed), and a partial molt of the head and body extends from September to December. Primary molt begins with the innermost quills in February or March, when the Arctic tern has nearly completed replacing the outer primaries.

Predation and mortality. Although adults usually band together to defend a colony, skuas and gulls occasionally take eggs or chicks from unattended nests in very small colonies. Egg cannibalism has been reported when colonies are disturbed by humans.

Ectoparasites. Feather lice, *Saemundssonia lockleyi, Quadraceps houri,* and *Q. sellatus.*

Habitat. Antarctic and sub-Antarctic zones, generally in coastal waters near breeding islands at all seasons. Some birds, however, are more pelagic and occur in warmer more temperate waters in the contranuptial season.

Distribution. In the Antarctic the Antarctic tern breeds on South

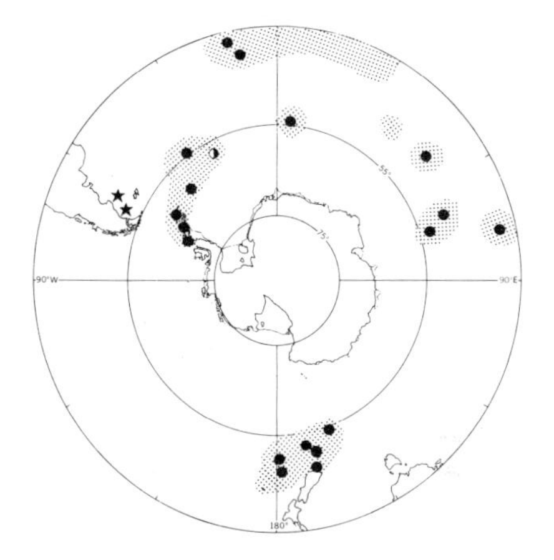

Antarctic tern

Georgia, South Sandwich (needs confirmation), South Orkney, and South Shetland islands; on the Antarctic Peninsula south to about 68°S; and on Bouvet, Crozet, Kerguelen, Heard, and Macquarie islands. Farther north it also breeds on the Tristan da Cunha group, Gough Island, Stag Island off the coast of South Africa, Ile Amsterdam, Ile Saint-Paul, and New Zealand sub-Antarctic Islands (Campbell, Auckland, Snares, Antipodes, and Bounty islands). It also occurs on Marion Island, but there is no proof of its ever having bred there [*Rand,* 1954]. In the austral summer it stays near the breeding grounds, but in the winter it probably moves northward as is suggested by several at-sea records of adults and first-year birds off southern South America, off South Africa, and in the central South Atlantic Ocean [*Courtnay-Latimer,* 1957]. Erroneous early reports of breeding at Gaussberg on Wilhelm II Coast in East Antarctica are based on misidentified Arctic terns [*Murphy,* 1938].

KERGUELEN TERN *Sterna virgata*

Resident

Identification. Plate 10. 13 in. (33 cm)/28 in. (71 cm). This insular Indian Ocean tern is more heavily pigmented than either of the other two terns in antarctic waters and is the only one with a gray underwing and largely gray tail. The inner web of the outermost primary is predominately gray (Figure 9). The dark gray body coloration greatly enhances the white moustachial streak and cheeks below the black cap, and the white rump contrasts with the dark tail. The outer webs of the tail feathers are gray, and the inner webs, which show only when the tail is spread, are white (Figure 9). When the bird is perched, the tail does not extend beyond the folded wing tips. This species assumes a pale gray or white forehead in its nonbreeding dress by February, a month before the Antarctic tern begins molting, but the underparts apparently remain dark gray throughout the year (needs confirmation). In Iles Crozet birds the gray forehead is less extensive than it is in the Iles Kerguelen population and occupies only one third of the black cap. Iles Crozet birds (*S. v. mercuri*) also differ slightly in bill and foot color from birds in the Iles Kerguelen and possibly on Marion Island (*S. v. virgata*). The bill of the Iles Crozet birds is reddish vermilion with irregular black blotches in summer and fades to dark red in winter rather than black like the bills of the Iles Kerguelen birds. Iles Crozet birds have bright red orange feet during breeding rather than dull red feet like those of the Iles Kerguelen birds. Both have black feet in winter.

The juvenile is similar to the young Antarctic tern in being strongly barred with black on the wings and back, but each feather is broadly tipped with tan, and the wings and tail are darker gray. The underwing coverts are conspicuously white, not gray like those of the adult. Presumably, the subsequent molts in this species parallel those of the Antarctic tern, but information is lacking.

The downy chick is tan above with an irregular pattern of black spots and speckling. The chin and throat are dark brown, whereas the rest of the underparts are light tan.

Flight and habits. Similar to the Antarctic tern in flight, but in this species the tail appears shorter. The Kerguelen tern is very bold, showing little fear of man.

Voice and display. A high-pitched scream and a chattering 'like the gritting of teeth.' The display has not been studied.

Food. Marine amphipods and isopods in coastal waters and Diptera, other insect larva, and spiders inland. The Kerguelen tern feeds with the rising tide on the coast and at dusk inland. Winter feeding habits have not been recorded.

Reproduction. Loosely colonial. Most sites are on high sloping broken ground with good drainage, usually not far from the sea. The Kerguelen tern sometimes breeds on shell or pebble (not sand) beaches a few meters from the high-water mark. The nest is a shallow scrape skimpily lined with dried plants, seaweed stalks, or tufts of *Azorella.* The reproductive schedule is about 2 months ahead of that of the Antarctic tern, which may later occupy the same breeding sites.

Arrival: Nest sites are reoccupied and pairing begins in early September on Iles Crozet and late September on Iles Kerguelen and continues to early October.

Eggs: Mid-October to mid-November. Later broods are probably replacement clutches. Clutch, 2 eggs on Iles Kerguelen, usually 1 on Marion Island, and invariably 1 on Iles Crozet. They are brownish green spotted or blotched with reddish brown or black and violet; average 45.6 × 31.3 mm (Iles Crozet) and 22 g.

Hatching: By mid-November. Incubation and brood periods unknown.

Fledging and departure: Late December. The young possibly leave the islands to feed at sea after fledging, since there are no winter records of immatures (needs confirmation).

Molt. Complete molt begins in November during incubation and is completed in January. A partial molt involving at least the crown takes place before breeding (information is needed on the timing and extent).

Predation and mortality. Little information is available, but adults drive skuas away from breeding areas.

Ectoparasites. Feather louse, *Saemundssonia lockleyi.*

Habitat. Freshwater, inland ponds, marshy terraces, and coastal beaches. In the winter the Kerguelen tern is probably more coastal, especially in Iles Crozet.

Distribution. Resident and breeding on Prince Edward Island, Marion Island (presence in winter on both islands needs confirmation), Iles Crozet, and Iles Kerguelen.

ARCTIC TERN *Sterna paradisaea*

Nonbreeding Migrant

Identification. Plate 10. 15 in. (38 cm)/30 in. (76 cm). This migratory northern tern that spends the austral summer in the Antarctic is very similar to the Antarctic tern but has distinctive dark gray margins to the outermost tail streamers. The underwing coverts are white with only a narrow dark streak on the inner web of the outer primary (Figure **9**). The back and underparts are pale gray, slightly lighter than those of any Antarctic tern population. In breeding dress the white moustachial streak and cheeks are less well defined. The northern species is also slimmer and has a more deeply forked tail and a shorter tarsus (15.5-17.5 mm) than all but the South Georgia population of the Antarctic tern (Figure **9**). The Arctic tern stands noticeably closer to the ground, and when the tail streamers are fully grown, they project considerably beyond the folded wing tips.

The molting schedule may also be of use in identification of adults. During most of the austral summer, Arctic terns are in nonbreeding dress with white forehead and underparts and are molting wings and tail, which appear ragged and short, especially if the outer feathers are in growth. Conversely, the Antarctic tern at this time is in breeding dress with fully grown wings and tail. Beginning in late February, Arctic terns molt the head and body again and assume the all-black cap and gray underparts of the breeding dress before starting their northward migration. They are thus usually out of phase with the Antarctic tern, which begins a complete molt into its nonbreeding dress in March and has white incoming feathers on the forehead and underparts (see the section on molt).

The juvenile resembles the nonbreeding adult except that each feather of the back and wing coverts is edged with buff, brown, or white, imparting a faint scaled appearance to the upperparts. The tail is shorter than that of the adult, and the outer web on each of the outer three quills is dark gray. Much of the body feathering is molted during southward migration, but first-year birds may always be identified because they retain some juvenile feathers, especially in the wings and tail, and they always have the dark cubital bar.

Flight and habits. Similar to the flight of the Kerguelen tern but flies in a more direct less undulating fashion. This difference may only be valid when Arctic terns are molting their flight feathers.

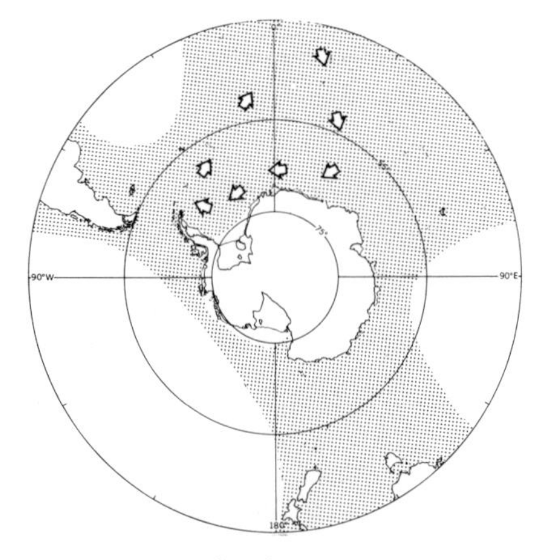

Arctic tern

Although Arctic terns occasionally swim, they more frequently rest on pieces of flotsam at sea.

Voice. A harsh drawn-out grating 'kee-yaah' with the emphasis on the second syllable and also a whistled 'kee-kee' rising in pitch and various chattering calls.

Food. Euphausiids and fish. Krill may constitute a more important item in the diet of this species than in that of the Antarctic tern (needs confirmation) and in some areas may even be the exclusive food.

Molt. Twice per year. A complete molt begins on the head and body during migration in September, and wings and tail complete growth while the bird is in the Antarctic. Arctic terns are therefore completing regrowth of the outer primaries in January and February when Antarctic terns are just beginning to molt the inner primaries. Another partial molt of the head and body but not wings and tail takes place from February to March.

Ectoparasites. Feather lice, *Saemundssonia lockleyi, Quadraceps houri,* and *Austromenopon atrofulvum.*

Habitat. Although the Arctic tern is found in inshore waters with Antarctic terns during the austral summer, it also ranges farther out to sea, and pelagic terns are most likely to be this species.

Distribution. The Arctic tern breeds in the subarctic and arctic zones of the northern hemisphere from May to September, migrating south to climatically similar areas of the southern hemisphere from September to April. The greatest breeding concentrations are in the North Atlantic Ocean, and only a small number breed in the North Pacific Ocean. Most Atlantic Ocean birds cross from southern Africa to antarctic waters in a southeasterly direction and may drift farther eastward to the south Indian Ocean islands, Australia, and New Zealand before reaching the pack ice off the continent. Most records are from the Weddell Sea east to about 150°E. These birds may take advantage of the East Wind drift near the continent to move west into the Weddell Sea, where they complete molting of wings and tail before returning north in March via Africa. The species is probably quite rare in the western Atlantic Ocean sector. Arctic terns recorded in West Antarctica, where they may be more common than was previously thought, may be wind-drifted first-year birds. These may belong to the Pacific Ocean population, but a few of them may be Atlantic Ocean birds that circumnavigated the continent before returning north to spend their first 'summer' in the Pacific Peru current [*Salomonsen,* 1967]. See map for distribution in the Antarctic.

Bridled tern

BRIDLED TERN *Sterna anaethetus*

Vagrant

A dark gray-backed tern (14 in., 36 cm/30 in., 76 cm) with white underparts, nape, forehead, and eyebrows. The wings are nearly black; the gray tail has variable amounts of white. The bill and feet are black. Atlantic Ocean birds are pure white below; Pacific and Indian ocean birds have the underparts washed with gray. In the nonbreeding season the back and crown are mottled and streaked with white. The juvenile resembles the nonbreeding adult but has buff, rather than white, mottling on the back. A weakened adult of this cosmopolitan tropical and subtropical species was collected near Islas Diego Ramírez in the Drake Passage at approximately 56°50′S, 68°45′W in late January 1969 [*Peterson and Watson,* 1971].

BROWN NODDY *Anous stolidus*

Breeding Migrant

Identification. Plate 10. 16 in. (41 cm)/33 in. (84 cm). A dark brown tern with a grayish white cap and a long rounded tail. The bill and feet are black. The brown noddy is more slender than any of the dark shearwaters, gadfly petrels, or jaegers with which it might be confused.

The juvenile usually has the pale area on the head restricted to the forehead.

Downy chicks vary in color from nearly pure white to dark gray with only a white crown and belly. Most are gray above with white peppering, heaviest on the forehead, crown, and nape; the belly is white. The bill and feet are black.

Flight and habits. The brown noddy is less graceful and buoyant in flight than black-capped terns, but it has strong steady wingbeats. It generally flies closer to the surface than other terns. In feeding it snatches fish from the surface, only occasionally splashing rather than plunging deeply.

Voice and display. The brown noddy has a soft throaty courtship note and, when it is disturbed at the nest, 'ka-r-rk' like a crow but in a rather muted and sweet tone. In display at the nest two birds face each other, fence with bills, show orange mouth linings, and nod, bow, and jerk heads. The male may feed the female.

Food. Around the Tristan da Cunha group the brown noddy eats only small fish.

Reproduction. Mostly colonial. On Tristan Island the brown noddies gather in small groups to nest on offshore rocks and on the main island, but usually on Inaccessible Island and always on Gough Island, solitary pairs nest in dense *Phylica* woods. Nest sites are in rock crevices in low steep cliffs near the coast, in open caves, or in trees 4-5 m up. The nest is composed of twigs, grass, leaves, or algae.

Arrival: September at islands, occupying nest sites in October.

Eggs: Mid-October to late January, mostly in November. Clutch, 1 chalky white egg with chestnut blotching; averages 52.4 × 35.3 mm.

Hatching: Mid-November to mid-January. Incubation period 32-35 days.

Fledging and departure: Mid-January to mid-March. Departure takes place by late March.

Molt. Wings and tail are molted during breeding, from December to March.

Predation and mortality. No predation has been recorded in the Tristan da Cunha group or on Gough Island.

Ectoparasites. None recorded for the Tristan da Cunha group and Gough Island populations.

Habitat. Subtropical waters, generally feeding at no great distance from the breeding grounds but migrating over a great expanse of open ocean.

Distribution. Breeds on numerous islands in the tropical and subtropical zones of the Atlantic, Pacific, and Indian oceans including the Tristan da Cunha group and Gough Island. These southernmost birds migrate to an unknown area from April to August.

BLACK NODDY *Anous tenuirostris*

Rejected Breeding Record

The black noddy, which is smaller (14 in., 36 cm/28 in., 71 cm) and blacker and has a whiter forehead, a relatively longer bill, and a shorter tail than the brown noddy, has also been recorded breeding in the Tristan da Cunha group. The only proof of its occurrence, however, is a specimen of a nearly fledged young bird reportedly taken from a nest on Inaccessible Island in mid-October 1873 [*Saunders,* 1877]. The date is remarkably early in the season for a tropical bird, and the record should be viewed with skepticism because no later visits have substantiated the occurrence of this species there [*Watson,* 1969].

LAND BIRDS

Only one true land bird, a pipit, now occurs naturally in the Antarctic, whereas three finches and a thrush breed farther north in the Tristan da Cunha group and on Gough Island. On the other hand, at least 16 other species representing a variety of families occur as vagrants or established introductions. A population of the New Zealand red-crowned parakeet (*Cyanorhamphus novaezelandiae erythrotis*) became extinct on Macquarie Island by 1911 [*Falla,* 1937]. This dearth of resident or breeding land birds in the Antarctic is probably due to the year-round lack of open ice-free ground, severe winters, and remoteness of antarctic landmasses.

PIGEON *Columba livia*

Introduction

'Introduced' pigeons, presumably feral rock doves, *Columba livia,* were reported at the whaling station on South Georgia in January 1968 (D. K. Bailey, unpublished report, 1969).

TURTLEDOVE *Streptopelia* sp.

Vagrant

A small dove tentatively identified as a Madagascar turtledove, *Streptopelia picturata* (breeds on Madagascar and Seychelles and was introduced on Mascarene Islands), was found dead on Marion Island in 1965-1966 [*van Zinderen Bakker,* 1971*a*]. No specimen was preserved, and it may also have been a migratory dove, such as the Eurasian turtledove, *S. turtur,* or a widespread African species rather than a sedentary insular species.

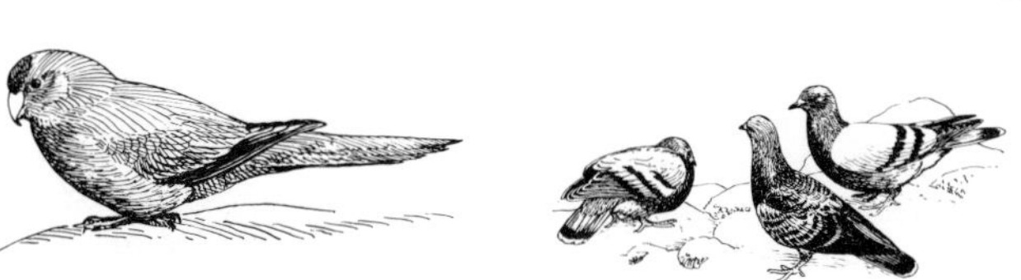

Red-crowned parakeet Pigeon

FORK-TAILED SWIFT *Apus pacificus*

Vagrant

A white-rumped fork-tailed swift (7 in., 18 cm/16 in., 41 cm) with brownish black back, white chin and throat, and barred underparts. The very long wings of swifts seem to flicker in flight. The fork-tailed swift is an east Asian migrant that occurs as a very rare straggler to New Zealand and was recorded on Macquarie Island in December 1958 [*Gibson,* 1959] and possibly also in December 1957 [*Keith and Hines,* 1958].

SPINE-TAILED SWIFT *Chaetura caudacuta*

Vagrant

A large (8 in., 20 cm/18 in., 46 cm) swift with a short square tail, white undertail coverts, and a light tan interscapular region. The

Fork-tailed swift Spine-tailed swift

crown, wings, and tail are glossy bluish black; the back, breast, and belly are brown; and the forehead and throat are white. Each tail feather has a spine on the tip. The spine-tailed swift is a central and northeast Asian species migrating regularly to Australia and Tasmania and occasionally invading New Zealand. It straggled to Macquarie Island in January 1960, when a specimen was taken [*Warham,* 1961].

UNIDENTIFIED SWIFTS

Vagrant

Other swifts of undetermined species were recorded on Ile Saint-Paul in early December 1957 [*Vanhöffen,* 1912] and on Macquarie Island in December 1957 [*Keith and Hines,* 1958].

BROAD-BILLED ROLLER *Eurystomus glaucurus*

Vagrant

A large-headed brown bird (12 in., 30 cm/27 in., 69 cm) with magenta underparts and striking purple wings. The rump and base of the slightly forked tail are blue. The legs are very short. The bill is yellow. A vagrant from the Madagascar population of this widespread African species was killed by a skua on Iles Kerguelen in November 1961 [*Milon,* 1964; *Falla,* 1964; *Prévost and Mougin,* 1970, pp. 128-129], and an individual was seen on Ile de la Possession, Iles Crozet, in late April 1969 [*Barrat,* 1974*b*].

BARN SWALLOW *Hirundo rustica*

Vagrant

A small (7.5 in., 19 cm /13.5 in., 34 cm) highly aerial bird with long streamers accentuating a deeply forked tail. The upperparts are dark metallic blue; the underparts are tan or rust colored. Swallows super-

Broad-billed roller

Barn swallow

Bank swallow

ficially resemble swifts but have triangular wings and appear to fly with more controlled and predictable wingbeats. The barn swallow is a holarctic species that was collected on Marion Island in 1965-1966 [*van Zinderen Bakker,* 1971*a*], on Iles Crozet in mid-May 1971 (European race [*Barrat,* 1974*b*]), on Tristan Island in late March 1938 and late October 1952 (both the American race [*Elliott,* 1953]), and at sea at 60°10′S, 61°15′W in late November 1963 and 55°20′S, 44°50′E in May 1934 [*Holdgate,* 1965*a*].

BANK SWALLOW *Riparia riparia*

Vagrant

A very small (4.75 in., 12 cm/9 in., 23 cm) brown swallow with a dark band across the white underparts. A specimen of this holarctic migrant was collected at sea between the Falkland Islands and South Georgia (53°S, 50°30′W) in mid-November 1967 (J. R. Beck, personal communication, 1968).

TRISTAN THRUSH *Nesocichla eremita*

Resident

Identification. Plate 11. 9.5 in. (24 cm)/13 in. (33 cm). A heavyset brown-backed songbird with strong dark spotting on the pale buff underparts. The neck and cheeks are orange buff heavily streaked with brown. The wings, particularly the shoulders, are marked with orange, and the whole underwing is orange. The smallest and palest race (*N. e. eremita*), which occurs on Tristan Island, is sepia and buff with a little tawny red. The bill is black; the feet are very dark brown. A darker more rufous bird intermediate in size and color occurs on Inaccessible Island (*N. e. gordoni*) and has a lighter bill with a pale tip. The Nightingale thrush (*N. e. procax*) is largest and darkest with a grayish tinge to the buff underparts, heavier spotting, and even more rufous on the wings. The bill is like that of *N. e. gordoni*, but the feet are flesh brown.

The juvenile has orange spots and streaks on the back and more pronounced smaller spots on the underparts.

No information is available about the chick.

Flight and habits. Flight is short, swirling, and sparrowlike. Behavior is typically thrushlike. The Tristan thrush climbs in tussock and moss, where it stays well hidden, although it is generally tame, inquisitive, and 'perky.' It tends to be most social on Nightingale Island but shy and silent on Tristan Island.

Voice and display. One call is a soft chirp accompanied by a flitting of wings; the other is a wheezing sibilant note. The song is a combination of the two, 'chissik, chissik, trrtkk, swee, swee, swee' or 'pseeooee, pseeooee, pseeooee, psee—ptee.'

Food. Almost omnivorous, foraging for seeds, fruit, amphipods, insects, spiders, and worms. The Tristan thrush also feeds on carrion such as stranded fish and bird carcasses left by skuas. Its tongue is adapted to egg sucking, and the thrush can break open all Tristan Island eggs except those of rockhopper penguins.

Reproduction. Territorial in pairs. The nest is a rough cup about 10 cm in diameter made of tussock fronds, grass stalks, green leaves, and moss scraps and placed a few centimeters to 30 cm off the ground at the base of a tussock. Nests are occasionally placed in a *Phylica* tree or on a ledge or other sheltered spot, even inside a shack. Only the female incubates.

Pairing: No information.

Eggs: Dates are variable, i.e., early September and October on Nightingale Island, November on Tristan Island, and as late as February (second clutch?) on Inaccessible Island. Clutch, 2-3 pale turquoise eggs thickly covered with brownish red speckles and blotches; average 29.5 × 22 mm on Inaccessible Island and 33.5 × 22.7 mm on Nightingale Island.

Hatching: No information. Incubation period unknown.

Fledging: Flying young on Inaccessible Island in February.

Molt. Recorded in January but needs study.

Predation and mortality. The Tristan thrush appears most abundant on Nightingale Island and rare and local on Tristan Island, where the total population is only 200-400 birds, possibly because of predation by rats and feral cats.

Ectoparasites. No information is available.

Habitat. Occurs at all elevations, from the beaches (Inaccessible Island, where it is absent from the plateau) to the higher slopes and moors (to 1200 m on Tristan Island, where it is absent from beaches and rare in shore cliffs), in tussock (Nightingale Island) and dock and tree ferns (Tristan Island).

Distribution. Resident on Tristan, Inaccessible, and Nightingale islands.

Blackbird

Song thrush

BLACKBIRD *Turdus merula*

Vagrant

The male (10 in., 25 cm/16 in., 41 cm) is wholly black with a yellow bill and eye ring. The female is dark gray, brown above and gray spotted with brown below; the bill is dark brown. A Palaearctic species introduced on New Zealand was reported as early as 1951 on Macquarie Island and finally confirmed in March 1960 [*Warham,* 1969].

SONG THRUSH *Turdus philomelos*

Vagrant

A medium-sized (9 in., 23 cm/15 in., 38 cm) brown terrestrial thrush with strongly spotted white underparts. Both sexes are alike, but juveniles have white spotting on the upperparts. A female of this Palaearctic species that has been introduced on New Zealand and has reached as far south as Campbell Island was collected from a group of three on Macquarie Island in late August 1967 [*Merilees,* 1971*b*].

SOUTH GEORGIA PIPIT *Anthus antarcticus*

Resident

Identification. Plate 8. 6.5 in. (17 cm)/9.5 in. (24 cm). A slim ground-dwelling songbird with a streaked brown body, grayish white outer tail feathers, and a slender bill. The upperparts are reddish tan heavily marked with dark brown and some light buff. The underparts are pale buff with heavy dark brown streaking on the breast and flanks and light streaking on the throat and belly. The straw-colored legs are long with a remarkably elongated claw on the hind toe.

The juvenile is even more heavily marked on the underparts than the adult.

The chick is naked at hatching but within 1 week has a plentiful pale buff down over most of its body. It remains in the nest until it is fully fledged.

Flight and habits. Flies in an undulating fashion, usually low over the ground and generally not for any great distance. The South Georgia pipit is extremely tame, even allowing capture by hand. It has a characteristic habit of running for a short distance, stopping, flicking its tail, and running again.

Voice and display. Twittering song, somewhat like that of an eastern North American song sparrow, *Melospiza melodia,* but lasting longer and much softer in tone. The flight song has been likened to that of a wagtail, *Motacilla.*

Food. Adult and larval springtails, beetles, flies, and marine copepods found in the grass or in algae cast up on the beach around tidal pools.

Reproduction. From mid-November to December the adults retire to offshore islets to breed. These islets become snow free sooner than the mainland and are covered with abundant tussock grass, in which the nest is hidden. The nest is a deep cup built of fine roots and dried grass on the ground and sometimes partly domed over. Nesting has also been recorded in rock crevices on the main island, but the actual nests have not been found. The only egg described was dull gray green thickly speckled and streaked with red brown; 22 × 17 mm. Four naked young were found in one nest on January 6, and fledged young were seen on January 2.

Predation and mortality. No predation has been recorded, but rats may be predators on the main island.

Ectoparasites. No information is available.

Habitat. Grassy glacial meadows near sea level. The South Georgia pipit feeds along the shore or beside glacial streams, using nearby thick grass for protective cover.

Distribution. Resident all year on South Georgia. The South Georgia pipit was probably derived from the South American and Falkland Island species *A. correndera,* but it is now quite distinct.

PIPIT *Anthus* sp.

Vagrant

An unidentified 'pipit' recorded on Macquarie Island in 1956 and 1957 [*Keith and Hines*, 1958] may have been the New Zealand pipit, *A. novaeseelandiae,* which breeds on New Zealand and some of its subantarctic islands.

STARLING *Sturnus vulgaris*

Resident

Identification. Plate 8. 8.5 in. (22 cm)/13 in. (33 cm). A stocky songbird with glossy greenish or purplish black plumage, short pointed wings, and a short square tail. The strong legs and feet are reddish brown. From April to August the plumage is speckled with white, especially on the underparts, and the long pointed bill is brownish gray. The spots abrade, so that by the breeding season the bird is uniformly dark and glossy. The bill is then bright yellow.

The juvenile is grayish brown, whitish on the throat, and lightly streaked with buffy white on the belly. The bill and feet are dark brown. This plumage is molted from February to April, when the bird assumes a first-year plumage similar to that of the adult. The first-year bird is still recognizable, however, because the white area of each feather is rounded and larger than it is in the adult, so that the young bird appears more spotted. The spots show less tendency to wear than those of the adult, and the metallic colors are less brilliant.

The chick at hatching is nearly naked and helpless. A wispy brown down later covers the head and parts of the body, but the bird is mostly bare until the juvenile feathers begin to develop.

Flight and habits. The rapid rustling beat of the wings, short tail, and direct flight are characteristic. On the ground the starling is very active, walking with a jerky strutting gait. Gregarious, especially in winter, when communal roosting in rocks, cliff crevices, and grass tussocks aids in protection from cold nights.

Voice and display. The call is a long descending whistled 'cheeoo'; the song is a variety of warbling, chirruping, gurgling, and whistled notes. The species is an excellent mimic.

Food. A ground-feeding omnivore, taking seeds, berries, and a variety of invertebrates including insects and worms. Its diet probably varies seasonally. The only food recorded on Macquarie Island is

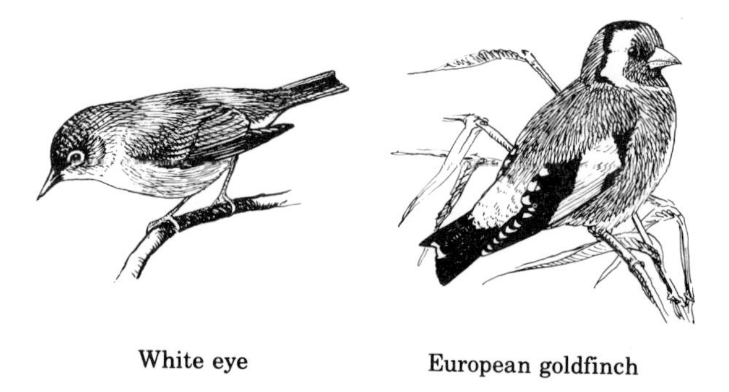

White eye European goldfinch

the red berries of taupeta, *Coprosma pumila.* The starling forages on the plateau in herb fields in *Festuca erecta* and on coastal terraces in small flocks, but during the breeding season, pairs scavenge along the beaches on moss and kelp jetsam.

Reproduction. Few data are available from Macquarie Island. Elsewhere the nest is an untidy mass of dry grass and feathers usually built in a protected cavity. On Macquarie Island the starling utilizes rock crevices in sea cliffs and caves. On New Zealand it breeds from September to November. Clutches range from 4 to 7 slightly glossed pale blue eggs; 30 × 21 mm. The incubation period is 13 days; fledging 21 days. Only one brood is raised. Young were heard calling from nests on Macquarie Island in November.

Molt. No information is available for the Macquarie Island population, but elsewhere adults begin molt after breeding

Predation and mortality. No information is available. See the section on distribution.

Ectoparasites. Nasal mites, *Sternostoma bruxellarum* and *Boydaia sturni;* feather lice, *Sturnidoecus sturni, Brueelia nebulosa, Menacanthus mutabilis,* and *Myrsidea cucullaris.*

Habitat. Found in dry tussock grasslands and wet meadows from sea level to the upper limits of vegetation.

Distribution. A western Palaearctic species that has been introduced in the New World, South Africa, Australia, and New Zealand. From New Zealand it colonized all the off-lying subantarctic islands and established itself on Macquarie Island early in the present century. A flock of 250 was recorded in 1960, and almost 1400 birds roosted at Green Gorge in 1967 [*Warham,* 1969; *Merilees,* 1971*a*].

WHITE EYE *Zosterops lateralis*

Vagrant

A tiny (4.7 in., 12 cm/7.5 in., 19 cm) bright yellow green bird with a gray back, brown sides, grayish white underparts, and a prominent white eye ring. The bill and feet are brown. The sexes are alike. The white eye is an Australian forest species that established itself naturally on New Zealand over 100 years ago and was recorded once on Macquarie Island (sight record in 1915 [*Falla,* 1937]).

EUROPEAN GOLDFINCH *Carduelis carduelis*

Vagrant

A brightly colored finch (5 in., 13 cm/9.5 in., 24 cm) with a red, white, and black head and a yellow band on the black wings. The back is brown, the rump and underparts are white, and the forked tail is black. Juveniles lack the head pattern and are streaked on the breast. The European goldfinch is a western Palaearctic species that has been introduced on New Zealand and has colonized all the New Zealand sub-Antarctic Islands. A male was collected on Macquarie Island in April 1956 [*Keith and Hines,* 1958], and another in early May 1967 [*Merilees,* 1971*a*].

REDPOLL *Acanthis flammea*

Resident

Identification. Plate 8. 5 in. (13 cm)/9 in. (23 cm). A small chunky streaked finch with red forehead, black chin, and small conical yellowish brown bill. The upperparts and flanks are tan or grayish white streaked with gray brown. The wings and forked tail are brownish black with a buff or white bar on the wing coverts. The male has a pink tinge on the breast.

The juvenile is similar to the female but has no red on the forehead and much less black on the chin and is more heavily streaked on the throat and breast.

The chick is covered with relatively long and plentiful dark gray down and remains in the nest until it is fully fledged.

Flight and habits. The redpoll usually flies at a considerable height, when it is best identified by its call and undulating flight. It is highly acrobatic, hanging from seed heads when it is feeding. It hops on the ground.

Voice and display. Metallic twitter in flight. The song is a rippling trill, either from the perch or in flight.

Food. Mainly seeds and also some insects and their larvae. The redpoll feeds both on the ground and in vegetation. On Macquarie Island the foods recorded are seeds from *Cotula, Pleurophyllum,* and Macquarie cabbage (*Stilbocarpa polaris*).

Reproduction. A nest in a clump of ferns, *Polystichum vestitum,* in February 1912 is the only Macquarie Island breeding record. On New Zealand, where the redpoll is abundant, it breeds in small aggregations from September to March. The nest is a cup of roots and grass thinly lined with feathers or hair on a firm twig foundation. The 4-6 eggs are dull bluish green spotted and streaked with light brown; 17 × 12.5 mm. The incubation period is 10-12 days; fledging 11-14 days. The species is probably double brooded.

Molt. No information is available for the Macquarie Island population. Elsewhere adults molt after breeding.

Predation and mortality. No information is available for the Macquarie Island population.

Ectoparasites. No information is available for the Macquarie Island population.

Habitat. Tussock hillsides and wet and dry meadows from sea level to the upper limits of vegetation. The redpoll is mainly coastal in winter.

Distribution. The redpoll is a northern holarctic species that was introduced into New Zealand and has naturally established breeding populations on Chatham, Snares, Auckland, Campbell, and Macquarie islands. Only small numbers occur on Macquarie Island, where it arrived early in the present century.

TRISTAN BUNTING *Nesospiza acunhae*

Resident

Identification. Plate 11. 7 in. (18 cm)/10 in. (25 cm). A stocky olive green backed finch with bright yellow underparts and eyebrow stripe. The breast and flanks are suffused with green, and the head has indistinct dark streaking. The grayish green wings and tail appear the same color as the back when they are folded. The conical straight bill and feet are brownish gray. Yellow pigments are reduced in the female, which appears more grayish and streaked above and washed out below. On Nightingale Island, *N. a. questi* is smaller and more yellow and has a less heavy bill than do birds from Inaccessible Island, probably *N. a. acunhae,* but see *Hagen* [1952].

The juvenile is strongly streaked with brown above and yellowish white below. It may spend 2 or more years in a more or less intermediate streaked plumage before becoming fully adult.

The chick is unknown.

Flight and habits. The Tristan bunting is cautious but not shy. It climbs to the highest stems of tussock, calls for awhile, and then flies a short distance to join other individuals. It does not flock but stays in pairs or family groups.

Voice and display. The song is a simple melodious twitter or chirping 'chickory chikky' followed by a wheezy 'tweeyer.' The male has a rattling call to lure the incubating female off the nest. Birds on Inaccessible Island have a greater variety of higher-pitched calls than those on Nightingale Island.

Food. Tussock and other grass seeds, insects, and spiders. Grit is also found in stomachs. On Inaccessible Island, birds are chiefly seed eaters, whereas on Nightingale Island they feed largely on insects found, in part, in lichens on *Phylica* trunks or low vegetation.

Reproduction. Nests, which have only been found on the plateau of Nightingale Island at 180 m, are built on or just above the ground at the base of a clump of sword grass or in bracken. The shallow cup is made of loosely woven sword grass and tussock stalks and lined with tussock fronds, grass heads, and fine water grass. Both sexes incubate.

Pairing: No information.

Eggs: December to January, possibly also earlier. Clutch, 1-2 (4-5 also reported but probably in error) pale turquoise eggs speckled pale chestnut with underlying mauve blotches; 23-26 × 16.5-18 mm.

Hatching: Recorded in early January. Incubation period unknown.

Fledging: Recorded in early February and early March.

Molt. On Inaccessible Island, adults begin molting wings by late January. Juveniles molt bodies at the same time but not wings or tails. Nightingale Island birds may molt earlier.

Predation and mortality. No information is available. The population on Inaccessible Island is several hundred. The species disappeared from Tristan Island before rats were introduced, possibly because of the disappearance of tussock.

Ectoparasites. No information is available.

Habitat. Generally found in dense tussock and dock but also occurs in *Phylica* woods, where it forages on tree trunks, in the lower branches, or on the ground.

Distribution. Resident on Nightingale and Inaccessible islands in the Tristan da Cunha group and extirpated from Tristan Island itself before 1873.

WILKINS' BUNTING *Nesospiza wilkinsi*

Resident

Identification. Plate 11. 8-8.5 in. (20-22 cm)/11-12 in. (28-30 cm). Similar to the Tristan finch in color but decidedly larger, particularly in the much heavier, stouter, and darker bill, which is strongly decurved rather than straight. Nightingale Island birds (*N. w. wilkinsi*) are larger, yellower, and paler below than Inaccessible Island birds (*N. w. dunnei*).

The juvenile is dull olive gray above and washed-out yellow green below with dark streaks all over. The bill is less heavy than that of the adult. Subadult birds may retain streaking on the back and breast during the second year.

The newly hatched chick has pale gray down.

Flight and habits. Climbs in branches of trees like a crossbill. Individuals tend to wander more on Inaccessible Island, where the habitat is patchy, than on Nightingale Island.

Voice and display. The call of the Nightingale Island bird is a clear 'flutelike note' similar to that of the Tristan bunting but with 'a more pure and pleasing tone' recorded as an almost whistling 'tweet tweeyeer, tweet tweeyer.' The Inaccessible Island bird has a similar, but harsher, call.

Food. On Nightingale Island, Wilkins' bunting eats *Phylica* nuts, which it cracks and husks before it swallows. It also eats 'cranberry,' *Nertera,* seeds and feeds young on *Empetrum* seeds. On Inaccessible Island, in addition to *Phylica* nuts, it takes hard seeds of several types and insects and also ingests sand particles.

Reproduction. Solitary pairs defend territories. Nests, which have only been discovered on Nightingale Island, are built on or near the ground in a *Spartina* tussock, water grass thicket, or weeds and bracken. The deep cup of tussock fronds is supported by *Phylica* twigs, occasionally partly domed and having a tunnel entrance. Both sexes incubate. The following schedule is from Nightingale Island; no information is available from Inaccessible Island.

Pairing: No information. Nests under construction in early December and early January.

Eggs: Presumably December to early February. Clutch, 2 pale or greenish blue eggs speckled with pale brown and blotched with pale mauve especially at the blunt end; 26.5-29 × 19.5-21 mm.

Hatching: Records in early January. Incubation period unknown.
Fledging: Fledged young being fed as late as April 4; independent young recorded in early February.

Molt. Adults start molting body and wings in January and February on Nightingale Island; the tail is molted later. The schedule on Inaccessible Island may be somewhat later.

Predation and mortality. No information is available. Total populations are 70-120 birds on Nightingale Island and 40-90 on Inaccessible Island, where the *Phylica* woods are sparse.

Ectoparasites. No detailed information is available, but Wilkins' bunting is said to be heavily parasitized by Mallophaga, hippoboscid flies, fleas, and ticks.

Habitat. *Phylica* woodland exclusively on Nightingale Island and also tussock on Inaccessible Island.

Distribution. Resident on Nightingale and Inaccessible islands in the Tristan da Cunha group.

GOUGH BUNTING *Rowettia goughensis*

Resident

Identification. Plate 11. 7.5 in. (19 cm)/10.5 in. (27 cm). A bright olive green finch with yellow belly, eyebrows, and moustache and a black throat and eye mask. The bill and feet are dark brown. The female is drab olive green with yellow buff on the belly and indistinct dark markings on the face.

The juvenile is orange buff streaked all over with blackish brown. Intermediate (first-winter?) birds are richer brown with bolder streaking on the back, upper breast, and sides; still older (second-year?) birds resemble the female but retain the rich brown back, streaking being restricted to the interscapular region and sides. The downy chick is unknown.

Flight and habits. Hops about on the ground. The Gough bunting is very tame. Its behavior is said to be quite unlike that of the Tristan bunting but rather suggestive of that of the Tristan thrush. It is highly territorial, but (family?) groups of five to seven birds occur together.

Voice and display. The call is a penetrating 'tissik'; the feeding note is a soft 'pseeping.'

Food. Small seeds, fruits, insects, and amphipods collected in dead seaweed on the beach or in vegetation and on rocks elsewhere.

Reproduction. Little information is available. One old nest has been found in a tussock, but the eggs and schedule are unknown. The young are already fledged in mid-November.

Molt. No information is available.

Predation and mortality. No information is available, but the species is abundant.

Ectoparasites. No information is available.

Habitat. Generally distributed all over the island, especially common along the coast among boulders, stranded wood, and seaweed a little above high-water mark and on open ground in the mountains.

Distribution. Resident on Gough Island. The Gough bunting is very closely related to the South American mountain finch genus *Melanodera.*

HOUSE SPARROW *Passer domesticus*

Unsuccessful Introduction

A small (5.7 in., 14 cm/10 in., 25 cm) brown bird with a streaked back, grayish underparts, and a light wing bar. Males have a black bib on the throat and upper breast, a gray crown, light gray or white cheeks, and a black bill (breeding season only). The female and immature lack any distinctive markings and have a brown bill. The house sparrow is larger, heavier, and browner than the redpoll. It is a Palaearctic symbiont of man that has been widely introduced around the world including southern South America and the Falkland Islands. It was apparently introduced on South Georgia, where there is one undated short-lived record (W. L. N. Tickell, unpublished report, 1960).

UNIDENTIFIED VAGRANT BIRDS

'A dark bird, larger than the giant fulmar, *Macronectes giganteus,* with long out-stretched neck, was seen flying at about 1000 feet on Macquarie Island in September 1957' [*Keith and Hines,* 1958]. Possi-

House sparrow

bly, this was the same individual identified as a great cormorant in October of that year.

'A small sparrowlike bird' was seen on Iles Kerguelen by Comer [*Verrill,* 1895], who suggested that it was not rare. No specimen was collected, and no other observer has reported a resident passerine bird on Iles Kerguelen.

A bird 'like a sparrow with a brown spotted breast' was collected on Macquarie Island in 1915, but the specimen cannot be located [*Falla,* 1937].

The lands frequented by the birds in this handbook range from small temperate subantarctic islands with stunted trees and lush tussock grass to the icy wastes of the vast Antarctic continent. This section presents brief discussions of the appearance, location, physical environment, climate, vegetation, exploration, and present status of knowledge of the birdlife for each of the landmasses. Table 4 lists the birds that breed in each area, and Table 5 shows the distribution of seabirds in various sectors of the ocean and north and south of 55°S, roughly the sub-Antarctic and Antarctic zones. These tables can be used for generating local checklists of the species that a visitor might expect to see.

The major landmasses and sea areas discussed in this section are shown on a polar projection to 35°S in the frontispiece. Figures 10 and 11 show in larger scale the South Orkney and South Shetland islands and the Antarctic Peninsula, but many geographic features mentioned in the text do not appear in these three maps. Greater detail, particularly for the continent, is shown on the map of Antarctica by the *American Geographical Society* [1965] and in maps in various local avifaunal accounts (see the analysis of references). Geographic names in this section and elsewhere in this handbook conform to the standards set forth by the U.S. Board of Geographic Names [*Geographic Names Division*, 1969].

SHAG ROCKS

Shag Rocks is a group of approximately six guano-covered islets that lie at 55°33′S, 42°02′W, 1500 km east of Tierra del Fuego and 240 km west of South Georgia. The highest point is about 75 m above sea level. The rocks rise sheer from the sea, and landing is probably impossible. Black Rock and another low rock lie 18.5 km southeastward of the main group.

In recent years, observers on small boats from a cruise ship found 'three species dominant on the rocks. Shags [blue-eyed?] were present in enormous numbers, as were great numbers of prions and quite a few wandering albatrosses. Undoubtedly other species occur on the rocks' (R. T. Peterson, personal communication, August 1974).

TABLE 4. Distribution of Birds

	South Georgia	South Sandwich Islands	South Orkney Islands	South Shetland Islands	Antarctic Peninsula	Antarctic Continent	Scott Island
Emperor penguin					X	X	
King penguin	X	?					
Adélie penguin		?	X	X	X	X	
Chinstrap penguin	X	X	X	X	X		
Gentoo penguin	X	X	X	X	X		
Rockhopper penguin							
Macaroni penguin	X	X	X	X	X		
Wandering albatross	X						
Black-browed albatross	X						
Gray-headed albatross	X						
Yellow-nosed albatross							
Sooty albatross							
Light-mantled sooty albatross	X						
Northern giant fulmar	X						
Southern giant fulmar	X	X	X	X	X	X	
Southern fulmar		X	X	X	X	X	
Antarctic petrel						X	
Cape pigeon	X	X	X	X	X	X	
Snow petrel	X	X	X		X	X	X
Narrow-billed prion							
Antarctic prion	X	X	X	X		X	X
Broad-billed prion							
Fulmar prion							
Fairy prion							
Blue petrel			?				
Great-winged petrel							
White-headed petrel							
Atlantic petrel							

Breeding on Islands and Landmasses

Balleny Islands	Peter I Island	Tristan da Cunha Group and Gough Island	Bouvetøya	Marion and Prince Edward Islands	Iles Crozet	Ile Amsterdam and Ile Saint-Paul	Iles Kerguelen	Heard Island	Macquarie Island
				X	X		X	X	X
X	X		X						
X	X		X					X	
				X	X		X	X	X
		X		X	X	X	X	X	X
			X	X	X		X	X	X
		X		X	X	?	X		X
							X	X	X
				X	X		X		X
		X		X		X			
		X		X	X	X			
				X	X		X	X	X
		X		X	X		X	?	X
			?	X	X		X	X	X
?	X		X						
?									
X	X		X		X		X	X	X
X			X						
					X		X		
				?	X		X	X	X
		X		X	X	X			
							?	X	
				X	?				?
				X	X		X		?
		X		X	?	?	X		
					?		X		X
		X							

TABLE 4.

	South Georgia	South Sandwich Islands	South Orkney Islands	South Shetland Islands	Antarctic Peninsula	Antarctic Continent	Scott Island
Kerguelen petrel							
Soft-plumaged petrel							
White-chinned petrel	X						
Gray petrel							
Sooty shearwater							
Flesh-footed shearwater							
Greater shearwater							
Little shearwater							
Wilson's storm petrel	X	X	X	X	X	X	X
Black-bellied storm petrel	X		X	X			
Gray-backed storm petrel	X						
White-faced storm petrel							
South Georgia diving petrel	X						
Kerguelen diving petrel	X						
Blue-eyed shag	X	X	X	X	X		
King shag							
Gray duck							
Yellow-billed pintail	X						
Kerguelen pintail							
Speckled teal	X						
Inaccessible Island flightless rail							
Weka							
Gough moorhen							
American sheathbill	X		X	X	X		
Lesser sheathbill							
South polar skua				X	X	X	
Brown skua	X	X	X	X	X		

(continued)

Balleny Islands	Peter I Island	Tristan da Cunha Group and	Gough Island	Bouvetøya	Marion and Prince Edward Islands	Iles Crozet	Ile Amsterdam and Ile Saint-Paul	Iles Kerguelen	Heard Island	Macquarie Island
		X			X	X		X		
		X			X	X				
		X			X	X	?	X		?
		X			X	X	?	X		?
										X
							X			
		X								
		X					?			
?	?			?		?		X	X	
		X		?		X	?	X		
		?				X		X		?
		?					?			
					X	X		X	X	?
		X			X	X	?	X	X	?
									X	
					X	X		X		X
										X
						X	I	X		
		X								
										I
		X								
					X	X		X	X	
X	?									
?		X		X	X	X	X	X	X	X

TABLE 4

	South Georgia	South Sandwich Islands	South Orkney Islands	South Shetland Islands	Antarctic Peninsula	Antarctic Continent	Scott Island
Southern black-backed gull	X	X	X	X	X		
Antarctic tern	X	?	X	X	X		
Kerguelen tern							
Brown noddy							
Tristan thrush							
South Georgia pipit	X						
Starling							
Redpoll							
Tristan bunting							
Wilkins' bunting							
Gough bunting							

Known to breed in area, X; introduced and breeding in the area, I; breeding

(continued)

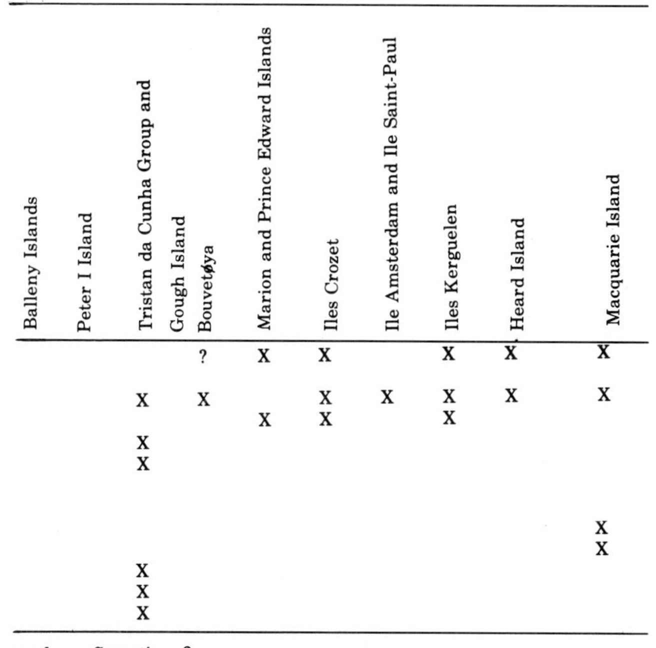

Balleny Islands	Peter I Island	Tristan da Cunha Group and Gough Island	Bouvetøya	Marion and Prince Edward Islands	Iles Crozet	Ile Amsterdam and Ile Saint-Paul	Iles Kerguelen	Heard Island	Macquarie Island
			?	X	X		X	X	X
		X	X		X	X	X	X	X
				X	X		X		
		X							
		X							
									X
									X
		X							
		X							
		X							

needs confirmation, ?.

TABLE 5. Distribution of Birds at Sea in Various

	South America and Scotia Ridge, 80° to 30°W		Atlantic Ocean and Weddell Sea, 50°W to 20°E	
	North	South	North	South
Emperor penguin	v	X		X
King penguin	X	x		
Adélie penguin	v	X	x	X
Chinstrap penguin	x	X	x	X
Gentoo penguin	X	X		v
Rockhopper penguin	X	v	X	
Macaroni penguin	x	X	x	x
Wandering albatross	X	X	X	X
Black-browed albatross	X	X	X	
Gray-headed albatross	X	X	X	x
Yellow-nosed albatross	x		X	
Sooty albatross		v	X	v
Light-mantled sooty albatross	X	X	X	X
Giant fulmars*	X	X	X	X
Southern fulmar	X	X	x	X
Antarctic petrel	v	X	v	X
Cape pigeon	X	X	X	X
Snow petrel	x	X	x	X
Narrow-billed prion	X	X		
Antarctic prion	X	X	X	X
Broad-billed prion	v		X	
Fulmar prion				
Fairy prion	x			
Blue petrel	X	X	X	X
Great-winged petrel		v	X	
White-headed petrel	v	v	X	x
Atlantic petrel	X	v	X	
Kerguelen petrel	v	v	X	X
Soft-plumaged petrel	X		X	x
Mottled petrel				
White-chinned petrel	X	X	X	X
Gray petrel	x		X	
Sooty shearwater	X	v	X	x

Sectors of Antarctic and Subantarctic Seas

Indian Ocean, 20° to 100°E		Australian, New Zealand, and Ross Seas, 100°E to 175°W		Pacific Ocean, 175° to 80°W	
North	South	North	South	North	South
v	X	v	X		X
X	x	X	x		
v	X	v	X		X
x	x	v	x		x
X		x	x		
X		X	x		
X	x	x	x		
X	X	X	X	X	X
X	x	X	X	X	X
X	x	X	X	X	X
X		X			
X					v
X	X	X	X	X	X
X	X	X	X	X	X
v	X	v	X	X	X
v	X	v	X	v	X
X	X	X	X	X	X
v	X		X		X
X	X	X	X	X	X
X	X	X	X	X	X
X		X		x	
X	x	X			
x		X	x		
X	X	x	X	X	X
X	v	X		X	v
X	X	X	X	X	X
X					
X	X	v	X		X
X		v			
		X	X		x
X	X	X	X	X	X
X		X	x	X	x
v	x	X	X	X	X

TABLE 5.

	South America and Scotia Ridge, 80° to 30°W		Atlantic Ocean and Weddell Sea, 50°W to 20°E	
	North	South	North	South
Flesh-footed shearwater				
Greater shearwater	x		X	
Little shearwater	v		X	
Wilson's storm petrel	X	X	X	X
Black-bellied storm petrel		X	X	X
Gray-backed storm petrel	X	x	x	
White-faced storm petrel	v		x	
South Georgia diving petrel	X	x	x	
Kerguelen diving petrel	X	x	X	x
Blue-eyed shag	X	X		
King shag	X			
American sheathbill	X	X		
Lesser sheathbill				
South polar skua	x	X	x	X
Brown skua	X	X	X	x
Southern black-backed gull	X	X	x	v
Antarctic tern	x	X	X	x
Kerguelen tern				
Arctic tern	X	X	X	X
Brown noddy			x	

Occurs abundantly in the sector, X; occurs sparingly in the sector, x; and
north of 55°S; the south entry refers to the area south of 55°S. These two
*The at-sea ranges of the two giant fulmars *Macronectes giganteus* and

(continued)

Indian Ocean, 20° to 100°E		Australian, New Zealand, and Ross Seas, 100°E to 175°W		Pacific Ocean, 175° to 80°W	
North	South	North	South	North	South
x		X			
x					
x		X		X	
X	X	X	X	x	x
X		X	x		x
X		X	x		
x		X		x	
X	x	X	X		x
X	x	X	x		x
x					
X		x			
X					
x	X	x	X	x	X
X	x	X	x		
X	v	X	v		
X	x	X	x		
X					
X	X	X	X	x	x

known only as a vagrant in the sector, v. The north entry refers to the area
areas correspond roughly to subantarctic and antarctic waters.
M. halli are not distinguished in this table.

SOUTH GEORGIA

South Georgia, the largest and highest island on the Scotia ridge, lies between 53°30′ and 55°S and 35°30′ and 38°30′W. The Falkland Islands (Islas Malvinas) are 1350 km WNW. The island is crescent shaped, its long axis running northwest to southeast and its concave side facing the southwest. It is 170 km long and 40 km wide at its widest. The spine of the island is a long high mountain ridge, Mount Paget being its highest point, at 2950 m. Although much of the coast is formed of high sea cliffs, the north side of the island is indented with numerous deep bays, fjords, and glacial valleys. There is no flat coastal plain, but here and there can be seen traces of a narrow raised beach at about 7.5 m above present sea level.

Along the warmer north side the summer snow line ranges between 450 and 600 m, but elsewhere it is much lower. The island is glaciated, and moraines are present in most wider valleys. Glacial ponds and swift cascading streams are everywhere, some of them quite large.

Numerous small rocky and mountainous islands lie off the coast. Among the largest and most important are Bird Island and Willis Islands at the northwest end, Annenkov Island on the southwest coast, and Cooper Island at the east end.

Clerke Rocks, a group of 15 or more granite islets and rocks that attain an elevation of 331 m, lie about 74 km southeast of Cooper Island.

Most of the rocks are folded metamorphosed acid sediments consisting of slates, silts, and graywackes with occasional thin limestone strata. A few igneous intrusions occur on the south and southeast coasts.

The climate is cold, cloudy, and windy with little seasonal variation. Local topography, especially elevation, affects meteorological measurements. High hills and mountains near Cumberland Bay on the north coast, where most of the weather records are taken, make the area warmer in summer, wetter, and much less windy than more exposed locations on the island or nearby waters would be. Although Bird Island is partly protected from the full force of the winds, it may give a better indication of true summer temperatures and wind directions. Mean monthly temperatures are below freezing from May to September, but the mean temperature in August, the coldest month, is only −2°C, and the minimum temperature near sea level in winter is only about −15°C. February, the warmest month, has a mean temperature of about 5°C at Bird Island and 7°C at Cumberland Bay. Summer temperatures seldom go above 9.5°C but have reached as high as 21°C at Cumberland Bay. Precipitation shows little seasonal variation, although it is slightly higher in winter. The average annual precipitation is about 1500 mm, mostly as rain in summer and snow in winter. Permafrost is present only near glaciers and at high elevations. Prevailing winds with a mean velocity of 16 km/h on the north side of the island are from the north and west. These warm moist winds are cooled when they pass over the sea surface and thus bring low clouds and much fog to South Georgia. Winds at Bird Island are predominately from the southwest and northwest. Severe storms and cold weather accompany winds from the southwest.

South Georgia is about 350 km south of the Antarctic convergence and is under the influence of cold water from the Weddell Sea gyre. Except for the extreme southeast end of the island, which experiences some pack ice in winter, its bays and harbors are ice free all year. The northern limit of pack ice is about 180 km south of the island in early spring and 900 km south in fall.

The flora consists of 20 species of flowering plants with several other species of ferns, mosses, and lichens. The lushest vegetation is near the coast. The tussock grass *Poa flabellata,* which grows up to 1.5 m high, is confined to coastal areas and reaches an elevation of 225 m on some slopes. It occurs in nearly pure closed stands on flat or gentle slopes especially on Bird Island and on the northwest tip of South Georgia away from penguin colonies or elephant seal grounds. A more open tussock of pronounced *Poa* stools and other secondary plants occurs where the grass is subject to disturbance by birds and mammals or the ground is less stable.

The climax vegetation along the well-drained flats or moderate slopes of the northeast coast is a grass heath or tundra meadow. Depending on local conditions of stability, water, and elevation it varies from a closed heath rich in grasses and sedges to a moss and lichen carpet, in which phanerogams are of little importance. The wettest areas are marshes and bogs, in which rushes and wet mosses are dominant. Low-growing feldmark of moss cushions and only scattered phanerogams occurs on bare rocky ground above 225-300 m and on unstable rocky slopes, screes, and moraines.

The island was discovered by Captain Cook in 1775 and claimed for Great Britain. It was a center for British and American sealers from 1778 to the early nineteenth century and was the site of several year-round whaling stations from 1904 until 1965. Human occupation has continued up to the present for weather observations and scientific studies. With man came domestic animals and some plants. Of the animals, only reindeer (several thousand) and innumerable rats have thrived, but horses, sheep, rabbits, mice, cats, and dogs have at times also lived on the island or still persist. Predators affect birds on coastal grasslands of South Georgia. Most of the offshore islands, especially Bird Island, are free from introduced pests and have large populations of birds.

Many expeditions have studied the extensive bird life of South Georgia. The most comprehensive recent reports are by *Matthews* [1929] and by *Rankin* [1951]. Some species monographs are based on fieldwork conducted at South Georgia or on Bird Island [*Stonehouse*, 1956, 1960; *Tickell*, 1962, 1968].

SOUTH SANDWICH ISLANDS

The 11 South Sandwich Islands form a 340-km long curving chain of volcanic peaks between 56°18′ and 59°28′S and 26°14′ and 28°11′W. They lie on the east rim of the Scotia ridge, the northernmost island being about 530 km ESE of South Georgia and the southernmost being 960 km east of the South Orkney Islands. The Antarctic convergence is more than 1000 km north. The islands from north to south are Zavodovski, Leskov, Visokoi, Candlemas, Vindication, Saunders, Montagu, Bristol, Bellingshausen, Cook, and Thule.

The entire arc is basaltic rock of fairly recent volcanic origin. Craters or fumaroles are known to have been active during the present century on all islands except Cook, Montagu, and Vindication, and some of the islands, most notably Zavodovski, have been active continuously. They range in size from Montagu Island, about 10.4 km in diameter, down to tiny Leskov Island, which is about 2.4 km in circumference. The highest peak is 1375 m, Mount Belinda on Montagu Island. Several other islands have peaks 600-1100 m high. Some are regular cones; others are more irregular in shape with several jagged peaks. Permanent snow and glacial ice cover all the islands; Montagu, Bristol, Cook, and Thule islands are heavily glaciated, the ice averaging about 60 m thick. Only Leskov, Candlemas, Vindication, and Bellingshausen islands have extensively ice-free areas. In spite of their more northerly position, the South Sandwich Islands are far more barren than the South Orkney and South Shetland islands owing to the influence of cold-water currents, ice, and winds from the Weddell Sea. The south islands have heavy pack ice and a severe maritime Antarctic climate from May to November, whereas the waters about the more northern islands may be free of pack ice all year, and the islands have a moderate cold sub-Antarctic climate. The islands are almost continually cloudy, and heavy sea swells make landing dangerous. Vegetation is limited almost entirely to algae and encrusting lichens except on Leskov, Visokoi, Candlemas, and Bellingshausen islands, where ground warmed by volcanic fumaroles supports luxuriant moss cushions with occasional hepatics, and on Candlemas Island, which has the grass *Deschampsia antarctica*.

High rocky cliffs and glaciers make Leskov, Bristol, Cook, and Thule islands inhospitable for landing by penguins, but they abound on most of the other islands especially where volcanic activity keeps the rocks warm and therefore free of snow and ice. Fulmars and other petrels throng the cliffs. The eight southern South Sandwich Islands were discovered by British Captain James Cook in 1775, and the three northern islands were sighted in 1819 by Russian Captain Thaddeus Bellingshausen, who landed on Zavodovski Island. Most of the few later nineteenth century landings were by sealers, and scientific knowledge is scant. Present knowledge of the birds is based mainly on a *Discovery* expedition survey in 1930 [*Kemp and Nelson,* 1931] and brief visits ashore in the 1950's and 1960's [*Wilkinson,* 1956, 1957; *Holdgate,* 1963*b; Baker et al.,* 1964]. The birds are still imperfectly known but warrant further study especially on the north islands, where the status of king penguins needs to be determined.

SOUTH ORKNEY ISLANDS

The South Orkney Islands are a group of four main ice-covered rocky islands and a number of smaller satellite islands that lie in the Scotia Sea between 60°30′ and 60°50′S and 44°15′ and 46°15′W (Figure 10). They are about 600 km northeast of the north end of the Antarctic Peninsula and 1440 km southeast of Tierra del Fuego. The Antarctic convergence is about 800 km to the north.

The largest and highest island is Coronation Island, which is about 48 km from east to west and 12 km at its widest. Its highest summit is Mount Nivea, 1265 m, and several other peaks in the central and east portions exceed 1000 m. It is largely covered with a permanent

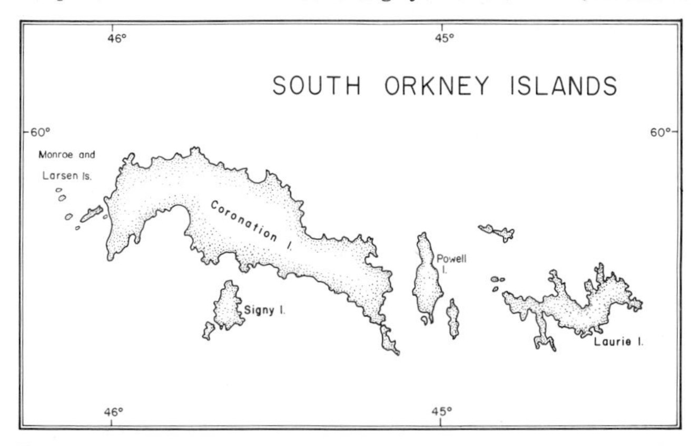

Fig. 10. South Orkney Islands showing locations of major islands.

ice cap and several glaciers that reach the sea. Most of the headlands and small islands off the south coast are ice free in summer, however, and are frequented by birds. The north coast is precipitous and has little ice-free habitat suitable for birds.

Low-lying Signy Island, 1.5 km south of the central part of Coronation Island, is the most biologically important of the group. Although it has a small ice cap and ice cliffs of a glacier on the south coast, about half its surface, especially on the east and west sides, becomes snow and ice free for at least 3 months during summer. It is roughly triangular in outline, 8 × 5 km (18 km^2), and rises to an elevation of only 281 m. There are several small freshwater lakes near the coast, and lowland moss banks are usually waterlogged.

Mountainous Laurie Island, the easternmost in the group, is irregularly shaped with several long peninsulas. It is about 24 km long east to west, and its highest peak is 940 m. Although glaciers cover most of its surface, the lowland coastal areas become snow free in summer.

Narrow mountainous Powell Island, immediately east of Coronation Island, is about 11 × 3 km and 620 m high. It is largely ice covered, especially on its east side, but a few headlands on the west are ice free, and low-lying Michelsen Island, joined to Powell Island by a narrow isthmus that is submerged at high tide, has important colonies of penguins and other birds.

Other smaller islands in the group include the Larsen Islands and Monroe Island in the extreme west; Moe Island off the south coast of Signy Island; the Robertson Islands, including Matthews, Skilling, and Atriceps islands, at the southeast end of Coronation Island; and Fredriksen and Saddle islands between Powell and Laurie islands. These islands are mostly ice free, and although they are high and steep, most have important bird colonies.

The majority of the rocks on Coronation and Signy islands are metamorphosed sediments, mainly quartz-mica schists and amphibolites with local marble outcrops. The extreme east end of Coronation Island, including the small offshore islands and most of Powell Island, is younger conglomerates with some shale beds. Parts of Powell Island and nearly all of Fredriksen and Laurie islands are graywacke sandstone and shales. All the islands show signs of glacial erosion, and the ground is covered with talus, moraines, and unconsolidated glacial detritus.

The vegetation of the South Orkney Islands is typical of the maritime Antarctic. Low cushion mosses and bushy fruticose lichens are the dominant plants and form thick peaty cushions at low elevations and more open communities above 100 m, reaching as high as

500 m on Coronation Island. In a few sites the grass *Deschampsia antarctica* or the pearlwort *Colobanthus quitensis* predominates but neither is of any significance for birds.

The South Orkney Islands experience a cold oceanic climate. Mean monthly temperature is about −4°C, but at least 1 and normally all 3 summer months have a mean temperature a little above freezing. From early December to late March, air temperatures are frequently above 0°C, occasionally to 9°C during north winds. Ice-free rocks and moss cushions absorb much radiant heat and increase summer snow and ice thaw. Mean winter temperature is about −10.3°C from June to August on Laurie Island and 1°-2°C warmer on Signy Island. Some precipitation falls on almost every day but averages only about 40 cm/yr, slightly more falling in February to April than falls in other months. Rain predominates in January to early March, and snow at other times. Evaporation is low owing to generally heavy cloudiness. Daily sunshine averages only about 1½ hours a day. The prevailing winds are from southwest, west, and northwest and average 26 km/h on Signy Island and 18 km/h on Laurie Island. September and October are the windiest months. Mean annual surface water temperature around the islands is −1°C and varies little seasonally. Sea ice surrounds the islands in late winter in most years.

The South Orkney Islands were discovered by American and British sealers in 1821. A British and a French expedition visited Laurie Island during the next 20 years, and three late nineteenth century sealing expeditions commercially extirpated fur and elephant seals from the islands. Scottish and English biological expeditions visited Laurie Island in 1903 and 1932-1933 [*Clarke,* 1906; *Ardley,* 1936], and an Argentine meteorological station, Orcadas, was established in 1904 and has continued in operation. Since 1947 the British have maintained a permanent base on Signy Island, which has been the major site for many bird species studies [*Beck,* 1969; *Beck and Brown,* 1971, 1972; *Burton,* 1968*a, b,* 1970; *Conroy,* 1972; *Jones,* 1963; *Pinder,* 1966; *Sladen,* 1958; *Tickell,* 1962].

Three localities in the South Orkney Islands have been designated specially protected areas under the Antarctic Treaty [*Conference on Antarctica,* 1959]. Moe Island (but not its offshore islets) is preserved as a representative base line sample of the maritime Antarctic ecosystem in case intensive research alters nearby Signy Island. Lynch Island, off the south coast of Coronation Island, is preserved for its extensive grass cover. In the central South Orkney Islands, southern Powell Island, Fredriksen, Michelsen, Christoffersen, and Grey islands, and their offshore islets are essentially undisturbed and have an expanding fur seal population.

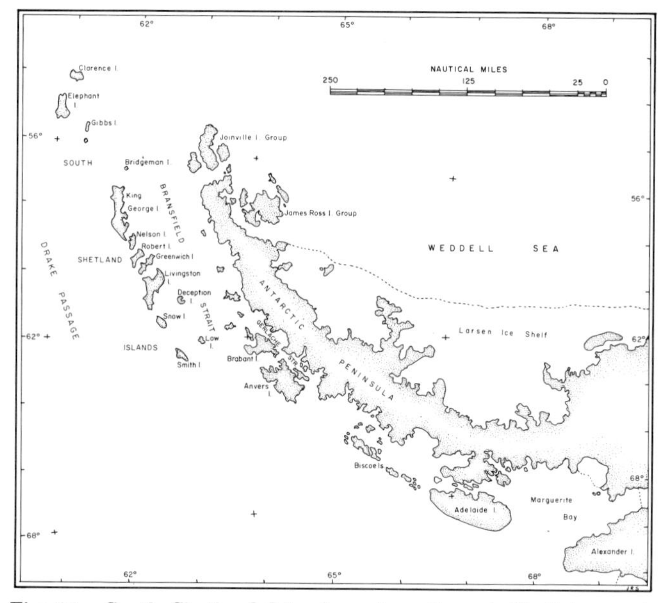

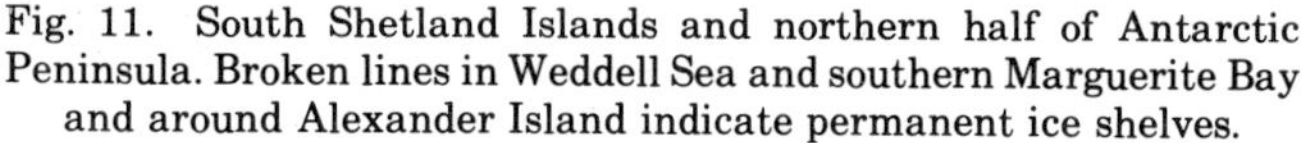
Fig. 11. South Shetland Islands and northern half of Antarctic Peninsula. Broken lines in Weddell Sea and southern Marguerite Bay and around Alexander Island indicate permanent ice shelves.

SOUTH SHETLAND ISLANDS

The South Shetland Islands form a 539-km-long chain of 11 main islands and a number of smaller islands and rocks that lie between 61° and 63°30′S and 53°30′ and 62°45′W (Figure 11). The chain trends northeast to southwest and parallels the coastline of the Antarctic Peninsula about 160 km to the south across the Bransfield Strait. Cape Horn is about 770 km to the north across the Drake Passage, and the South Orkney Islands are 480 km to the east. The South Shetland Islands divide naturally into two groups. The two major islands of the Elephant or northeast group are Clarence and Elephant, and those of the main or southwest group are King George, Nelson, Robert, Greenwich, Livingston, Snow, Deception, Smith, and Low islands. The total land area is about 4700 km^2.

The north approaches to the South Shetland Islands are shallow (the 200-m contour is more than 35 km away) and marked by many islets, reefs, and rocks. The south coasts, on the other hand, drop off rapidly to the great depths of the Bransfield Strait and are essentially free of rocks and islands. All of the southwest islands except Smith and Low islands lie on the same 200-m shelf.

Each of the major islands supports an active ice dome or several glaciers. Ice cliffs fringe much of the coastline. Nunataks project through the ice here and there in the interior, but in general, the only ice-free areas available to breeding birds are lowland peninsulas, headlands, coastal cliffs, screes, and boulder beaches. The easternmost island in the chain, Clarence Island, is 21 km north to south and 14.5 km east to west. The highest point is a rounded ice cone of 1924 m. Two small islets lie within 2 km of the east coast, and a string of 5 small islets extends 1.8-16 km to the north. Cornwallis Island, 24 km northwest of Clarence Island, is 3.8 × 2 km and rises sheer to sharp crests about 470 m high. Elephant Island, the largest and best known of the northeast group, is 9.5 km WSW of Cornwallis and about 30 km west of Clarence Island. It is 40 km east to west and 24 km at its greatest width. The coastline alternates glaciers and high steep cliffs with occasional narrow rocky beaches. Landing is difficult. The highest peak, Mount Pendragon, on the south end, is 973 m. More than 95% of the island is covered with ice and snow; only a few headlands, beaches, and precipitous inland cliffs are snow free in summer. Off its north, west, and southwest coasts lie a number of rocks and small islets. The Gibbs Island group lies 28 km to the SSW of Elephant Island. Narrow and steep, Gibbs Island itself is in two parts connected by a low boulder-strewn isthmus 80 m long. Both portions are about 300 m high. Three small high islands, Aspland (734 m high), Eadie, and O'Brien (420 m), are about 8 km to the west of Gibbs Island. Nearly circular Bridgeman Island, which stands alone 61 km further SSW, is a truncated volcanic cone less than 1 km in diameter and 240 m high.

King George Island, the largest and northeasternmost of the main group, is about 100 km southwest of the Gibbs Island group and 45 km west of Bridgeman Island. It is about 65 km long and 5-40 km wide. The central portion is an ice-capped dome that reaches about 610 m. The south coast is deeply indented by three bays, King George, Admiraltv. and Maxwell bays, Potter Cove being an inlet of the last. Glaciers feed into the bays, which are separated by largely ice-free peninsulas. The north coast is heavily glaciated, but the Fildes Peninsula on the west end of the island is almost ice free. The next two islands in the chain are Nelson Island (22 × 13 km), immediately southwest of Fildes Peninsula, King George Island, and Robert Island (18 × 12 km), 10 km southwest of Nelson Island. Both islands have ice caps but are of moderate elevation with no conspicuous peaks. Ice-free areas with significant bird populations are on Harmony Point (3.2 × 1.6 km), Nelson Island, and Coppermine

Peninsula, Robert Island. Although Greenwich Island, 4 km southwest of Robert Island across English Strait, is larger (26 × up to 11 km) and higher (at least two peaks over 620 m) than Nelson and Robert islands, it has no known important bird concentrations. On the other hand, the numerous ice-free islands in the approaches to English Strait, especially Heywood Island, may have important penguin colonies. Livingston Island, 4 km to the southwest of Greenwich Island, has about the same dimensions as King George Island. Its southeast coast is mountainous, rising to 1600 m at Mount Friesland; a central ice cap is 350-450 m; and the low-lying Byers Peninsula on its west end is the largest (14.5 × 5 km) ice-free area in the South Shetland Islands. Three other little-known islands lie farther southwest. Ice-capped Snow Island, 6 km southwest of Livingston Island, is 19 × 11 km. Smith Island, 40 km southwest of Snow Island, is 35 × 10 km and remarkably high. Its sheer black coastal cliffs rise from 600-1000 m to three peaks 1280, 1860, and 2105 m high. Low-lying Low Island, 26 km southeast of Smith Island, is about 13 × 8 km and almost entirely mantled with snow.

Deception Island, best known of the South Shetland Islands, is a horseshoe-shaped flooded caldera. It lies 20 km south of a promontory on the southeast end of Livingston Island. The island, the largest of the three recent volcanic centers in the South Shetland Islands, is 14 × 13 km. Its large central harbor, Port Foster, is accessible from the sea through Neptunes Bellows on the southeast. The rim has a mean elevation of 300 m, its highest points being Mount Pond (542 m) in the east and Mount Kirkwood (467 m) in the south. The land is composed of lava and cinders, but above 100 m it is mostly glaciers of ash-covered ice, which reach the sea at many places along the coast and on the east side of Port Foster. Owing to numerous active fumaroles at sea level the water in Port Foster is warmer than the surrounding sea and except for the immediate vicinity of the entrance is largely devoid of birdlife. The outer coast, on the other hand, is fringed with penguin rookeries. The largest are on the southwest near Vapour Col, on the northeast at Macaroni Point, and on the southeast at Rancho Point (Baily Head). A remarkably straight stretch of black sandy beach, which provides easy landing for penguins and is backed by high ice cliffs, connects Macaroni Point and Rancho Point. The 100-m-high cliffs in Neptunes Bellows, elsewhere about the coast, and on offshore stacks harbor Cape pigeons.

The geologic history of the South Shetland Islands has been long and diverse. Only King George, Livingston, and Deception islands have been well studied, and many of the rocks have not been precisely identified and dated. The islands of the Elephant Island group are composed mostly of old, highly folded metamorphic schists, amphibolites, and marbles. Although the rocks of the main group are predominately volcanic, Precambrian and Cambrian schists occur in southwest Livingston Island, and the core of the islands is made up of Carboniferous and Jurassic sedimentary rocks. Tertiary and later lavas were extruded during several stages of activity along a fault zone on the flanks of the sedimentary rocks. The major fault lies parallel to the southwest boundary of the islands and represents an uplift of 2 km and lateral displacement from the floor of the Bransfield Strait before or during the Tertiary. Local parallel and cross faulting can be identified on each of the islands. Volcanic activity was particularly intense during the Pliocene and Pleistocene and continues at three centers today. Penguin Island is a fresh cone; Bridgeman Island erupted in 1821, 1829, and 1880; and Deception Island erupted most recently in 1842, 1967, and 1969.

Glaciation has played a major role in eroding the landscape and building moraines. Sea level changed repeatedly during glaciation, and raised beaches are conspicuous features of the coastal landscape. In the past, glaciation was far more extensive than it is at present. At one time in the Pleistocene a single continuous ice cap 250 × 65 km covered all the southern islands that lie on the 200-m shelf. The main axis of the ice cap was then on the submarine shelf north of the present islands, which was exposed during a low stand of sea level. Ice flow was largely southeast through and over the present islands into the deep Bransfield Strait. The narrower straits between the present islands plus the deep bays of the south coast of King George Island are the result of ice gouging across the axis of the islands during this period. At present, marine erosion is severe on the northwest coasts during summer storms and in mild winters, when there is little fast ice to protect the shoreline.

The maritime Antarctic climate of the South Shetland Islands is cold, wet, and windy. The islands are subject to warm moist winds from the north and west over the Drake Passage and have the greatest frequency of storm depressions in the Antarctic. As a result, overcast skies and fog prevail, and clear days are the exception. Meteorological observations have been recorded for several years at a number of sites in the islands. The following account is based mainly on records from King George Island.

Mean annual temperature is −2.7°C. The means for December to March are at or above freezing from 0.0°C to 1.3°C, but those for April to November range from −0.7°C (November) to −8.0°C (July). Summer maximums can range above 15°C on infrequent calm sunny days. Temperatures that birds experience near ground level in moss banks or rocks may be up to 20°C warmer than the air above them owing to warming of dark surfaces by the sun. Precipitation, measured near sea level on the south coast of King George Island, is about 40 cm/yr, most falling as snow, but drifting blizzards make accurate measurement difficult. At higher elevations on ice caps, up to 100 cm may fall, and lee slopes may receive as much as 150 cm. Rain can fall at sea level in any month. Cloud cover is heavy, especially in summer, and relative humidity often ranges over 90%. Mean wind velocity is 21 km/h, gales with gusts over 100 km/h are frequent, and calms are rare.

Sea temperature is about 1°C in summer and below freezing in winter. The shores of the islands are ice free about 6 months in average years, and fast ice may persist for only short periods in mild winters.

The terrestrial flora of the South Shetland Islands consists of numerous species of mosses and lichens, a few liverworts, two phanerogams, and two algae. Vegetation is restricted to areas that become snow and ice free in summer and is far more luxurious on the south coasts, which are less exposed to spray and strong winds, than on the north. Thick moss carpets grow on deep peat where water accumulates in lowland flat areas and depressions; drier moss turf occurs on well-drained slopes. Small patches of the grass *Deschampsia antarctica* and the pearlwort *Colobanthus quitensis* are found in sheltered sites below 50 m. An open feldmark of fruticose lichens and moss cushions grows on rocks up to 300 m that are away from salt spray, guano, and wind, whereas crustose lichens occur on more exposed rocks. Algae sheets grow on wet ground near bird colonies.

The South Shetland Islands were discovered in 1819 by the British merchant captain William Smith when his ship was blown southward while it was en route from Buenos Aires to Valparaiso and were explored in 1820 by American sealer Nathaniel B. Palmer. New England sealers used the islands, mainly Deception Island, as bases during the 1820's and 1870's. In 1906 a British whaling station was established at Whalers Bay near the entrance to Port Foster, Deception Island, and shore operations continued there until 1931. Elephant Island gained notoriety as the overwintering site for part of the crew of Shackleton's wrecked *Endurance* in 1916.

Eights [1833] of Albany Institute, New York, was the first scientist to visit anywhere in the Antarctic. In 1829 he studied penguins and other animals and geology in the South Shetland Islands on an expedition exploring for new sealing grounds. Aside from Deception Island the South Shetland Islands remained largely unknown and unstudied until the late 1940's, when Argentine, Chilean, and British scientific stations were built on King George, Greenwich, and Deception islands. A Russian base was built on King George Island in 1968.

The avifauna was studied by *Racovitza* [1900] and *Gain* [1914], and reports on the birds of Harmony Point, Nelson Island, Coppermine Peninsula, Robert Island, and Elephant Island have appeared recently [*Araya and Aravena*, 1965; *Pefaur and Murua*, 1972; *Furse and Bruce*, 1971], but much further work remains, particularly in the large penguin rookeries on Deception Island. Byers Peninsula and Cape Shirreff on Livingston Island and Coppermine Peninsula have been designated specially protected areas on the basis of their important animal and plant populations. Fildes Peninsula, King George Island, is specially protected because it has lakes that are ice free in summer.

ANTARCTIC PENINSULA

The mountainous ice-covered Antarctic Peninsula at the southwest end of the Scotia ridge includes some of the most spectacular scenery in the Antarctic. The peninsula, its fringing islands, and the South Shetland Islands make up three roughly parallel mountain chains that are a continuation of the Andes Mountains of South America, 1050 km to the north across the Drake Passage. The peninsula extends 1600 km in a narrow 'S' shaped curve 45-260 km wide between 63°15′S, 56°00′W and about 75°00′S, 60°00′ to 80°00′W (Figure 11). Among the larger islands that lie off the northeast tip of the peninsula are d'Urville, Joinville, and Dundee islands (Joinville Island group) and James Ross, Vega, Seymour, and Snow Hill islands (James Ross Island group). They are separated from the mainland by Antarctic Sound and by Prince Gustav Channel, respectively. Along the west coast from north to south the better-known and larger islands are Hoseason Island, Trinity Island, Brabant Island, Melchior Islands, Anvers Island, and Wiencke Island (collectively known as the Palmer Archipelago and separated from the Danco Coast on the Antarctic Peninsula by the Gerlache Strait); Argentine Islands; Biscoe Islands; Adelaide Island; Charcot Island; Latady Island; and Alexander Island. Larsen ice shelf and the Weddell Sea mark the east limit of the Antarctic Peninsula; Bransfield Strait lies between the Trinity Peninsula at the north end of the Antarctic Peninsula and the South Shetland Islands; and the South Pacific Ocean and the Bellingshausen Sea border the more southern islands and coast. Marguerite Bay separates Adelaide Island from Alexander Island, and permanently frozen George VI Sound lies between Alexander Island and the mainland.

The peninsula consists of a gently convex ice-capped narrow central plateau 60 m to 40 km wide. The plateau is more than 2000 m high in the south, but its average elevation drops gradually to the deeply dissected north tip. Exposed peaks on Anvers Island and near 70°S rise to 2800 m, and the highest peak, Mount Jackson, at 72°S, is nearly 3500 m. Most of the other island and mainland peaks are between 1000 and 2000 m high. Glaciers flow east and west over the steep plateau scarps and across the featureless lower slopes to the sea. The west coast is a succession of ice-cliffed fjords, whereas most of the east coast is fringed with the Larsen ice shelf. Glaciation is far more pronounced on the northwest coast, which is exposed to the moisture-laden onshore winds, than on the more protected east and south coasts, where more rock is exposed.

The larger islands, Charcot, Latady, Alexander, Adelaide, Biscoe, and Anvers islands, have extensive ice domes and active glaciers. They are thus similar to the mainland in offering only scattered ice-free headlands and low-lying peninsulas as nesting habitat for birds. Even many of the much smaller islands such as those in the Argentine Islands have permanent ice caps. Their north coasts rise to rocky crests, and the bulk of the ice is on the south side, which ends abruptly in ice cliffs. Snow- and ice-free areas are now restricted to a coastal fringe and low-lying wave- or ice-cut platforms on the north side or to coastal cliffs and steep inland outcrops. Some very small islands, especially those in Arthur Harbor immediately south of Anvers Island, become largely ice free in summer and provide numerous breeding sites for birds.

The geology of the Antarctic Peninsula is relatively well known. The majority of the rocks that make up the core of the peninsula are remarkably thick volcanic deposits of late Cretaceous or early Jurassic dioritic intrusives. They are both contemporaneous and petrographically similar to the rocks that make up the bulk of the Andes Mountains. Widespread on the east side of the peninsula from Joinville Island to 75°S are highly metamorphosed graywackes, shales, and mica schists that are of late Paleozoic, most probably Carboniferous age. In the same general area, especially on the Joinville and James Ross island groups, are a number of fossil beds of mid-Jurassic, Cretaceous, lower Cenozoic, and Pliocene ages. On the west side of the peninsula, Alexander Island has Cretaceous or early Jurassic fossil-bearing sediments. On the Joinville and James Ross island groups are Pliocene and possibly Miocene volcanic rocks, cone-shaped Paulet Island possibly representing a recent eruption.

The entire Antarctic Peninsula area has been gouged and otherwise eroded by Pleistocene glaciation, and the lowlands are littered with moraines and other glacial deposits. The present ice cover, however, is only a remnant of the far thicker ice caps that formerly existed, but some of the nunataks and coastal headlands have apparently never been entirely covered with glacial ice.

The vegetation is typical of the maritime Antarctic zone, but owing to greater aridity, shorter growing season, and lower temperature it is more scattered and less luxuriant than the vegetation in the South Orkney and South Shetland islands. Mosses and lichens predominate, but algae, two hepatics, the grass *Deschampsia antarctica,* and the pearlwort *Colobanthus quitensis* are found sparingly south to Marguerite Bay. Climatic conditions, especially amount of precipitation and temperature, are of great importance in determining the nature of the vegetation at any locality. Much of the vegetation on rocks and unstable soil consists of open communities of lichens and moss cushions. Dry stable slopes and cliff ledges have extensive moss turf, whereas lush moss carpets or hummocks grow where water tends to accumulate. Deeper moss banks provide nesting sites for surface-nesting skuas and for burrowing Wilson's storm petrels. Algae sheets occur on damp ground near bird colonies.

The climate of the peninsula is the coldest and driest in the maritime Antarctic zone. Isotherms virtually parallel the coastline of the peninsula with the 0°C isotherm for the warmest month bisecting the north part of the mainland. Most of the islands, except those inshore from Biscoe Islands and all of Alexander Island, have mean temperatures above freezing for at least 1 month (usually January) in summer. The mean of the coldest month ranges from −13.6°C in the north to −16°C at 68°S. Precipitation, mostly in the form of snow, decreases from more than 400 mm annually in the north to less than 200 mm in the south and also decreases from west to east. The Melchior Islands may receive up to 800 mm annually. The area south of 65°S lies in the East Wind drift (mean annual velocity less than 20 km/h), whereas the area to the north is subject to stronger winds from the west (mean annual velocities 23-29 km/h), which bring heavy clouds and precipitation. Nevertheless the entire Antarctic Peninsula area is less cloudy than the more maritime South Shetland and South Orkney islands. Protected Hope Bay at 63°24′S and Marguerite Bay at 68°S have less than 200 cloudy days a year, whereas the more exposed Argentine Islands have about 240 cloudy days annually.

The north limit of pack ice is 240-280 km northwest of the Antarctic Peninsula in winter, and permanent sea ice extends from Marguerite Bay southward. The Antarctic convergence is 550-925 km to the north and west.

The east coast of the Antarctic Peninsula is far colder and less hospitable and consequently far less well known than the west side. It is bathed by the ice-laden waters of the Weddell Sea, and much of the coastline is permanently fringed by the Larsen ice shelf or pack ice. Only the islands of the Joinville Island and James Ross Island groups become ice free enough during most summers to serve as nesting habitat for penguins and most other birds.

The portion of the Antarctic Peninsula south of the north end of Adelaide Island lies south of the Antarctic circle and thus receives 24 hours of daylight in summer and a like period of darkness in winter. The northernmost portion of the peninsula has a short night in midsummer and enough daylight in midwinter to allow some birds to overwinter.

The discovery of the Antarctic Peninsula followed 1 year after the sealers' rush to the newly discovered South Shetland Islands. British Captain Edward Bransfield is formally credited with first sighting 'Trinity Land' in January 1820. On a rare clear day some New England sealers may have seen the mainland mountains from Deception Island earlier, but there are no unequivocal reports. Russian Admiral Thaddeus Bellingshausen sighted Alexander Island (Alexander I Land) in 1821. American sealing captains Nathaniel B. Palmer and John Davis, looking for places to land crews, explored the islands and west coast. Davis was the first to land on the Antarctic Peninsula (and thus strictly speaking on the Antarctic continent) at Hughes Bay on the Danco Coast. Antarctic sealing and American interest in the peninsula area declined by the mid-1820's, but French and British expeditions under Dumont d'Urville in 1839 and Sir James Ross in 1842-1843 explored the north tip of the peninsula. Biological information on the Antarctic Peninsula remained scant until about the turn of the century, when Norwegian whaling revived interest. A Belgian expedition under Adrien de Gerlache explored the west coast in 1897-1899 [*Racovitza*, 1900], a Swedish expedition under Otto Nordenskjöld explored both coasts in 1902-1903 [*Andersson*, 1905], and important French expeditions under Jean B. Charcot in 1903-1905 and 1909-1910 produced an outstanding report on the birds [*Gain*, 1914]. British and American ornithologists studied birds in the Gerlache Strait and on Stonington Island during the 1920's and early 1940's [*Bagshawe*, 1938, 1939; *Eklund*, 1945; *Friedmann*, 1945].

During and after World War II, and particularly during and after the International Geophysical Year in 1957, permanent or summer stations were established on the peninsula and islands by the Argentines, British, and Chileans at Hope Bay, Melchior Islands, Danco Island, Paradise Harbor (Danco Coast), Anvers Island, Doumer Island, Wiencke Island, Argentine Islands, Prospect Point, Detaille Island, Stonington Island (Marguerite Bay), Horseshoe Island, Adelaide Island, and Fossil Bluff (Alexander Island) and near Seal Nunataks on the Larsen ice shelf. In 1965 the United States took over the abandoned British station at Arthur Harbor, Anvers Island, which was given the name Palmer Station. A new Palmer Station was erected in 1967 on nearby Gamage Point and has become an important center for biological studies. Only a few recent reports, however, deal with the avifauna of the Antarctic Peninsula [*Araya,* 1965; *Holdgate,* 1963*a*; *Sladen,* 1958; *Stonehouse,* 1953].

Two island sites near the Antarctic Peninsula have been designated specially protected areas. The Dion Islands in Marguerite Bay are the site of the only emperor penguin colony on the west side of the Antarctic Peninsula. Green Island in the Berthelot Islands near Argentine Islands has exceptionally luxuriant vegetation.

ANTARCTIC CONTINENT

Most of the 14.3 million km^2 surface of the Antarctic continent is covered with glacial ice 2000-4000 m thick, which obscures land contours and makes much of the continent unfit as bird habitat. Only about 8000 km^2 of antarctic rock and soil are exposed, of which a significant fraction is on the Antarctic Peninsula and most of the rest lies along the periphery of the continent proper, where the ice sheet is considerably thinner than it is inland. Although long stretches of coast are continuous ice cliffs and calving glaciers, in some areas, strong offshore winds and the warming effect of the sun on dark rocks allow mountain ranges, headlands, beaches, and small islands to become snow and ice free in summer and therefore to be attractive to birds for breeding. A few of the highest inland mountain ranges and peaks project above the ice as nunataks, but most of them are too remote from the sea to be used by breeding birds.

The heavy ice sheet, which includes nine tenths of the total glacial ice in the world, depresses the land beneath it, and possibly half of the continent now lies below sea level. The continental shelf is also depressed to 400-500 m, at least 200 m deeper than it is in the more temperate areas of the world. There are traces that even more extensive glaciers and a thicker ice cap were present in the past. Some of the coastal areas rebounded upward when they were relieved of their former heavy ice burdens and left raised ice-free beaches on which Adélie penguins now nest.

The center of the roughly circular continent is offset to the east with respect to the geographic south pole. Consequently, although most of the coast from Enderby Land to Victoria Land (45° to 170°E) is about 66°S, much of the coast of Queen Maud Land (20° to 45°E) is about 70°S, and that of Marie Byrd Land averages 75°S. The Weddell Sea and its ice shelves penetrate the periphery beyond 82°S, and the Ross Sea and Ross ice shelf extend inland to 85°30′S, but the 60-m-high ice cliffs of the Ronne, Filchner, and Ross ice shelves bar navigation in either sea beyond about 78°S. These two great embayments possibly meet under the ice sheet in a deep subglacial basin perhaps lying below sea level that divides the continent into two parts of unequal area. East Antarctica, the larger and higher portion, which fronts on the Atlantic and Indian oceans, has an ice dome up to 4200 m thick; that of West Antarctica, which fronts on the Pacific Ocean, is up to 2000 m thick.

The most striking land feature of Antarctica is the more than 3000-km-long escarpment of the Transantarctic Mountains, which runs across the continent from Victoria Land at least to Coats Land and possibly to Queen Maud Land. Its central portion is submerged by ice, so that its undoubted connection with the Pensacola Mountains, Shackleton Range, and Theron Mountains in Coats Land and its possible continuation into the inland and coastal mountains of Queen Maud Land are obscured. The 3000- to 4000-m peaks of the Transantarctic Mountains partly hold back the flow of the East Antarctic ice sheet in Victoria Land, but innumerable glaciers flow between the mountains and coalesce to form the floating Ross ice shelf.

The highest elevation on the continent is 5140-m Vinson Massif in the Ellsworth Mountains, which lie between the Transantarctic Mountains of East Antarctica and the mountains of Marie Byrd Land in West Antarctica.

The geology of the readily accessible exposed protions of Antarctica is well known, but because the bulk of the rocks of the continent are covered with a mass of glacial ice, there are vast gaps in knowledge, and much of the following account is highly simplified. The rocks of East and West Antarctica differ to such an extent in age and composition that they are considered separate geologic provinces. A shield of Precambrian and early Paleozoic metamorphic rocks (gneisses, amphibolites, and schists) and intruded granites is uplifted and exposed as the 'basement complex' in Queen Maud Land and along the rest of the Indian Ocean coast and as the 'Ross system' in the Transantarctic Mountains. It probably also underlies most of East Antarctica. Lying exposed on top of this base in the Transantarctic Mountains are fossil-bearing marine, estuarine, and freshwater sediments and coal of Devonian to Jurassic age ('Beacon system'), which are interbedded with broad bands of hard volcanic dolerite ('Farrar system'). These later sediments were uplifted in the mid-Tertiary to late Tertiary to form the high Transantarctic Mountains.

The Ellsworth Mountains probably constitute a separate tectonic area. Although they resemble the Beacon system of the Transantarctic Mountains, the sediments are considerably thicker and more distinctly folded.

Similar rock formations and fossil remains in other southern hemisphere continents and subcontinents provide striking evidence that East Antarctica, along with southern Africa, India, eastern Australia, and South America, once made up a vast southern continent, Gondwanaland. This large landmass underwent climatic vicissitudes from glaciation to desert desiccation and had a rich flora and fauna before it broke up beginning in the Triassic and its constituent parts drifted to their present positions.

Most of East Antarctica is tectonically stable at present. There are only two active volcanoes, 3794-m-high Mount Erebus on Ross Island and Mount Melbourne in north Victoria Land. Franklin and Beaufort islands, however, are the remains of two of several former craters that belong to this same volcanic complex in the west Ross Sea.

The mountains of West Antarctica, which rest on a base of Paleozoic granite and dolerite, are similar in age and structure to those of the Antarctic Peninsula. They are metamorphosed Paleozoic and Mesozoic sediments, especially graywackes, interbedded with volcanic rocks that were uplifted in the late Cretaceous or early Tertiary and that have subsequently undergone further volcanism. Several 1000-m-high peaks in the Ford Ranges in Marie Byrd Land are late Cenozoic volcanoes similar in age and structure to Mount Erebus and other volcanoes in the west Ross Sea.

The climate of the Antarctic continent is cold, dry, and windy. Although the world's lowest temperature (−88.3°C) has been recorded at Vostok Station at 3488-m elevation near the center of the Antarctic continental ice sheet, such extreme low temperature is hardly characteristic of the low coastal areas inhabited by birds. Mean annual temperatures at coastal stations from 66° to 69°S and 141° to 39°E range from −9.9°C to −11.8°C. Summer means are −0.9°C to −3.3°C, temperatures occasionally rising to +5°C to +8°C. Winter means are −16.1°C to −17.7°C with extreme temperatures near −40°C. Mean annual summer and winter temperatures farther south at 70° to 78°S in Queen Maud Land and the Ross Sea are 5°C to 10°C colder, as are temperatures at inland nunataks.

Precipitation is difficult to measure owing to blizzards. Virtually all of it falls as snow, and there is no marked seasonal variation in amount. More precipitation falls at coastal stations (36-42 cm/yr and possibly 85 cm/yr at Mirnyy Station) than at those in the Ross Sea (12 cm/yr). Inland mountains presumably receive still less, and some dry valleys in Victoria Land are complete deserts. Mean relative humidity is 60-75% at coastal stations.

Although the winds are light to moderate on the high continental ice dome, when they cascade down the snow and ice slopes of the plateau, they accelerate and carry snow and particles of ice with them. An area of particularly strong and constant winds is along the Adélie Coast, where mean wind velocity is 72 km/h and gusts of over 240 km/h are common. These winds, which compound the antarctic cold, have a favorable aspect for nesting birds that are able to avoid their full force by seeking sheltered sites at ground level. Winds sweep coastal rocks and lowlands free of snow early in the season and help move pack and sea ice away from the coast in summer and thereby permit nesting by penguins, several petrels, and skuas. Strong winds, coupled with sea currents and the constantly progressing Ross ice shelf, make the sea ice in McMurdo Sound unstable and subject to unseasonably early breakup for its latitude. Not only is resupply of the huge U.S. research station possible from the sea, but emperor and Adélie penguins occupy their southernmost breeding sites in the area.

The coastline of the continental Antarctic lies near or south of the Antarctic circle. Daylight is continuous or prolonged during the 4 months of high summer, and birds can therefore find food for young almost continuously. On the other hand, winter, when most birds are absent, can be a long period of near-total darkness.

Less than 1% of the exposed ground and rock in continental Antarctica has plant cover of any sort, and what vegetation there is is of virtually no significance to birds. It consists primarily of scattered open communities of lichens and, where sufficient moisture is available, mosses and algae.

Although British Captain James Cook circumnavigated Antarctica generally south of 60°S, crossed south of the Antarctic circle 3 times, and penetrated as far south as 71°10′S, 106°54′W on January 30, 1774, he failed to sight the mainland. Russian Admiral Thaddeus Bellingshausen likewise circumnavigated the continent, and although he discovered Alexander Island in 1821, he made no claims to have seen the mainland anywhere, nor did British sealer James Weddell, who finding ice conditions extraordinarily light in 1823 reached 74°15′S, 34°17′W in the Weddell Sea. Three captains employed by Enderby Brothers, a British whaling company, made the first landfalls on the continent during the 1830's. Captain John Biscoe first sighted Enderby Land on February 28, 1831; in December 1833, Captain Peter Kemp sighted land on the Kemp Coast; and in March 1839, John Balleny discovered the Sabrina Coast in Wilkes Land. In the early 1840's three national expeditions made important antarctic discoveries. A French expedition under Dumont d'Urville landed on an islet in the Géologie Archipelago, Adélie Coast, in January 1840; in January and February of the same year the United States Exploring Expedition under Lieutenant Charles Wilkes charted a series of landfalls between 160° and 98°E, revealing the continental nature of the landmass; and in the following year, British Captain James C. Ross sailed into the Ross Sea as far south as Ross Island and the Ross ice shelf and landed on Possession and Franklin islands.

For the next 50 years, although exploitation of seals continued on the subantarctic islands, no attention was given the larger south land.

In the early 1890's, however, whalers investigating whaling potential in the Antarctic made new discoveries that reactivated interest in the south and opened the way for a 'heroic' period of Antarctic exploration. Norwegian Captain C. A. Larsen discovered Oscar II and Foyn coasts in the west Weddell Sea in 1892-1893, and Captain Leonard Kristensen made the first landing on mainland Victoria Land at Cape Adare in 1894-1895.

In 1897 there began a series of large-scale government-sponsored continental landings and exploration and overwintering parties. Scientific investigations were an important component of these expeditions and included the first inventories and life history studies of continental birdlife [*Clarke*, 1907; *Hanson*, 1902; *Levick*, 1914, 1915; *Lowe and Kinnear*, 1930; *Reichenow*, 1908; *Wilson*, 1907]. English, German, Scottish, Norwegian, Japanese, and Australian expeditions under Borchgrevink (who had first been ashore at Cape Adare with Kristensen in 1895), von Drygalski, Bruce, Amundsen, Scott, Shackleton, Shirase, and Mawson explored the coasts of the Ross and Weddell seas and parts of Adélie Coast intensively and ventured far inland. Their efforts and the race for new discoveries culminated in separate expeditions led by Amundsen and Scott reaching the south pole in December 1911 and January 1912 (although Scott's party died on the return to the coast). This period of heroic exploration ended during World War I.

Aircraft and motorized land vehicles were first used to supplement shipboard support in antarctic exploration in the late 1920's and were extensively employed by the Americans Richard E. Byrd and Lincoln Ellsworth and the Australians Hubert Wilkins and Sir Douglas Mawson for exploration in the 1930's and 1940's. Bird studies of longer duration began from more or less fixed stations [*Eklund*, 1945; *Falla*, 1937; *Friedmann*, 1945; *Perkins*, 1945; *Siple and Lindsey*, 1937], but World War II interrupted plans to make the stations permanent.

After the war and particularly beginning with the International Geophysical Year in 1957, permanent bases were set up by most of the 12 nations that eventually signed the Antarctic Treaty [*Conference on Antarctica*, 1959] for cooperation in scientific research.

TABLE 6. Coastal Stations in Continental Antarctica

Country	Base	Location	Date Open	Ornithological References
United States and Argentina*	Ellsworth	77°43′S, 41°07′W Filchner ice shelf	1957-1962	*Luna Perez* [1963]
Argentina	General Belgrano	77°47′S, 38°14′W Filchner ice shelf	1955	
United Kingdom	Halley Bay Base Z	75°31′S, 26°42′W Brunt ice shelf at Halley Bay, Caird Coast	1956	*Ardus* [1964], *Brook and Beck* [1972], *Jarman* [1973], and *Novatti* [1959]
South Africa†	SANAE	70°19′S, 2°24′W Princess Martha Coast	1962	*La Grange* [1962]
USSR	Lazarev	69°58′S, 12°55′E Lazarev ice shelf, Princess Astrid Coast	1959-1961	*Konovalov* [1964]

Belgium‡	Roi Baudouin	70°26′S, 24°19′E Breid Bay, Princess Ragnhild Coast	1958-1961 1964	*Loy* [1962]
Japan	Shōwa	69°00′S, 39°35′E East Ongul Island, Prince Olav Coast	1956-1962	*Haga* [1961] and *Matsuda* [1964]
USSR	Molodezhnaya	67°40′S, 45°50′E Alasheyev Bight, Enderby Land	1962	*Korotkevich and Ledenev* [1970]
Australia	Mawson	67°36′S, 62°52′E Holme Bay, Mac. Robertson Land	1954	*Brown* [1966]
Australia	Davis	68°35′S, 77°58′E Vestfold Hills, Ingrid Christensen Coast	1957	*Brown* [1966]
USSR	Mirnyy	66°33′S, 90°0.1′E Mabus Point, Queen Mary Coast	1956	*Korotkevich* [1964] and *Pryor* [1965, 1968]
USSR	Oazis	66°16′S, 100°45′E Bunger Hills, Knox Coast	1956-1958	*Korotkevich* [1964]

TABLE 6. (continued)

Country	Base	Location	Date Open	Ornithological References
Australia and United States§	Wilkes	66°15′S, 110°32′E Clark Peninsula, Budd Coast	1957	*Eklund* [1961], *Orton* [1963, 1968], and *Penney* [1968]
France ‖	Dumont d'Urville	66°40′S, 140°01′E Géologie Archipelago, Adélie Coast	1956	*Arnaud* [1964], *Etchécopar and Prévost* [1954], *Guillard and Prévost* [1964], *Isenmann* [1970*a*, *b*, 1971], *Isenmann and Jouventin* [1970], *Isenmann et al.* [1971], *Jouventin* [1971], *Lacan* [1971], *Le Morvan et al.* [1967], *Mougin* [1967], *Prévost* [1953, 1961, 1963, 1964, 1969], *Prévost and Sapin-Jaloustre* [1964, 1965], *Prévost and Vilter* [1963], *Sapin-Jaloustre* [1960], and *Trawa* [1970]
United States and New Zealand	Hallett	72°18′S, 170°18′E Cape Hallett, Victoria Land	1957	*Maher* [1962] and *Reid* [1964, 1966]
United States	McMurdo	77°51′S, 166°37′E Hut Point Peninsula,	1956	*Ainley* [1972], *Ainley and Schlatter* [1972], *Austin* [1957],

		Ross Island, McMurdo Sound		*Douglas* [1968], *Emison* [1968], *Emlen and Penney* [1964], *LeResche and Sladen* [1970], *Müller-Schwarze* [1971], and *Sladen et al.* [1968]
New Zealand	Scott	77°51′S, 166°48′E Pram Point, Ross Island, McMurdo Sound	1957	*Spellerberg* [1969, 1970, 1971*a*, *b*], *Taylor* [1962], and *Young* [1963*a*, *b*]

*The United States opened Ellsworth Station in February 1957 and transferred it to Argentina in February 1959.

†Norway opened Norway Station on shelf ice about 30 km from the sea (70°30′S, 2°32′W) on Princess Martha Coast, Queen Maud Land, in December 1956 and transferred it to South Africa in 1960. The station was closed in February 1962, when the program of scientific observations was transferred to SANAE Station, also on shelf ice but about 10 km nearer the coast.

‡Belgium opened Roi Baudouin Station in January 1958 and closed it in February 1961. Operations resumed in February 1964 with joint Belgian-Netherlands personnel.

§Wilkes Station was originally opened by the United States in January 1957 but was transferred to Australia in January 1959. U.S. scientists continue to conduct research there.

‖ A French station was opened at Port Martin (66°49′S, 141°24′E), Adélie Coast, in February 1950. It was destroyed by a fire in January 1952, and operations were temporarily shifted until January 1953 to Pétrel Island, Géologie Archipelago, where the permanent station was subsequently built.

The stations are resupplied and personnel rotated annually, ice breaker and aircraft logistics thus being necessitated. These bases have been the sites or support centers for surveys of bird distribution and intensive studies of population ecology, life history, physiology, behavior, migration, and dispersal of penguins, petrels, and skuas (Table 6).

Some inland mountain ranges 100-250 km from the shore have sufficient exposed cliff face or scree to offer nest sites for Antarctic and snow petrels, south polar skuas, and possibly Wilson's storm petrels. Thus far, only the Theron Mountains in Coats Land [*Brook and Beck,* 1972]; the Tottan Hills, Mühlig-Hofmann Mountains, Wohlthat Mountains, Schirmacher Hills, and Sør Rondane Mountains in Queen Maud Land [*Ardus,* 1964; *Konovalov,* 1964; *Konovalov and Shulyatin,* 1964; *Loy,* 1962]; and the Fosdick Mountains in Marie Byrd Land [*Perkins,* 1945] are known to have such inland colonies, but *Dow and Neall* [1968] suggest that the Morozumi Range in Victoria Land may also have nesting birds. Many other nunataks are so poorly known by biologists that there may be other bird colonies yet to be discovered.

Six sites on the Antarctic continent have been designated specially protected areas largely on the basis of their concentrations of breeding seabirds. They are (1) the rock exposure on the east side of Taylor glacier in Mac. Robertson Land (for its emperor penguin rookery, the largest one wholly on land), (2) Rookery Islands in Holme Bay, Mac. Robertson Land, (3) Ardery and Odbert islands in Vincennes Bay, Budd Coast, (4) Beaufort Island in the Ross Sea, (5) Cape Crozier on Ross Island (including whatever nearby locality is occupied by the emperor penguin rookery), and (6) Cape Hallett, Victoria Land.

SCOTT ISLAND

Scott Island is an isolated volcanic island at 67°24'S and 179°55'W, north of the Ross Sea, about 560 km north of Cape Adare on the continent, and 2420 km southeast of New Zealand. It lies south of the Antarctic convergence but north of the normal winter limit of pack ice. It consists of two parts that represent the remains of a crater and a plug. The main island is about 400 × 200 m, its long axis running north to south. From the precipitous 50-m-high cliffs on the north end, rough basalt and lava rock slope irregularly to about 2 m above sea level near the south end. The island is entirely ice covered during winter, but much of the higher north portion becomes ice free in summer. Haggits Pillar, a column 62.5 m high, which lies 730 m west of the main island, is pierced by a large cavern.

Scott Island was only discovered in 1902. Its steep rocky sides and rough seas make landing difficult. Ornithologists have not been ashore, but the bird life has been investigated at least once from the sea [*Harper,* 1972]. Snow petrels, Antarctic prions, and Wilson's storm petrels nest abundantly in rock fissures of both the main island and the pillar.

BALLENY ISLANDS

The Balleny Islands consist of a 185-km-long chain of three main islands (Young, Buckle, and Sturge islands) and several smaller ones (including Row, Borradaile, and Sabrina islands) lying between 66°15′ and 67°35′S and 162°30′ and 165°E, northwest of Cape Adare, Victoria Land. Their total area is less than 400 km^2. The islands, which are high, steep, and presumably of volcanic origin, are largely ice covered. Glaciers descend through breaks in cliffs and project as ice tongues into the sea. The islands were sighted by British Captain John Balleny in February 1839. Landing is hazardous but has been accomplished by boat or helicopter on Young Island (once), Borradaile Island (4 times), Buckle Island (once), and Sabrina Island (3 times). Vegetation at the few sites visited is meager. Terrestrial algae are common, lichens are rare, and mosses are absent. Information on birds is scanty, being derived mainly from observations during short visits and aerial reconnaissance [*Hatherton et al.*, 1965; *Dawson et al.*, 1965].

Sabrina is a specially protected area under the Antarctic Treaty [*Conference on Antarctica,* 1959].

PETER I ISLAND

Peter I Island is the only oceanic island in the extreme south section of the Pacific Ocean. It lies south of the Antarctic circle at 68°55′S, 90°26′W, in the Bellingshausen Sea. The nearest point on West Antarctica is about 400 km away. The island is surrounded by thick impenetrable pack ice except during late summer and has therefore seldom been visited. Authorities differ on its dimensions and elevation. Estimates of length and width range from 18 × 8 km to 25 × 10 km. Dome-shaped Lars Christensen Peak, an extinct volcanic crater, is said to be 1200 or 1750 m high. The island is covered with ice and snow all year except where the slopes are precipitous or headlands are exposed to high winds. The east coast is steep, whereas the west is dominated by a high central piedmont plateau. Floating shelves of glacial ice project into the sea from the gradual slopes of the north and south sides and elsewhere around the island, but in a few places the glaciers terminate on narrow rocky beaches. Precipitous rock cliffs line the central east coast, and rocky headlands and capes bound two large bays on the west coast. Off the east coast lie the two flat-topped ice-free Tvistein Pillars.

The island was discovered by Russian Captain Thaddeus Bellingshausen in January 1821 but owing to the great difficulty in landing and heavy ice was not further investigated until the second *Norvegia* expedition landed in 1929 and claimed it for Norway.

Holgersen [1961] visited the island in February 1948 and recorded a few breeding penguins and several colonies of southern fulmars. Snow petrels, Wilson's storm petrels, and south polar skuas may also breed, but the numbers of birds observed in the nearby icy waters are not great.

TRISTAN DA CUNHA GROUP AND GOUGH ISLAND

The Tristan da Cunha group consists of three volcanic islands lying on the east slope of the mid-Atlantic ridge at 37°02′ to 37°24′S and 12°12′ to 12°42′W, approximately midway between South Africa and the coast of Brazil. Tristan Island, the largest and highest island, is the northernmost in the group. Inaccessible Island is 40.5 km to the WSW, and Nightingale Island is 38 km to the SSW. They are 22 km apart. Gough Island, a fourth volcanic island of similar geology, climate, and vegetation, lies 425 km SSE of Tristan Island. All four islands rise abruptly from the ocean floor, more than 3500 m deep.

Tristan Island, the most recent (circa 1,000,000 years old) and least eroded, is roughly circular with a diameter of 11.2-12.8 km and an area of 95 km^2. From its steeply inclined conical peak 2060 m high the land falls away abruptly on all sides to a more gently inclined 'base' area between 900 and 600 m. Cliffs of up to about 600 m bound most of the island but are broken here and there by low-lying coastal strips, which afford landing and access to the island.

The remaining three islands are the highly eroded remains of large volcanic cones that were formerly about the same size as Tristan Island. Nightingale Island, the oldest (18 million years) and most eroded island, is now the smallest and lowest. Adjacent to Nightingale Island (about 2 km^2) are Middle and Stoltenhoff islands plus several smaller offshore islets and rocks. The highest point is about 337 m. This island has the densest seabird colonies in the Tristan da Cunha group. Irregularly square Inaccessible Island is about 18 km^2 and consists of a tilted flattop ranging from 275 m in the east to 550 m in the west and entirely surrounded by nearly vertical cliffs.

Gough Island is elongated northwest to southeast, 13 × 5-6 km, and is about 57 km^2. The central portion of the island is a 600-m-high plateau with several mountain peaks 800-910 m high. Eight deep canyons or 'glens' cut back into the central highlands up to 2 km and are separated by steep-sided ridges. The coasts are almost entirely steep cliffs with narrow boulder beaches, particularly on the west side, where they are nearly continuous and 300-450 m high. The eroded south end of the island is the only extensive area below 300 m.

All four islands are composed entirely of volcanic lavas, rocks, cinders, and ash, indicating multiple eruptions. Although Nightingale and Inaccessible islands have probably been long dormant, Tristan Island has undergone frequent volcanic activity. The most recent eruption, in 1961-1962, extended a lava platform on the north coast near the settlement. Gough Island last erupted about 2400 years ago.

The climate is oceanic and mild in comparison to the climates of the more southern islands and has a narrow range of temperature, strong westerly winds, heavy rainfall and cloud cover, and high humidity. Mean annual temperature at sea level on Tristan Island is 14.7°C. December through March average near 18°C; June through October average about 12.5°C. The extreme temperatures that have been recorded are 24.9°C in January and February and 3.0°C in August. Winds fluctuate between northwest and southwest and are strongest in winter. Southerly winds bring cold squalls and sunny periods; northerly winds bring warm but overcast and rainy conditions. From March to December, depressions are accompanied by high rainfall, but in summer the South Atlantic Ocean anticyclones migrate southward and bring fine weather, especially with southwest winds. Rainfall and cloud cover are strongly influenced by local orography. At sea level on the north coast of Tristan Island, mean annual rainfall is about 1675 mm, spread fairly evenly through the year, although November to March are the wettest months. Rain falls an average of 250 days a year. The base (600-900 m) receives about 2500 mm, and the highlands about 5000 mm. Snow covers the peaks from May to October and occasionally reaches down to 900 m. Relative humidity at sea level averages about 80%, and during northwest winds the semipermanent orographic clouds on the average cover seven eighths of the sky.

Gough Island is cooler and wetter with an annual mean temperature at sea level of 11.3°C and a range from 0°C to 24.6°C. Mean annual rainfall is about 3400 mm at sea level, and precipitation falls almost 300 days a year. Snow occurs as low as 450 m in winter.

The mean position of the Subtropical convergence lies near Tristan Island but occasionally may shift considerably to the north. Sea temperatures range from 18°C in summer to 14°C in winter off Tristan Island and from 13°C in summer to 11°C in winter off Gough Island. Sea currents in the nearby waters are generally northeast.

Owing to its size and elevation, Tristan Island has the most varied vegetation in the group, but in the lowlands the natural vegetation has been affected by grazing animals, fire, woodcutting, and introduced plants. Tussock grassland of *Spartina arundinacea* or *Poa flabellata* (the latter only on Gough Island) grows into a closed canopy 2-3 m high in undisturbed areas. Tussock occurs up to 600 m over the lowland coasts, cliffs, and slopes that are exposed to winds and sea spray. It is particularly luxuriant near bird colonies on Nightingale Island and on the west cliffs and slopes of Gough Island. Whether *Poa* ever occurred in the Tristan da Cunha group is unknown, but it now dominates the grasslands of Gough Island below 300 m.

In protected lowland flat areas and slopes away from sea spray and bird and seal colonies but where deep peat accumulates, fern bush communities replace the tussock. They grow up to an elevation of 600 m on Tristan Island and to 300 m on Gough Island. Dense evergreen thickets of *Phylica arborea* trees form a canopy up to 5 m high. Each tree sends up numerous suckers from its roots or semiprocumbent main trunk buried in the peat. The forest understory consists of tree and terrestrial ferns. Epiphytes hang densely from the branches, whereas mosses, liverworts, and lichens grow on the trunks or as a ground carpet. The tree ferns *Histiopteris incisa* and *Blechnum palmiforme* dominate some fern bush communities, the latter especially above 300 m.

Above the tussock and fern bush from 600 to 1200 m on Tristan Island and from 300 to 600 m on Gough Island, wet heaths of *Empetrum rubrum, Acaena,* and various grasses, sedges, and ferns are widespread on Tristan, Gough, and Inaccessible islands. On Tristan Island, introduced *Rumex acelosella* and *Holcus lanatus* dominate some heaths.

Feldmark of low-growing small cushion plants, grasses, mosses, and lichens occurs on stable ground in exposed sites of the uplands of both Tristan and Gough islands. On the peaks of Tristan Island, where the only soil is loose cinders, or on fresh lava, plants are absent.

In the damp oceanic conditions of the islands, peat formation is widespread, but the deepest accumulations are in the poorly drained sphagnum bogs. Owing to the restricted coastal lowlands and steep lower slopes most of the bogs are at elevations above 500 m. Deep valley bogs occur at 500-700 m on Gough Island and at 750 m on Tristan Island and perhaps on Inaccessible Island. Shallower plateau bogs with some standing surface water occur on Gough Island and at 200 m in south central Nightingale Island.

A broad belt of giant kelp *Macrocystis* surrounds each of the islands in places, extending 2-300 m offshore. The kelp harbors considerable numbers of food organisms and is a favored forage ground for birds.

Portuguese Admiral Tristão d'Acunha discovered the three north islands in 1506, and at about the same time another Portuguese captain, Gonçalo Alvarez, discovered Gough Island, which at one time bore his name. The latter island was erroneously renamed for a British captain who may not have even seen it in 1731. The islands were visited periodically by sealers from 1790 to 1820 and by whalers throughout most of the nineteeth century. Tristan Island was first settled in 1810, and a short-lived British garrison was established in 1816-1817. The settlement on the north coast fluctuated during the nineteeth century with periodic immigration and emigration owing to visits by passing sailing ships. From 1908 to 1948, however, the settlement was largely isolated. The human population has ranged up to 275. Since 1948 and 1950, when a local crawfish industry was established, British officials and South African meteorologists have been in residence. All persons were evacuated during the eruption of October 1961, but most had returned by November 1963. Although a South African meteorological station with rotating personnel has been maintained on Gough Island since 1958, neither it nor the other two islands has ever had a permanently resident staff.

Cattle, goats, sheep, pigs, donkeys, and other domestic animals including poultry, geese, dogs, cats, rats, and mice have been introduced on Tristan Island and have affected the vegetation and nesting birds near the settlement, but except for mice and a few sheep on Gough Island, none of the other islands now has introduced animals, although pigs and goats were on Inaccessible Island in 1873 and a few sheep were present in 1938.

The land birds of the Tristan da Cunha group and Gough Island are of great biogeographic interest [*A. L. Rand,* 1955], and they and the seabirds have been studied at length during several recent expeditions [*Hagen,* 1952; *Elliott,* 1953, 1957; *Holdgate,* 1958, 1965*b*; *Swales,* 1965]. The summer-breeding seabirds are reasonably well known. The winter-breeding petrels (great-winged and gray petrels and little shearwater) and land birds warrant further attention.

BOUVETØYA

Bouvetøya, a Norwegian possession, is the southernmost of the volcanic islands on the mid-Atlantic ridge, at 54°26'S, 3°24'E. It lies 3330 km south of Capetown, South Africa, and 2035 km east of the South Sandwich Islands. Bouvetøya is about 1000 km south of the Antarctic convergence and somewhat north of the usual limit of winter pack ice. The island is roughly rectangular in outline, measuring 9.6 km east to west and 7 km north to south. It totals about 50 km². The highest points are two peaks about 780 m high (935 m high according to *Holdgate et al.* [1968]), surrounding an ice-filled inactive volcanic crater, Wilhelm II plateau. Most of the island is covered by an ice cap about 100 m thick. The slopes on the east are gentle and uniform, but in other directions they are steeper and less regular. The south and east coasts are generally bounded by glaciers, but steep rocky cliffs and headlands up to 500 m high occur on the north, west, and southwest and make landings difficult. A low lava platform about 400 m wide and 800 m long, which appeared on the west coast between 1955 and 1958, is the only extensive ice-free lowland. Most of the island's meager moss and lichen cover and bird colonies are on the platform. A single rocky islet, Larsøya, lies off the southwest coast, and small offshore rocks are found all around the island. Most of the year the island is continuously cloud and fog bound. The climate is severe with a mean temperature of −1.5°C and a mean monthly maximum of about 2°C in summer. Pack ice from the Weddell Sea gyre reaches the island occasionally in winter but rarely in summer.

Bouvetøya was originally discovered by Captain Jean-Baptiste Bouvet de Lozier on January 1, 1939, but ice and poor weather conditions prevented his obtaining a good fix or even determining its insular nature. It was resighted and identified as an island in 1808 by British Captains James Lindsay and Thomas Hopper on Enderby Brothers whalers, but until the *Valdivia* expedition in 1898 its position was not precisely determined. Sealers visited it sporadically in the nineteenth century, but the first scientists to land were those of the *Norvegia* expeditions in November 1927 and again in December 1928 [*Holgersen*, 1945] when the island was also claimed for Norway. Bouvetøya has been visited briefly by scientists since then [*Holdgate et al.*, 1968; *Solyanik*, 1964], but the inventory of breeding bird species is still incomplete. More information is needed on their numbers, which may fluctuate according to volcanic activity and the consequent availability of nest sites.

PRINCE EDWARD ISLANDS

Marion and Prince Edward islands (referred to collectively as the Prince Edward Islands) are the most recently formed volcanic islands near the mid-Indian ridge, at 46°54′S, 37°45′E and 46°38′S, 37°57′E, respectively. They lie 2300 km southeast of South Africa and 925 km west of Ile aux Cochons, Iles Crozet. The Antarctic convergence is about 220 km to the south. Marion Island, the larger of the two, is roughly oval with a coastline of 27 km and an area of 290 km^2. It is 24 km east to west and 17 km north to south. Its highest point is Jan Smuts Peak, 1230 m.

Prince Edward Island, lying 22 km NNE of Marion Island, is also roughly oval, about 10 km long in its northwest to southeast axis and 44 km^2 in area. The coastline is 29 km, and its highest peak is 672 m. The two islands lie on a submarine platform about 183 m deep and probably were connected during the Pleistocene.

Both islands have undergone essentially the same geologic history. They have been in existence about 500,000 years and show two distinct stages of volcanism, each representing several eruptions. Older gray lavas predominate in the east. On Marion Island at least, they were eroded and smoothed by glaciation between 100,000 and 15,000 years ago. They are generally overlain by layers of fresher unworn black lavas, which began erupting about 15,000 years ago. Both islands have a prominent escarpment on the west, below which is a flat coastal plain underlain by the later black lava. Tectonic movements produced horsts and section grabens in the older gray lavas on Marion Island, but there is no unequivocal evidence of faulting on Prince Edward Island.

Marion Island has a low domelike profile broken by about 130 conical hills up to 200 m high, which mark eruption centers of the second stage. The island consists of three distinct regions: the 1000-m-high central highland plateau with several volcanic cones; the slopes, which occupy the greater part of the island from the highland to the coast on the north, east, and southeast but only to the edge of the coastal escarpment on the west; and the flat marshy 'coastal plain' 50 m above sea level. The 200- to 300-m-high escarpment generally parallels the west coast 1 to 2 km inland. Black lava has overflowed the escarpment, its steepness on the north and west sides thus being reduced. The west coastal plain terminates in irregular vertical sea cliffs 15 m high. Although small lakes are common on Marion Island, there is only one perennial stream, and much of the rainfall undoubtedly reaches the sea by underground flow through the porous volcanic rock and ash.

Prince Edward Island has considerable vertical relief. It consists of an older central and eastern block of horizontal lava flows dominated by three prominent cones and sloping upward from the lower east coast to end abruptly at the 400-m-high west escarpment. The nearly vertical west escarpment is continuous with the 400- to 500-m-high cliffs of the southwest coast and the lower grassy slope of the north coast. The west coastal plain, which is about 50 m above sea level, shows several cones including the semicircular remains of a 1.5-km-wide crater. Lakes are rare on the island.

The summer snow line on Marion Island is about 1000 m, so that much of the central highland is snow covered all year except the windward sides of the easternmost peaks, which are swept bare briefly in summer. Permanent ice occurs in a depression in the central part of the plateau. There is evidence of extensive glaciation on older gray lavas of Marion Island but not on Prince Edward Island or on the more recent black lavas of both islands. No traces of glacial moraines have been found.

The islands have a typical cold sub-Antarctic climate characterized by low mean temperature with little diurnal or seasonal variation, strong westerly winds often of gale force, abundant rain and snow, and heavy cloud cover. Average daily temperatures range between 10.5°C and 5°C in summer (February mean is 7.8°C) and 6°C and 1°C in winter (August and September means are 3.6°C). The maximum temperature recorded was 22.3°C in February, and the minimum was −6.8°C in September. Average sea surface temperature measured along the north coast varies from 6.1°C in February and March to 4°C in August and September. Average monthly wind velocity is 32 km/h. Gales of 55 km/h or more are most frequent in July to September and least frequent in February and March. Precipitation falls, on the average, 25 days per month and averages about 2.5 m/yr; slightly more falls in April through July than falls at other times of the year. Northwesterly storms bring rain; southwesterly winds bring snow. Cloud cover varies little throughout the year. On the average, the sky is three-quarters or more obscured by clouds.

Weather data were recorded on the east coast of Marion Island near sea level. The west coast should have somewhat more wind, precipitation, and cloudiness, and Prince Edward Island may be drier than Marion Island.

The native flora of Marion Island consists of 22 species of which 15 are phanerogams. Tussock grass *Poa cookii* occurs on well-drained coastal or inland slopes that are exposed to salt spray or fertilized by petrel droppings. More protected well-drained inland slopes, where the influence of salt spray and birds is minimal, have mixed herb fields of ferns, grasses, sedges, dwarf shrubs, and cushion plants. The dominant species on warm north-facing slopes is the fern *Blechnum ponnamarina*, on east-facing slopes protected from the wind *Acaena adscendens*, and on higher more exposed slopes *Azorella selago* and *Poa cookii*. Poorly drained valleys have swampy ground covered with a carpet of mosses, sedges, and rushes. *Azorella* cushions and several species of mosses form feldmark communities on lava flows exposed to high winds. Those above 300 m are composed almost entirely of mosses and lichens. Thirteen species of exotic weeds are present on Marion Island, but only one of these, the grass *Poa annua*, occurs on Prince Edward Island. Introduced mice and cats occur, and the latter prey on birds.

Discovered in 1772 by Nicholas Marion-Dufresne and possibly more than 100 years earlier by the Dutch, Marion and Prince Edward islands were inhabited by sealers during the late eighteenth and early nineteenth centuries. Because of the great difficulty in landing, however, Marion Island had been visited by few scientists until 1947, when both were annexed by South Africa and later proclaimed nature reserves. A meteorological station was established in 1948. The birds of Marion Island have been studied by *Crawford* [1952], *La Grange* [1962], *R. W. Rand* [1954, 1955, 1956], and *van Zinderen Bakker* [1971*a, b, d*]. Prince Edward Island is known from only one visit by an ornithologist [*van Zinderen Bakker*, 1971*a*].

ILES CROZET

Iles Crozet constitute an archipelago of five volcanic islands, several rocks, and a reef on an isolated shallow bank between 46° and 46°30′S and 50° and 52°30′E. They lie 2400 km southeast of the Cape of Good Hope, 2400 km north of the Antarctic continent, 925 km east of Marion Island, and 1800 km west of Iles Kerguelen. The Antarctic convergence is 400 km to the south. Two small islands, Ile aux Cochons and Ile des Pingouins, a cluster of rocks, Ilots des Apôtres, and the reefs Brisants de l'Héroine make up the west group. The two higher east islands, Ile de la Possession and Ile de l'Est, lie about 100 km to the east. The total land surface of Iles Crozet is about 500 km^2.

Twelve rocky peaks and monoliths make up Ilots des Apôtres. Only the easternmost, Grande Ile, which is about 3 km in diameter and 289 m high, affords any possibility of landing. Fifty-three km to the south is Ile des Pingouins, a jumble of virtually inaccessible cliffs 3 km long and up to 360 m high. Neither of these two islands has been visited by an ornithologist.

Eighteen kilometers SSE of Ilots de Apôtres and 35 km NNE of Ile des Pingouins is Ile aux Cochons, a cloud-capped volcanic cone about 9 km in diameter and about 800 m high. The island is the least eroded and was probably the last to erupt in the archipelago. Its steep west slopes are covered with gray boulders and terminate in high sea cliffs. A few adventitious cones interrupt the regular slopes of the northwest and south sides.

Ile de la Possession, about 100 km farther east, the largest island in the archipelago, is roughly rectangular, 30 km long from northwest to southeast and about 15 km wide. Its high interior is a snow-capped chain of mountains, the principal summits of which range from 500 m to the 934-m-high Pic du Mascarin. Deeply eroded valleys lead from the interior to sandy beaches on the north and east coasts. Between the valleys on these two sides and almost continuously on the west and south coasts, rocky headlands and steep cliffs meet the sea.

Ile de l'Est is 20 km to the east of Ile de la Possession. It is roughly oval, 18 × 10 km, and has an area of 130 km^2. The 55-km-long coastline is broken by only two important bays, one on the north and one on the south. The high but deeply dissected island is dominated by several crests and ridges from 700 to 900 m high, and the two highest peaks are 1050 and 1012 m. Two broad flat valleys, bounded by rocky ridges, on the south side are separated from two similar valleys on the north by low saddles and provide relatively easy means of crossing the island. The floors of these valleys and of three other lesser valleys have broad stream beds and lush vegetation. Although the valley shores have extensive sand or shingle beaches, most of the rest of the coastline consists of steep cliffs, which range as high as 700 m on the west.

The plant life of the three larger islands is relatively uniform. Luxuriant tussock grass *Poa cookii* grows near the coast; Kerguelen cabbage, *Acaena* clumps, and *Azorella* cushions dominate undisturbed moors. Vascular plants reach altitudes of 200-300 m, but only sparse mosses and lichens occur on the higher slopes. The numerous introduced rabbits have grazed extensively in some areas of the two east islands, but in spite of the presence of rabbits on the east side of Ile aux Cochons, its lush vegetation is virtually virgin. Algae, particularly the giant kelp *Laminaria*, are common in coastal waters.

Climatic information is scant. Iles Crozet may be slightly warmer than either Marion Island or Iles Kerguelen, but like those islands they are cold, wet, windy, and cloudy. Based on the measurements of one summer, the mean temperature is about 7°C, ranging from 17°C to 1.5°C; rain falls on 80% of the days; winds with a mean velocity of 35 km/h blow mostly from the west and northwest; and gale winds, up to a maximum of 115 km/h, occur on 80% of the days. Winters are not severe, however. Although snow falls, it does not persist for long at sea level. No glaciers are present, but there is some permanent snow and ice on the peaks of Ile de la Possession and Ile de l'Est. Traces of past glaciation exist as low as 100 m.

The islands were discovered by Nicholas Marion-Dufresne in 1772 and occupied sporadically by American and other sealers during the nineteenth century, when cats, rabbits, rats, and mice were introduced. Pigs and goats were introduced on Ile aux Cochons and Ile de la Possession, respectively, but failed to establish themselves. Iles Crozet are claimed by France and since 1938 have been a national park. Because of difficulty in landing, only Ile de la Possession had been visited briefly by scientists until a meteorological station was

TABLE 7. Distribution of Birds Breeding in Iles Crozet

	Ile aux Cochons	Ile de la Possession	Ile de l'Est
King penguin	X	X	X
Gentoo penguin	X	X	X
Rockhopper penguin	X	X	X
Macaroni penguin	X	X	X
Wandering albatross	X	X	X
Gray-headed albatross		X	X
Sooty albatross		X	X
Light-mantled sooty albatross	X	X	X
Northern giant fulmar	?	X	X
Southern giant fulmar	X	X	X
Cape pigeon		X	
Narrow-billed prion			X
Antarctic prion			X
Broad-billed prion	X	X	
Fairy prion			?
Blue petrel			X
Great-winged petrel			probably
Kerguelen petrel		X	X
Soft-plumaged petrel			X
White-chinned petrel		X	X
Gray petrel		X	probably
Wilson's storm petrel			probably
Black-bellied storm petrel			X
Gray-backed storm petrel			X
South Georgia diving petrel		X	X
Kerguelen diving petrel		X	
King shag	X	X	X
Kerguelen pintail	X	X	X
Lesser sheathbill		X	X
Brown skua	X	X	X
Southern black-backed gull	probably	X	X
Antarctic tern		X	X
Kerguelen tern		X	X

Known to breed on island, X; breeding needs confirmation, ?.

established there in 1962. French ornithologists have visited the three larger islands and have begun publishing studies of the bird populations in recent years [*Dreux and Milon,* 1967; *Despin et al.,* 1972; *Mougin,* 1969*a, b, c,* 1970*a, b, c*] (Table 7).

ILE AMSTERDAM AND ILE SAINT-PAUL

Two volcanic islands, Ile Amsterdam and Ile Saint-Paul, lie on a narrow submarine ridge within 100 km of each other in the southern Indian Ocean at 37°50′S, 77°31′E and 38°43′S, 77°31′E, respectively. They are 1300 km northeast of Iles Kerguelen and just north of the Subtropical convergence.

Ile (Nouvelle) Amsterdam is a compact oval volcanic cone 10 × 7 km and 55 km^2 in area. Its two highest points, 911 and 829 m, are the remains of the wall of the principal crater, whose floor is a 600-m-high uneven plateau, which dominates the southwest central part of the island. Other secondary craters, cones, and vents are scattered on the plateau and flanks of the island. Lava flows radiate down from the plateau to the coast, where they end in almost uninterrupted cliffs usually above a narrow rock or boulder beach. The cliffs are generally 20-40 m high except on the west side, where the island has been eroded severely and the cliffs are up to 700 m high with high rock piles at their bases. These rocks near Pointe d'Entrecastreaux, which are accessible only from the sea, provide the main habitat for nesting seabirds on the island. A long gently sloping lava flow that meets the sea in the northeast affords the only landing point from which the island can be scaled and is also the site of a base and small settlement, La Roche Godon. Although its volcanism is quite recent and erosion anywhere other than along the coast is minimal, Ile Amsterdam shows no present activity. The volcanic soil is porous, and in spite of ample rainfall most of the year, there is little standing water except in the peat bogs of the high plateau, and only a few persistent streams exist in the southwest.

Ile Saint-Paul, 100 km to the south, is a 270-m-high volcano, the eastern third of which has been exploded or eroded away. The crater thus breached by the sea forms a 1-km-wide circular bay on the northeast coast. The 7-km^2 island is shaped roughly like a right triangle, its hypotenuse and crater bay facing northeast and its 3- and 5-km sides on the west and south. The walls of the crater and the cliffs of the eroded east side are about 200 m high and steep, whereas the lava flows of the north, west, and south sides slope more gently down toward eroded coastal cliffs 30 m high. The continuous cliffs are virtually unclimbable, so that the only ready means for penguins or boats to land on Ile Saint-Paul is for them to cross the shallow 0.8- to 2.5-m sill into the crater. Current volcanic activity is limited to vapors from salty thermal springs at two points around the crater. There may be some freshwater springs and small ponds on the west slopes.

Both islands appear relatively verdant from a distance in spite of some bare cliffs and lava flows. Drier lowland slopes up to about 250 m are covered with meadows of high tussock grass *Poa novarae.* The wetter slopes of Ile Saint-Paul and the higher slopes of Ile Amsterdam up to about 600 m have a dense nearly impenetrable growth of sedges. Formerly more widespread, *Phylica arborea* trees up to 7 m high and tall ferns grow in protected areas of Ile Amsterdam, but they are disappearing owing to browsing by cattle. The highland plateau of Ile Amsterdam above 600 m is covered by sphagnum bogs or a wind-contoured feldmark of *Acaena* and mosses.

The climate is subtropical, markedly oceanic, warm, windy, and humid. The summer, October to March, is warmer, drier, and less windy than the winter, April to October. Mean annual temperature is about 12.9°C; January and February, the warmest months, average 15.2°C, and June, the coolest month, averages 9.9°C. Frost occurs in winter in the highlands of Ile Amsterdam, where snow falls occasionally but does not persist. Rain is frequent, at least every other day in summer and 80-90% of the days in winter, but the total precipitation of about 1100 mm is not excessive. Winds blow consistently from the northwest, west, or southwest with an average monthly wind speed of about 19 km/h in March to 33 km/h in September. Clouds generally obscure the highlands of Ile Amsterdam above 600 m. Mean monthly sea surface temperatures range from a low of 12.3°C in September to 17.3°C in February and average about 0.3°C warmer near Ile Amsterdam than at Ile Saint-Paul.

Although Ile Amsterdam was discovered in 1522 by Magellan's shipmates after his death and Ile Saint-Paul was discovered probably about 100 years later, they were not landed on until 1696. They were visited sporadically by explorers, scientists, castaways, and lobster fishermen from the eighteenth through early twentieth centuries and were claimed by France in the mid-1800's. Attempts at colonization by fishermen from Madagascar and La Réunion in the 1840's and 1930's failed, but a meteorological station was established on Ile Amsterdam in 1950. Cattle were established in 1871 and persist semiwild on Ile Amsterdam today. Rats, mice, and rabbits are also established on both islands. Little remains of what may have once been reasonably extensive seabird populations, and there are no native land birds [*Jouanin and Paulian,* 1960; *Segonzac,* 1972].

ILES KERGUELEN

Iles Kerguelen, the largest landmass in the south Indian Ocean, make up a compact archipelago of one very large island, Ile Kerguelen (Grande Terre), and about 300 smaller islands, islets, and rocks between 48°27′ and 50°S and 60°27′ and 70°35′E. The entire archipelago extends 200 km from Ilot du Rendez-Vous (Ilot Réunion) in the north to Rochers du Salamanca in the south and 150 km from Iles de la Fortune in the west to the east end of Péninsule Courbet on Ile Kerguelen. The surface area is 7000 km^2, of which Ile Kerguelen comprises 6000 km. Iles Kerguelen lie about 2000 km north of the Antarctic continent, 1800 km WSW of Australia, 3300 km SSE of Madagascar, and 4200 km ESE of the Cape of Good Hope. Heard Island, 500 km to the southeast, lies on the same extensive submarine platform. A 200-m-deep submarine sill extends many kilometers to the northwest and northeast and demonstrates that the volcanic Iles Kerguelen once formed a far more significant landmass.

Roughly triangular, Ile Kerguelen is 120 km north to south and 140 km northwest to southeast. Its 1350-km-long shoreline is deeply dissected by gulfs and bays into a number of major peninsulas linked to the mainland by low narrow isthmuses. These peninsulas are themselves further dissected by fjords into smaller peninsulas. The sea thus penetrates deeply into the land, of which no point is more than 20 km from the coast. Ile Kerguelen displays sharp relief, high mountains and extensive 200- to 800-m-high plateaus being broken up by deep glacial valleys and lakes. Several snow- and ice-capped peaks are over 1000 m in elevation, and double-peaked Mont Ross rises to 1850 m near the south coast. Travel in the interior is rendered difficult by the many rivers, streams, and torrents that flow from the mountain snowfields and glaciers and are augmented by abundant rain. Only the east end of the island has extensive flat lowlands.

Steep rocky cliffs up to 600 m high alternate with long breaker-washed sand and pebble beaches along the south and west coasts. Although these portions of the coast are inaccessible from the sea, elsewhere fjords provide safe anchorage and landings.

Ile Kerguelen has a profusion of lakes and ponds of various kinds: glacial lakes gouged out by ice movement and dammed by moraines, narrow deep fault lakes, coastal lagoons shut off from the sea by uplift, and crater lakes. Large lowland lakes subject to strong winds seldom freeze, but small ponds may be covered with ice after a calm night, and lakes above 500 m may stay frozen for extended periods.

The cloud-covered ice cap, Calotte Glaciaire Cook, in the west is the still impressive remains of an ice sheet that formerly covered the entire landmass. It is 50 km from north to south and about 20 km wide and has a total area of about 600 km. It covers a flat to undulating plateau at a mean elevation of 1000 m. The higher slopes are gentle, but at lower elevations the outflow glaciers are steep and deeply crevassed. Two ice tongues reach the sea in the west, but the terminal moraines in the lowlands in the north, east, and south are distant from the sea. Other smaller mountain glaciers occur on Pic Guynemer (1088 m) on Péninsule Loranchet, Mont Richards (1049 m) on Presqu' île de la Société de Géographie, Mont Crozier (979 m) in the west part of Péninsule Courbet, Mont Ross, and the Péninsule Rallier du Baty (1262 m) in the southwest. Only that of Mont Ross reaches the sea.

A dozen significant off-lying islands have areas of 10-180 km. The largest and highest are Ile Foch (687 m) in the north and Ile de l'Ouest (617 m) in the west. Most are near Ile Kerguelen, but Iles Nuageuses and Iles Leygues (Iles Swain) are more than 10 km out in the open sea to the north, and some islets to the south are farther still. Although the islands are small in area, they are important rabbit-free refuges for vegetation, and rat- and cat-free sanctuaries for breeding birds.

Iles Kerguelen are entirely volcanic in origin. Four eruptive phases can be differentiated. The earliest basalts are of limited occurrence and undetermined age. Later intrusive rocks are probably of Eocene age. The bulk of the islands and in particular the plateaus are composed of thick layers of basaltic lavas separated by beds of tuffs and soft agglomerates that are probably of Miocene age. Lignite formed during this period of great volcanic activity indicates that trees similar to those now in New Zealand were present and that the climate was milder and less windy than it is at present. The high mountains are the result of still later Pliocene or early Pleistocene volcanism that occurred before glaciers covered the entire archipelago, eroded the highlands, gouged out cirques and valleys, and left moraines and massive alluvial conglomerates. A few thermal springs on the west coast of the Péninsule Rallier du Baty are the only current traces of volcanic activity. Many of the glaciers seem to be retreating.

Iles Kerguelen have a typical sub-Antarctic oceanic climate with cold summers and moderately severe winters. Temperatures vary little in different seasons, precipitation is abundant and frequent, cloud cover is heavy, and strong westerly winds blow almost constantly.

Temperatures have possibly been warming over the past 70 years. Scanty records in 1902 and 1929 show a mean annual temperature of about 3.2°C; since 1952, on the other hand, the mean has been above

4.5°C. The summer mean is now about 7.4°C, and the winter about 2.6°C. In summer the amplitude of diurnal variations is often as high as 10°C to 12°C. The extreme temperatures recorded at sea level are −9.4°C in June and 23°C in January, but most temperatures are 3°C to 11°C from December to April and −2°C to 5°C for the rest of the year.

Relative humidity, which averages between 70 and 80%, varies from 100% during rains brought by north and northwest winds and by warm fronts to 40% with cold dry southwest winds. Cloud cover is heavy, especially in summer; orographic clouds obscure most mountaintops above 700 m, but foggy days are exceptional.

Precipitation occurs on 250-300 days a year and varies little from season to season. It averages over 1100 mm in the protected southeast, but the mountains in the west may receive twice as much. Most rain is light drizzle or short-lived showers; heavy downpours of long duration are rare. Snow falls on less than 60 days a year at sea level and melts rapidly. On Calotte Glaciaire Cook, on the other hand, it snows almost daily owing to the high condensation of moisture-laden air over cold ice. The snow line on Ile Kerguelen is now from 900 to 1000 m, whereas in 1902 and 1929 it was between 650 and 700 m.

Eighty percent of the winds are from the northwest, west, or southwest. Mean monthly velocity varies from 28 to 45 km/h, the strongest winds occurring in the spring. Gales of 100-200 km/h occur almost every month, and usually, less than 5 days per month are calm.

Sea temperature has also warmed during the past 70 years. In 1902 it ranged from 1.4°C in September to 5.8°C in February (annual mean of 3°C); in 1952 the coldest month was July, at 2.1°C, and the warmest was February, at 7°C (annual mean, 4.7°C). Except in shallow brackish bays the sea never freezes. Although icebergs were reported frequently in the past, a sighting is now a rare event.

The flora of Iles Kerguelen is poor in native vascular species. It is composed of 28 phanerogams, 4 ferns, 2 club mosses, and some hepatics plus a few introduced grasses and other weed species. Mosses and lichens occur in profusion and probably number more than 200 species. Introduced rabbits have ravaged the native vegetation on Ile Kerguelen. Only on the Péninsule Rallier du Baty in the extreme southwest and on most of the offshore islands is the vegetation unaltered. Severe climatic conditions limit the altitudinal distribution of lush phanerogamic vegetation to below 400 m. Some hardier species occur to 700 or 800 m, but only mosses and lichens occur from there to the snow line.

The vegetation of Iles Kerguelen is harder to classify than that of other subantarctic islands, partly because of disturbance and partly because of the extreme ecological tolerance of individual species. The four major species, the Kerguelen cabbage *Pringlea antiscorbutica,* the tussock *Poa cookii,* the mosslike cushion *Azorella selago,* and the mat *Acaena adscendens,* occur from sea level up to 500 m or even higher and in differing relative abundances and different forms in a wide range of communities. The first three species have been radically affected by rabbits, so that *Acaena* has spread.

Closed tussock grassland is virtually absent on lowland slopes of Ile Kerguelen, where rabbits abound, although scattered *Poa cookii* tussocks are widespread. Tussock still occurs undisturbed on some smaller islands and in wet mountain uplands near 500 m. Herb fields of Kerguelen cabbage, *Poa cookii, Azorella selago,* and *Acaena adscendens* are widespread over much of Iles Kerguelen. The cabbage still dominates in steep rocky cliffs and some moist habitats from sea level to the limit of phanerogamic growth. *Azorella* is typical of the moist moraine slopes and wind-swept plateaus of the interior, where it forms deep peat, but either it is being replaced in many localities by *Acaena* or where it has been eaten by rabbits, the underlying peat may have eroded to bedrock or infertile moraine. Highland moors are dominated by *Acaena* mixed with *Azorella* in dry situations and with rushes under wetter bog conditions. The two grasses *Deschampsia antarctica* and *Agrostis magellanica,* which also occur in bogs, are more usually found along banks of streams. Lakes and ponds teem with minute green algae. Above the moors, either an open feldmark of dwarf grasses, ferns, mosses, and lichens or areas totally devoid of vegetation occur.

Marine habitats are particularly rich about Iles Kerguelen, where more than 100 species of algae have been recorded. Close inshore a kelp *Durvillaea utilis* dominates rocky shores. The giant kelp *Macrocystis pyrifera* grows many meters long in deeper protected bay and coastal waters.

Along with the native flora and fauna of Tristan Island, Ile Amsterdam, and Ile Saint-Paul the native flora and fauna of Iles Kerguelen have suffered disastrously from man's deliberate and accidental introductions of mammals. The most important were the rabbits liberated in 1874 by the British Transit of Venus Expedition, which destroyed much of the lowland vegetation of Ile Kerguelen and thus altered the landscape. The ensuing erosion has reduced nesting habitat for burrowing petrels. The house mouse and black rat arrived with sealers in the nineteeth century, and although the former does not harm birds, the latter preys on the eggs and chicks of petrels. Cats, which were unsuccessfully introduced to control rodents in the early 1800's, died out by 1850 but were reintroduced about 100 years later and are now an abundant scourge in petrel colonies. Dogs were present from 1903 until about 1928, and an introduction of mink in 1959 proved unsuccessful. About 1000 head of sheep were introduced in 1909 but died out in 1932. They were reintroduced and allowed to roam freely on Ile Kerguelen and some of the islets in 1949. They now number about 800 head. In 1957 a small flock of wild mouflon was added to the sheep, and by 1968 they numbered at least 42 head. A small herd of reindeer was established on Ile Kerguelen in 1955-1956. Mules, ponies, pigs, and cattle have been kept for short periods in recent years near the scientific station.

Iles Kerguelen were discovered and claimed for the king of France by Captain Yves Joseph de Kerguelen-Trémarec in February 1772 during his search for a south continent. He revisited the islands 2 years later and determined their insular nature. In 1776, British Captain James Cook explored the islands, named them for Kerguelen, and gave them the widely used alternate name 'Island of Desolation.' American and British sealers began visiting the islands in the early 1800's, and they and a few whalers used them as a base sporadically throughout much of the rest of the century. Several British, German, American, and French scientific expeditions visited the islands between 1840 and 1939 and produced descriptive studies of the flora and fauna, including important bird publications [*Kidder and Coues,* 1875; *Cabanis and Reichenow,* 1876; *Sharpe,* 1879; *Studer,* 1889; *Loranchet,* 1915-1916; *Falla,* 1937].

France formally annexed Iles Kerguelen in 1893. Except for sealers and shipwrecked sailors, however, the islands remained uninhabited until 1909, when Frères Bossière established a whaling station manned by Norwegians at Port Jeanne d'Arc at the head of Golfe du Morbihan. They also simultaneously established an inland sheep station. Both enterprises were interrupted in 1914-1921 and ceased altogether in 1932. Iles Kerguelen were a dependency of Madagascar from 1924 to 1955 and since then have been autonomous within the Terres Australes et Antarctiques Françaises. In 1949-1950 the French government constructed a meteorological station at Port-aux-Français, which they later expanded to include a biological laboratory. The year-round population of the station is 90-110. Scientists at the new station have carried out a few studies on the rich avifauna, but the birds need much further attention [*Paulian,* 1953; *Milon and Jouanin,* 1953; *Bauer,* 1964; *Derenne et al.,* 1972]. In spite of its remoteness it is remarkable that no true land bird exists on an archipelago as large and as well vegetated as Iles Kerguelen. Possibly, Captain George Comer [*Verrill,* 1895] did see a small resident passerine, but there were no earlier reports, and it has vanished without a trace.

HEARD ISLAND

Heard Island is the highest and southernmost island in the Indian Ocean, situated at 53°05′S, 73°30′E, about midway between South Africa and Australia and 1100 km north of the Antarctic continent. The Antarctic convergence is about 180 km to the north in summer. Heard Island lies on the same submarine plateau as Iles Kerguelen, 500 km to the northwest. The roughly circular central part of the island, about 24 km in diameter, is dominated by the dome-shaped 2400-m-high massif Big Ben, which culminates in the active volcano Mawson Peak, 2745 m high. A mountainous headland, Laurens Peninsula, 715 m high extends 10 km to the northwest, and 10-km-long Spit Point extends to the southeast, Heard Island thus having an overall length of about 43 km.

The volcano is intermittently active; vapors issue from the summit and from other vents as low as 1200 m. Seven extinct craters are present on the older portion of the island, Laurens Peninsula. Geologically, the island is formed of diverse lavas and volcanic ash with a few limestone deposits on Laurens Peninsula.

Glacial ice, in places as much as 150 m deep, covers approximately 90% of the island. Over half of the coastline is bounded by ice cliffs. Streams cascade down from melting glaciers and help form sand and gravel flats in some valleys. Glacial moraines provide soil for lowland vegetation and thus indirectly nesting sites for birds.

The surrounding waters are treacherous with rocks, reefs, and shoals. The largest are the three Shag Islands (Sail and Drury rocks and Shag Island) 12 km off the north coast, and Wakefield reef, 5 km off the southwest coast.

The flora consists of seven or eight vascular plants, some mosses, and lichens. Short tussock grass (*Poa cookii*), Kerguelen cabbage, and *Azorella* cushions occupy established moraines and valleys up to 200 m, *Acaena adscendens* occurring at protected localities such as Spit Point. Mosses and lichens occur on bare ground from 200 m to the snow line. Giant kelp is present along the coast but not as abundantly as it is at Iles Kerguelen. No exotic plants or animals are known to have been established.

The climate is cold, windy, and wet. Mean annual temperature at sea level is about 1°C. The winter mean monthly temperature is about −2°C, and the summer mean is about 2.5°C. The extremes vary from −9.5°C to 12.5°C. Winds blow almost constantly from the northeast, east, or southeast with an average velocity of about 30 km/h. Gale force winds are frequent, and velocities occasionally reach 200 km/h. Rain, snow, or sleet falls on about 300 days per year. Snow remains at sea level from April to mid-November. The summer snow line lies near 400 m.

Seawater temperature around the island varies from 3°C in summer to −1°C in winter. Although some ice forms in protected bays in winter, the northern limit of pack ice is usually 550 km south of Heard Island in August but might reach the island in some years.

Heard Island, possibly seen as early as 1833, was discovered by Captain J. J. Heard in 1853 and first landed on in 1855. Sealers occupied the island continuously for 20 years and sporadically thereafter until at least 1929. A scientific station was established in 1947, when the island was ceded by Great Britain to Australia. The station was closed in 1955, but Heard Island has been investigated frequently by biologists since then. The birds are the subject of one comprehensive report [*Downes et al.,* 1959] and other studies.

MCDONALD ISLANDS

McDonald Islands are exposed peaks on the Kerguelen submarine plateau at 53°03′S, 72°36′E, 38 km west of Heard Island. They were discovered in 1854 but only first landed on in 1971, for less than 1 hour [*Budd,* 1972]. There are three islands in the McDonald Islands. McDonald Island, the largest, consists of two parts, each about 1.3 km², bounded by steep cliffs and joined by a narrow central isthmus. The north half is a sloping plateau, 30 m high in the southeast to 120 m high on the northwest. The south half is a steep-sided hill about 230 m high. Beaches on either side of a promontory on the northeast end and along the east coast possibly give access to the plateau and from there to the hill.

Flat Island, 100 m north of McDonald Island, is a plateau about 55 m high bounded by steep cliffs except in the southeast corner, where penguins can land in a small rocky cove and scramble up a slope to the plateau. Meyer Rock, 1 km northwest, is apparently barren and 170 m high.

The islands are basalt, lava, and tuff, presumably the result of two or more eruptions from wet vents near sea level. The southern hill of the main island is the eroded cone of one eruption, and the plateau is the remains of another crater wall and dike.

Extensive tussock (*Poa cookii*) occurs on the east slopes of the hill and on the lower parts of the plateau, which is dotted with small tarns. An *Azorella selago* cushion covers the higher part of the plateau, and Kerguelen cabbage is present on the east slopes above the cliffs.

Birds that were seen or were presumed to be breeding included the macaroni penguin, the most abundant species, the southern giant fulmar, the lesser sheathbill, the brown skua, and a species of diving petrel.

MACQUARIE ISLAND

Macquarie Island is the southernmost of the subantarctic islands south of New Zealand. It lies just north of the Antarctic convergence at 54°30′S, 158°57′E, about 1280 km southeast of Tasmania, 1000 km southwest of South Island, New Zealand, and 1440 km from Cape Adare on the Antarctic continent. Although Macquarie Island is faunistically and botanically allied with the New Zealand sub-Antarctic Islands (Bounty, Snares, Antipodes, Auckland, and Campbell islands), it is not on the same submerged New Zealand 'continental' plateau as they are but represents an exposed mountain range on a large separate ridge.

Macquarie Island is about 33.6 km long with a maximum east to west width of 4.8 km and an area of 112 km^2. The island consists of a central plateau 250 m high in the north to 300 m high in the south. From the plateau, steep slopes descend to a narrow raised coastal terrace varying from 10 m wide on the east coast to 800 m wide on the south. The slopes are dissected locally by a few glacially scoured valleys and many cascading streams. In the southwest and south, steep cliffs rise directly from the sea. The central plateau is rolling and hilly, the highest peak, Mount Hamilton, being only 433 m above sea level. Small deep ice-formed lakes, ponds, and bogs punctuate the landscape and have a total area of about 2.5 km^2. Some are in closed basins and drain by soakage, whereas others feed streams that cascade to the sea.

The rocks are all igneous and alkaline.

Although glaciers once covered the entire island and gouged out small lakes and rounded even the highest points, no permanent ice or snow now persists.

The cold temperate climate is marked by small seasonal and diurnal variation in temperature, high humidity, and strong winds. Mean temperature is 4.4°C with a difference of only 6°C between the mean temperatures of the warmest and coolest months and with a diurnal range of only 5.7°C. Temperature also drops about 5.7°C/300 m, so that the mean annual temperature at the highest point is near freezing. Annual precipitation, mostly misty rain, is about 1 m, distributed fairly evenly throughout the year. Evaporation is slight, and relative humidity averages 92%. Low flat ground is perpetually wet, if not frozen. The mean wind velocity is 35 km/h near sea level, and gusts of 100 km/h are frequent all year. Wind velocity may be higher on the plateau. Winds are predominately from the west and northwest, and occasional south and north winds bring storms. Cloud cover is heavy most of the year. Only from November to March does sunshine average more than 2 h/d.

From offshore the island appears as a long irregular green mountain range. Although the flora consists of only 35 native species, Macquarie Island is extensively vegetated. Tussock grassland is common on the well-drained coastal terraces, on the slopes up to 300 m, and even on protected upland flat areas. Lush herb fields dominated by rosettes of *Pleurophyllum hookeri* and by Macquarie cabbage occur on flats and slopes where winds are moderate and where there is a high water table. Fens of rush and sedge occupy wetter flats, where the alkaline water table is at the surface or above ground level. Only at Handspike Point on the raised coastal terrace at the north end is there an extensive acid water peat bog. Low-growing feldmark, especially *Azorella* cushion, covers most of the island uplands above 180 m that are exposed to high winds, and feldmark occasionally occurs as low as 90 m.

Giant kelp grows luxuriantly in the surrounding waters, and the beaches are heavily littered with its remains, which afford foraging ground for wekas.

Wild and introduced animals have had a profound effect on the natural vegetation. Penguins and elephant seals moving through and over tussock and depositing large amounts of guano have flattened or obliterated much of the grass in the lowlands. Three species of weeds are well established, but only a grass, *Poa annua,* is widespread and abundant.

Rabbits introduced in 1880 to provide fresh meat for sealers have ravaged much of the lush native tussock and herb field vegetation and thus have indirectly caused extensive erosion on slopes. They may also compete with some petrels for burrows. Introduced cats, rats, and wekas are important predators on eggs and young of many birds. Goats, sheep, cattle, horses, and dogs have also been introduced but have not become established.

The island was discovered by the Australian sealer Frederick Hasselburg in 1810. During the nineteenth century it was inhabited by bands of sealers, who extirpated the fur seals and reduced elephant seals and king penguins drastically. Since 1933 the island has been declared a sanctuary. It is administered politically from Tasmania. A permanent scientific station has been established on a dry shingle beach at the north end of the island to support long-term studies. The birds have been studied for many years, but although *Warham* [1956, 1962, 1963, 1967, 1971*a*] has reported on several species, there is no comprehensive monograph on the avifauna.

The Judge and Clerk Islands and Bishop and Clerk Islands, lying 17 km and 37 km south of Macquarie Island, are small groups of barren rocks. The Bishop and Clerk Islands have been surveyed briefly from the air [*MacKenzie*, 1968]. The main rock is about 45 m above sea level, but the others are low and subject to waves. Royal and gentoo penguins molt and may breed; black-browed albatrosses, cormorants, gulls, and possibly skuas nest.

ANALYSIS OF REFERENCES

This analysis of references that are cited in full in the reference list that follows is intended as a guide to further readings on antarctic and subantarctic birds. The sequence of listings is general references, bird references (mostly by families), study methods, land geography, and sea geography. Accounts that deal primarily with a single species or family of birds are generally listed only under the appropriate systematic section and not by geography even if the work was conducted at only one site. Accounts of vagrants that are cited in the text to document records do not appear in this analysis unless they include several species (e.g., *Keith and Hines* [1958] and *Barrat* [1947*b*]). Literature has been consulted comprehensively through September 1974 and selectively into early 1975.

GENERAL

Alexander, 1954
Andersson, 1905
Arnaud et al., 1967
Barrat and Mougin, 1974
Bierman and Voous, 1950
Boswall and Prytherch, 1969
Carrick and Ingham, 1967, 1970
Cassin, 1858
Dixon, 1933
Falla, 1937, 1952, 1964
Falla et al., 1966
Gain, 1914
Holgersen, 1945, 1957
Lowe and Kinnear, 1930
Murphy, 1936, 1964
Oliver, 1955
Ozawa, 1956, 1967*a*
Peale, 1848
Prévost and Mougin, 1970
Reichenow, 1908
Roberts, 1941
Sladen and LeResche, 1970
Stonehouse, 1964, 1965
Voous, 1965
Watson et al., 1971
Wilson, 1907
Winterbottom, 1971

PENGUINS

Ainley, 1972
Ainley and Schlatter, 1972
Arnaud, 1964
Austin, 1957
Bagshawe, 1938
Bauer, 1964, 1967
Bougaeff, 1974
Budd, 1961, 1962, 1968, 1970
Budd and Downes, 1965
Carins, 1974
Caughley, 1960*a*, *b*
Conroy and Twelves, 1972
Conroy and White, 1973
Despin, 1972
Douglas, 1968
Downes, 1955
Emison, 1968
Emlen and Penney, 1964
Guillard and Prévost, 1964
Gwynn, 1953*a*
Isenmann, 1971
Isenmann and Jouventin, 1970
Jouanin and Prévost, 1953
Jouventin, 1971
Kooyman et al., 1971
LeResche and Sladen, 1970
Levick, 1914, 1915
Luna Perez, 1963

Matsuda, 1964
Mougin, 1972, 1974
Müller-Schwarze, 1971
Novatti, 1959
Ozawa, 1967*b*
Penney, 1968
Pettingill, 1960, 1964
Prévost, 1961
Prévost and Sapin-Jaloustre, 1964, 1965
Prévost and Vilter, 1963
Rand, R. W., 1955
Reid, 1964
Roberts, 1940*b*
Sapin-Jaloustre, 1960
Sladen, 1958, 1964
Stirling and Greenwood, 1970
Stonehouse, 1953, 1960, 1967, 1970*a*, *b*, 1975
Taylor, 1962
Trawa, 1970
van Zinderen Bakker, 1971*d*
Voisin, 1971
Warham, 1963, 1971*a*, 1972

PROCELLARIIFORMES

Bang, 1966
Beck, 1970
Dixon, 1933
Matthews, 1949, 1951
Murphy, 1964
Prévost, 1964
Warham, 1971*b*

ALBATROSSES

Bang, 1966
Bauer, 1964
Dixon, 1933
Gibson, 1963, 1967
Jameson, 1958
Mougin, 1970*a*, *b*
Rowan, 1951
Sorensen, 1950
Tickell, 1964, 1967, 1968
Tickell and Gibson, 1968
Tickell and Pinder, 1967
van Zinderen Bakker, 1971*b*
Warham and Bourne, 1974
Warham et al., 1966

FULMARINE PETRELS

Bang, 1965
Beck, 1969, 1970
Bourne and Warham, 1966
Brook and Beck, 1972
Brown, 1966
Conroy, 1971*b*, 1972
Hudson, 1966, 1968*b*
Isenmann, 1970*a*, *b*
Johnstone, 1974
Konovalov and Shulyatin, 1964
Løvenskiold, 1960
Maher, 1962
Mougin, 1967, 1968
Orton, 1968
Pinder, 1966
Prévost, 1953, 1964, 1969
Pryor, 1965
Shaughnessy, 1970, 1971
Stonehouse, 1958
Tickell and Scotland, 1961
Voisin, 1968
Voous, 1949
Warham, 1962

PRIONS

Bang, 1966
Beck, 1970
Falla, 1940
Fleming, 1941
Fullagar, 1972*a*, *b*
Harper, 1972
Richdale, 1944*a*, *b*, 1965
Strange, 1968
Tickell, 1962

GADFLY PETRELS

Dell, 1952
Harper, 1973
Harper et al., 1972
Imber, 1973
Mougin, 1969*c*
Murphy and Pennoyer, 1952
Richdale, 1964
Warham, 1956, 1967

SHEARWATERS

Bang, 1966
Barrat, 1974*a*
Elliott, 1971
Gibson-Hill, 1949
Mougin, 1970*c*, 1971
Richdale, 1954, 1963
Rowan et al., 1951
Segonzac, 1970
Southern, 1951
Watson, 1971
Woods, 1970

STORM PETRELS

Bang, 1966
Beck, 1970
Beck and Brown, 1971, 1972
Brodkorb, 1963
Huber, 1971
Lacan, 1971
Mörzer Bruyns and Voous, 1964
Mougin, 1968
Murphy, 1918
Roberts, 1940*a*
Serventy, 1952*b*

DIVING PETRELS

Bourne, 1968
Murphy and Harper, 1921
Richdale, 1945
Thoresen, 1969

CORMORANTS

Behn et al., 1955
Murphy, 1916*a*
Rand, 1956
Voisin, 1970

DUCKS

Delacour, 1956
Johnsgard, 1965
Kidder and Coues, 1875
Murphy, 1916*b*
Phillips, 1923
Spenceley, 1958
Weller, 1972, 1975
Weller and Howard, 1972

RAILS

Beintema, 1972
Eber, 1961
Wilson and Swales, 1958

SHEATHBILLS

Jones, 1963
Kidder and Coues, 1876

SKUAS

Brook and Beck, 1972
Burton, 1968*a*, *b*, 1970
Dalinger and Freytag, 1960
Eklund, 1961
Johnston, 1973
Kuroda, 1962

Le Morvan et al., 1967
Reid, 1966
Spellerberg, 1969, 1970, 1971*a*
Stonehouse, 1956
Young, 1963*a*, *b*

GULLS

Fordham, 1964
Kinsky, 1963

TERNS

Courtenay-Latimer, 1957
Downes, 1952
Gain, 1914
Kullenberg, 1946
Murphy, 1938
Salomonsen, 1967

LAND BIRDS

Murphy, 1923

STUDY METHODS, BANDING, AND PHYSIOLOGY

Anderlini et al., 1972
Austin, 1957
Bauer, 1964, 1967
Boswall and Prytherch, 1969
Chippaux et al., 1972
Douglas, 1968
Eklund, 1961
Emison, 1968
Emlen and Penney, 1964
Gibson, 1963, 1967
Hudson, 1966
Isenmann et al., 1971
Jarman, 1973
King et al., 1967
Kooyman et al., 1971
LeResche and Sladen, 1970
Margni and Castrelos, 1963
Mougin, 1972, 1974
Penney and Sladen, 1966
Prévost and Sapin-Jaloustre, 1964
Prévost and Vilter, 1963
Risebrough and Carmignani, 1972
Routh, 1949
Shaughnessy, 1970
Sladen, 1958
Sladen and LeResche, 1970
Sladen et al., 1968
Spellerberg, 1969
Stonehouse, 1967
Tickell, 1967
Tickell and Gibson, 1968
Tickell and Scotland, 1961
Trawa, 1970
Warham, 1971*b*
Watson and Amerson, 1967

SOUTH AMERICA, FALKLAND ISLANDS, AND TIERRA DEL FUEGO

Bennett, 1926
Cooke and Mills, 1972
Humphrey et al., 1970
Jehl, 1973, 1974
Johnson, 1965, 1967
Murphy, 1936
Olrog, 1963
Pisano V., 1972
Reynolds, 1935

SOUTH GEORGIA

Lönnberg, 1906
Matthews, 1929
Morris, 1962
Rankin, 1951

SOUTH SANDWICH ISLANDS

Baker et al., 1964
Holdgate, 1963*b*
Kemp and Nelson, 1931
Wilkinson, 1956, 1957

SOUTH ORKNEY ISLANDS

Ardley, 1936
Bennett, 1920
Clarke, 1906

SOUTH SHETLAND ISLANDS

Araya and Aravena, 1965
Bennett, 1920
Furse and Bruce, 1971
Gain, 1914
Pefaur and Murua, 1972
Racovitza, 1900
Stroud, 1953

ANTARCTIC PENINSULA

Araya, 1965
Bagshawe, 1939
Eklund, 1945
Friedmann, 1945
Gain, 1914
Holdgate, 1963*a*
Racovitza, 1900

ANTARCTIC CONTINENT

Ardus, 1964
Austin, 1957
Bougaeff, 1974
Brook and Beck, 1972
Caughley, 1960*b*
Cendron, 1953
Dow and Neall, 1968
Etchécopar and Prévost, 1954
Falla, 1937
Friedmann, 1945
Haga, 1961
Hanson, 1902
Isenmann et al., 1971
Konovalov, 1964
Korotkevich, 1959, 1964
Korotkevich and Ledenev, 1970
La Grange, 1962
Lowe and Kinnear, 1930
Loy, 1962
Orton, 1963
Perkins, 1945
Prévost, 1963
Pryor, 1968
Siple and Lindsey, 1937
Spellerberg, 1971*b*

SCOTT ISLAND

Harper, 1972

BALLENY ISLANDS

Dawson et al., 1965
Hatherton et al., 1965

PETER I ISLAND

Holgersen, 1951, 1959

TRISTAN DA CUNHA GROUP AND GOUGH ISLAND

Broekhuysen and Macnae, 1949
Elliott, 1953, 1957
Elliott, 1970
Hagen, 1952
Holdgate, 1958, 1965*b*
Rand, A. L., 1955
Swales, 1965

BOUVETØYA

Holdgate et al., 1968
Holgersen, 1960
Solyanik, 1964

MARION AND PRINCE EDWARD ISLANDS

Crawford, 1952
La Grange, 1962
Rand, 1954
van Zinderen Bakker, 1971*a*

ILES CROZET

Barrat, 1974*b*
Despin et al., 1972
Dreux and Milon, 1967
Milon, 1962
Mougin, 1969*a*, *b*
Prévost, 1970

ILE AMSTERDAM AND ILE SAINT-PAUL

Jouanin, 1953
Jouanin and Paulian, 1954, 1960
Paulian, 1953, 1956
Segonzac, 1972
Vanhöffen, 1912
Vélain, 1877

ILES KERGUELEN

Derenne et al., 1972, 1974
Hall, 1900
Kidder and Coues, 1875
Loranchet, 1915-1916
Milon and Jouanin, 1953
Paulian, 1953

HEARD AND MCDONALD ISLANDS

Budd, 1972
Downes et al., 1959
Ealey, 1954*a*, *b*

MACQUARIE ISLAND

Carrick, 1957
Gillham, 1967
Law and Burstall, 1953
MacKenzie, 1968
Merilees, 1971*a*
Warham, 1969

AUSTRALIA AND NEW ZEALAND

Falla et al., 1966
Oliver, 1955
Ornithological Society of New Zealand, 1970
Serventy et al., 1971
Slater, 1970
Warham and Keeley, 1969

AT SEA

Atlantic Sector

Bierman and Voous, 1950
Clarke, 1907
Cline et al., 1969
Dabbene, 1921-1926
Lathbury, 1973
Murphy, 1914
Novatti, 1962
Olrog, 1958
Tickell and Woods, 1972
van Oordt, 1939
van Oordt and Kruijt, 1953, 1954

Indian Ocean Sector

Falla, 1937
Gill, 1967
Loy, 1962
Paulian, 1953
Rand, 1962, 1963
Routh, 1949
van Zinderen Bakker, 1971*c*

Pacific Sector Including Australian and New Zealand Waters

Bierman and Voous, 1950
Darby, 1970
Dell, 1960
Fleming, 1950
Gain, 1914
Harrison, 1962
Holgersen, 1957
Norris, 1965
Spellerberg, 1971*b*
Szijj, 1967

REFERENCES

Ainley, D. G., Flocking in Adélie penguins, *Ibis, 114,* 388-390, 1972.

Ainley, D. G., and R. P. Schlatter, Chick raising ability in Adélie penguins, *Auk, 89,* 559-566, 1972.

Alexander, W. B., *Birds of the Ocean,* 2nd rev. ed., 306 pp., Putnam, New York, 1954.

American Geographical Society, Antarctica, 1:5,000,000, New York, 1965.

Anderlini, V. C., P. G. Connors, R. W. Risebrough, and J. H. Martin, Concentrations of heavy metals in some antarctic and North American sea birds, in *Proceedings of the Colloquium on Conservation Problems in Antarctica,* edited by B. C. Parker, pp. 49-61, Virginia Polytechnic Institute and State University, Blacksburg, 1972.

Andersson, K. A., Das hörere Tierleben im antarktischen Gebiete, *Wiss. Ergeb. Schwed. Südpolarexpedition 1901-1903, 5, Zool., 1,* 1-58, 1905.

Araya, B., Notas preliminares sobre ornitología de la Antártica chilena, *Rev. Biol. Mar., 12,* 161-174, 1965.

Araya, B., and W. Aravena, Las aves de Punta Armonía, Isla Nelson, Antártica chilena; Censo y distribución, *Publ. Inst. Antartico Chileno, 7,* 1-18, 1965.

Ardley, R. A. B., The birds of the South Orkney Islands, *Discovery Rep., 12,* 349-376, 1936.

Ardus, D. A., Some observations at the Tottanfjellà, Dronning Maud Land, *Brit. Antarctic Surv. Bull., 3,* 17-20, 1964.

Arnaud, P., Observations écologiques à la colonie de manchots empereurs de Pointe Géologie (Terre Adélie) en 1962, *Oiseau, 34,* no. spec., 2-32, 1964.

Arnaud, P., F. Arnaud, and J.-C. Hureau, Bibliographie générale de biologie antarctique et subantarctique, *Com. Nat. Fr. Rech. Antarctiques, 18,* 1-180, 1967.

Austin, O. L., Jr., Notes on banding birds in Antarctica, and on the Adélie penguin colonies of the Ross Sea sector, *Bird-Banding, 28,* 1-26, 1957.

Bagshawe, T. W., Notes on the habits of the gentoo and ringed or antarctic penguins, *Trans. Zool. Soc. London, 24,* 185-306, 1938.

Bagshawe, T. W., *Two Men in the Antarctic,* pp. 254-283, Cambridge University Press, London, 1939.

Baker, P. E., et al., A survey of the South Sandwich Islands, *Nature, 203,* 691-693, 1964.

Bang, B. G., Anatomical adaptations for olfaction in the snow petrel, *Nature, 205,* 513-515, 1965.

Bang, B. G., The olfactory apparatus of tubenosed birds (Procellariiformes), *Acta Anat., 65,* 391-415, 1966.

Barrat, A., Note sur le pétrel gris *Procellaria cinerea, Com. Nat. Fr. Rech. Antarctiques, 33,* 19-24, 1974*a*.

Barrat, A., Note sur les oiseaux visiteurs de l'Ile Crozet (46°25'S, 51°45'E), *Com. Nat. Fr. Rech. Antarctiques, 33,* 25-27, 1974*b*.
Barrat, A., and J.-L. Mougin, Données numeriques sur la zoogéographie de l'avifaune antarctique et subantarctique, *Com. Nat. Fr. Rech. Antarctiques, 33,* 1-18, 1974.
Bauer, A., Utilisation de la photographie verticale à l'étude ornithologique des îles australes; Dénombrement des manchotières de l'Ile aux Cochons (Archipel des Crozet) et de l'Ile de Kerguelen, *Terres Australes Antarctiques Fr., 25,* 34-38, 1964.
Bauer, A., Dénombrement des manchotières de l'Archipel des Crozet et des Iles Kerguelen à l'aide de photographies aériennes verticales, *Terres Australes Antarctiques Fr., 41,* 3-21, 1967.
Beck, J. R., Unusual birds at Signy Island, South Orkney Islands, 1966-67, *Brit. Antarctic Surv. Bull., 18,* 81-82, 1968*a*.
Beck, J. R., An early record of a pomarine skua in Marguerite Bay, Graham Land, *Brit. Antarctic Surv. Bull., 18,* 83, 1968*b*.
Beck, J. R., Food, moult and age of first breeding in the Cape pigeon, *Daption capensis* Linnaeus, *Brit. Antarctic Surv. Bull., 21,* 33-44, 1969.
Beck, J. R., Breeding seasons and moult in some smaller antarctic petrels, in *Antarctic Ecology,* vol. 1, edited by M. W. Holdgate, pp. 542-550, Academic, New York, 1970.
Beck, J. R., and D. W. Brown, The breeding biology of the black-bellied storm-petrel *Fregetta tropica, Ibis, 113,* 73-90, 1971.
Beck, J. R., and D. W. Brown, The biology of Wilson's storm petrel, *Oceanites oceanicus* (Kuhl), at Signy Island, South Orkney Islands, *Brit. Antarctic Surv. Sci. Rep., 69,* 1-54, 1972.
Behn, F., J. D. Goodall, A. W. Johnson, and R. A. Philippi B., The geographic distribution of the blue-eyed shags, *Phalacrocorax albiventer* and *Phalacrocorax atriceps, Auk, 72,* 6-13, 1955.
Beintema, A. J., The history of the island hen (*Gallinula nesiotis*), the extinct flightless gallinule of Tristan da Cunha, *Bull. Brit. Ornithol. Club, 92,* 106-115, 1972.
Bennett, A. G., Breves notas sobre las aves antârticas, *Hornero, 2,* 25-34, 1920.
Bennett, A. G., Notas sobre aves sub-antârticas, *Hornero, 2,* 255-258, 1922.
Bennett, A. G., A list of birds of the Falkland Islands and dependencies, *Ibis, 68,* 306-333, 1926.
Bierman, W. H., and K. H. Voous, Birds observed and collected during the whaling expeditions of the 'Willem Barendsz' in the Antarctic, 1946-1947 and 1947-1948, *Ardea,* extra no., 1-123, 1950.
Boswall, J., and R. J. Prytherch, A discography of bird sounds from the Antarctic, *Polar Rec., 14,* 603-612, 1969.
Bougaeff, S., Observations écologiques à la colonie de manchots empereurs de Pointe Géologie (Terre Adélie) en 1970, *Com. Nat. Fr. Rech. Antarctiques, 33,* 89-110, 1974.
Bourne, W. R. P., On the status and appearance of the races of

Cory's shearwater *Procellaria diomedea, Ibis, 97,* 145-149, 1955.

Bourne, W. R. P., Notes on the diving petrels, *Bull. Brit. Ornithol. Club, 88,* 77-85, 1968.

Bourne, W. R. P., and J. Warham, Geographical variation in the giant petrels of the genus *Macronectes, Ardea, 54,* 45-67, 1966.

Brodkorb, P., Catalogue of fossil birds, 1, Archaeopterygiformes through Ardeiformes, *Bull. Fla. State Mus., 7,* 179-293, 1963.

Broekhuysen, G. J., and W. Macnae, Observations on the birds of Tristan da Cunha islands and Gough Island in February and early March 1948, *Ardea, 37,* 97-113, 1949.

Brook, D., and J. R. Beck, Antarctic petrels, snow petrels and south polar skuas breeding in the Theron Mountains, *Brit. Antarctic Surv. Bull., 27,* 131-137, 1972.

Brown, D. A., Breeding biology of the snow petrel, *Pagodroma nivea* (Forster), *A.N.A.R.E. Rep., Ser. B, 89,* 1-63, 1966.

Budd, G. M., The biotopes of emperor penguin rookeries, *Emu, 61,* 171-189, 1961.

Budd, G. M., Population studies in rookeries of the emperor penguin, *Aptenodytes forsteri, Proc. Zool. Soc. London, 139,* 365-388, 1962.

Budd, G. M., Population increase in the king penguin (*Aptenodytes patagonica*) at Heard Island, *Auk, 85,* 689-690, 1968.

Budd, G. M., Further population growth in Heard Island king penguins, *Auk, 87,* 366-367, 1970.

Budd, G. M., McDonald Island reconnaissance, 1971, *Polar Rec., 16,* 64-67, 1972.

Budd, G. M., and M. C. Downes, Recolonization of Heard Island by the king penguin, *Aptenodytes patagonica, Emu, 64,* 302-316, 1965.

Burton, R. W., Stray birds at Signy Island, South Orkney Islands, *Brit. Antarctic Surv. Bull., 11,* 101-102, 1967.

Burton, R. W., Breeding biology of the brown skua, *Catharacta skua lönnbergi* (Mathews) at Signy Island, South Orkney Islands, *Brit. Antarctic Surv. Bull., 15,* 9-28, 1968*a*.

Burton, R. W., Agonistic behaviour of the brown skua, *Catharacta skua lönnbergi* (Mathews), *Brit. Antarctic Surv. Bull., 16,* 15-39, 1968*b*.

Burton, R. W., Biology of the great skua, in *Antarctic Ecology,* vol. 1, edited by M. W. Holdgate, pp. 561-567, Academic, New York, 1970.

Cabanis, J., and A. Reichenow, Uebersicht der auf der Expedition Sr. Maj. Schiff 'Gazelle' gesammelten Vögel, *J. Ornithol., 24,* 319-330, 1876.

Carins, M., Facial characteristics of rockhopper penguins, *Emu, 74,* 55-57, 1974.

Carrick, R., The wildlife of Macquarie Island, *Aust. Mus. Mag., 12,* 255-260, 1957.

Carrick, R., and S. E. Ingham, Antarctic sea-birds as subjects for ecological research, Proceedings of the Symposium on Pacific-

Antarctic Sciences, Tokyo, 1966, *JARE Sci. Rep., Spec. Issue 1,* 151-184, 1967.
Carrick, R., and S. E. Ingham, Ecology and population dynamics of antarctic seabirds, in *Antarctic Ecology,* vol. 1, edited by M. W. Holdgate, pp. 505-525, Academic, New York, 1970.
Cassin, J., *United States Exploring Expedition During the Years 1838, 1839, 1840, 1841, 1842 Under the Command of Charles Wilkes, U.S.N., Mammalogy and Ornithology,* J. B. Lippincott, Philadelphia, Pa., 1858.
Caughley, G., The Cape Crozier emperor penguin rookery, *Rec. Dominion Mus., 3,* 251-262, 1960*a*.
Caughley, G., The Adélie penguins of Ross and Beaufort seas, *Rec. Dominion Mus., 3,* 263-282, 1960*b*.
Cendron, J., Notes sur les oiseaux de la Terre Adélie, *Oiseau, 23,* 212-220, 1953.
Chippaux, A., et al., Enquête serologique chez des oiseaux des Iles Kerguelen vis à vis des certains arbovirus, in *Transcontinental Connections of Migratory Birds and Their Role in the Distribution of Arboviruses,* edited by A. I. Cherepanov, pp. 181-183, Nauka, Novosibirsk, USSR, 1972.
Clarke, W. E., Ornithological results of the Scottish National Antarctic Expedition, 2, On the birds of the South Orkney Islands, *Ibis, 48,* 145-187, 1906.
Clarke, W. E., Ornithological results of the Scottish National Antarctic Expedition, 3, On the birds of the Weddell and adjacent seas, Antarctic Ocean, *Ibis, 49,* 325-349, 1907.
Cline, D. R., D. B. Siniff, and A. W. Erickson, Summer birds of the pack ice in the Weddell Sea, Antarctica, *Auk, 86,* 701-716, 1969.
Conference on Antarctica, Antarctic Treaty, Washington, D. C., Dec. 1, 1959.
Conroy, J. W. H., Wilson's phalarope (*Steganopus tricolor*) in the Antarctic, *Brit. Antarctic Surv. Bull., 26,* 82-83, 1971*a*.
Conroy, J. W. H., The white-phase giant petrel of the South Orkney Islands, *Brit. Antarctic Surv. Bull., 24,* 113-115, 1971*b*.
Conroy, J. W. H., Ecological aspects of the biology of the giant petrel, *Macronectes giganteus* (Gmelin), in the maritime Antarctic, *Brit. Antarctic Surv. Sci. Rep., 75,* 1-74, 1972.
Conroy, J. W. H., and E. L. Twelves, Diving depths of the gentoo penguin (*Pygoscelis papua*) and blue-eyed shag (*Phalacrocorax atriceps*) from the South Orkney Islands, *Brit. Antarctic Surv. Bull., 30,* 106-108, 1972.
Conroy, J. W. H., and M. G. White, The breeding status of the king penguin (*Aptenodytes patagonica*), *Brit. Antarctic Surv. Bull., 32,* 31-40, 1973.
Cooke, F., and E. L. Mills, Summer distribution of pelagic birds off the coast of Argentina, *Ibis, 114,* 245-251, 1972.
Courtenay-Latimer, M., On the breeding of *Sterna vittata tristanensis* Murphy off the coast of Cape Province, South

Africa, *Bull. Brit. Ornithol. Club, 77,* 82-83, 1957.

Crawford, A. B., The birds of Marion Island, south Indian Ocean, *Emu, 52,* 73-85, 1952.

Dabbene, R., Los petreles y los albatros del Atlantico austral, *Hornero, 2*(3), 157-179; *2*(4), 241-254; *3*(1), 1-33; *3*(2), 125-158; *3*(3), 227-238; *3*(4), 311-348; 1921-1926.

Dabbene, R., Captura de un batitú (*Bartramia longicauda*) en las Islas Shetland del Sud, *Hornero, 3,* 197, 1923.

Dalinger, R. E., and O. Freytag, Observaciones sobre el skua polar del sur en Bahía Margarita, *Inst. Antartico Argent. Contrib., 51,* 1-19, 1960.

Darby, M. M., Summer seabirds between New Zealand and McMurdo Sound, *Notornis, 17,* 28-55, 1970.

Dawson, E. W., et al., Balleny Islands Reconnaissance Expedition, 1964, *Polar Rec., 12,* 431-435, 1965.

Delacour, J., *The Waterfowl of the World,* vol. 1, pp. 216-221; vol. 2, pp. 60-65, 124-135; Country Life, London, 1956.

Dell, R. K., The blue petrel in Australasian waters, *Emu, 52,* 147-154, 1952.

Dell, R. K., Sea-bird logs between New Zealand and the Ross Sea, *Rec. Dominion Mus., 3,* 293-305, 1960.

Derenne, P., J. Prévost, and M. van Beveren, Note sur le baguage des oiseaux dans l'Archipel de Kerguelen depuis 1951, *Oiseau, 42,* no. spec., 111-129, 1972.

Derenne, P., J. X. Lufbery, and B. Tollu, L'avifaune de l'Archipel Kerguelen, *Com. Nat. Fr. Rech. Antarctiques, 33,* 57-87, 1974.

Despin, B., Note préliminaire sur le manchot papou *Pygoscelis papua* de l'Ile de la Possession (Archipel Crozet), *Oiseau, 42,* no. spec., 69-83, 1972.

Despin, B., J. L. Mougin, and M. Segonzac, Oiseaux et mammifères de l'Ile de l'Est, *Com. Nat. Fr. Rech. Antarctiques, 31,* 1-106, 1972.

Dixon, C. C., Some observations on the albatrosses and other birds of the southern oceans, *Trans. Roy. Can. Inst., 19,* 117-139, 1933.

Douglas, D. S., Salt and water metabolism of the Adélie penguin, in *Antarctic Bird Studies, Antarctic Res. Ser.,* vol. 12, edited by O. L. Austin, Jr., pp. 167-190, AGU, Washington, D. C., 1968.

Dow, J. A. S., and V. E. Neall, Biological observations from the Rennick glacier region, Antarctica 1967-68, *Notornis, 15,* 117-119, 1968.

Downes, M. C., Arctic terns in the Subantarctic, *Emu, 52,* 306-310, 1952.

Downes, M. C., Size variation in eggs and young of the macaroni penguin, *Emu, 55,* 19-23, 1955.

Downes, M. C., E. H. M. Ealey, A. M. Gwynn, and P. S. Young, The birds of Heard Island, *A.N.A.R.E. Rep., Ser. B, 1,* 1-135, 1959.

Dreux, P., and P. Milon, Premières observations sur l'avifaune de l'Ile-aux-Cochons (Archipel Crozet), *Alauda, 35,* 27-32, 1967.

Ealey, E. H. M., Ecological notes on the birds of Heard Island, *Emu, 54,* 91-112, 1954*a*.

Ealey, E. H. M., Analysis of stomach contents of some Heard Island birds, *Emu, 54,* 204-210, 1954*b*.

Eber, G., Vergleichende Untersuchungen am flugfähigen Teichhuhn *Gallinula chl. choloropus* und an der flugunfähigen Inselralle *Gallinula nesiotes, Bonner Zool. Beitr., 12,* 247-315, 1961.

Eights, J., Description of a new crustaceous animal found on the shores of the South Shetland Islands, with remarks on their natural history, *Trans. Albany Inst., 2,* 53-69, 1833.

Eklund, C. R., Condensed ornithology report, East Base, Palmer Land, *Proc. Amer. Phil. Soc., 89,* 299-304, 1945.

Eklund, C. R., Distribution and life-history studies of the south-polar skua, *Bird-Banding, 32,* 187-223, 1961.

Elliott, C. C. H., Additional note on the seabirds of Gough Island, *Ibis, 112,* 112-114, 1970.

Elliott, C. C. H., Ecological considerations and the possible significance of weight variations in the chicks of the great shearwater on Gough Island, *Ostrich,* suppl. 8, 385-396, 1971.

Elliott, H. F. I., The fauna of Tristan da Cunha, *Oryx, 2,* 41-53, 1953.

Elliott, H. F. I., A contribution to the ornithology of the Tristan da Cunha group, *Ibis, 99,* 545-586, 1957.

Emison, W. B., Feeding preferences of the Adélie penguin at Cape Crozier, Ross Island, in *Antarctic Bird Studies, Antarctic Res. Ser.,* vol. 12, edited by O. L. Austin, Jr., pp. 191-212, AGU, Washington, D. C., 1968.

Emlen, J. T., and R. L. Penney, Distance navigation in the Adélie penguin, *Ibis, 106,* 417-431, 1964.

Etchécopar, R. D., and J. Prévost, Données oölogiques sur l'avifaune de Terre Adélie, *Oiseau, 24,* 227-247, 1954.

Falla, R. A., Birds, *Brit. Aust. N. Z. Antarctic Res. Exped. 1929-1931 Rep., Ser. B, 2,* 1-304, 1937.

Falla, R. A., The genus *Pachyptila* Illiger, *Emu, 40,* 218-236, 1940.

Falla, R. A., Antarctic birds, in *The Antarctic Today,* edited by F. A. Simpson, pp. 216-228, A. H. and A. W. Reed, Wellington, 1952.

Falla, R. A., Distribution patterns of birds in the Antarctic and high latitude Subantarctic, in *Biologie Antarctique, Proceedings of the First SCAR Symposium on Antarctic Biology,* edited by R. Carrick, M. W. Holdgate, and J. Prévost, pp. 367-376, Hermann, Paris, 1964.

Falla, R. A., R. B. Sibson, and E. G. Turbott, *A Field Guide to the Birds of New Zealand and Outlying Islands,* 254 pp., Collins, London, 1966.

Falla, R. A., C. A. Fleming, and F. C. Kinsky, *Notornis, 18,* 64-66, 1971.

Fleming, C. A., The phylogeny of the prions, *Emu, 41,* 134-155, 1941.

Fleming, C. A., Some South Pacific sea-bird logs, *Emu, 49,* 169-188, 1950.

Fordham, R. A., Breeding biology of the southern black-backed gull, *Notornis, 11,* 3-34, 110-126, 1964.

Friedmann, H., Birds of the United States Antarctic Service Expedition 1939-1941, *Proc. Amer. Phil. Soc., 89,* 305-313, 1945.

Fullagar, P. J., Notes on the races of prions, *Aust. Bird Bander, 10,* 35, 1972*a.*

Fullagar, P. J., Identification of prions—*Pachyptila* spp., *Aust. Bird Bander, 10,* 36-39, 1972*b.*

Furse, J. R., and G. Bruce, Ornithology report, in *Joint Services Expedition: Elephant Island 1970-71,* edited by M. K. Burley, Annex F, F1-F11, Ministry of Defence, London, 1971.

Gain, L., Oiseaux antarctiques, *Deuxieme Exped. Antarctique Fr. 1908-1910, 2,* 1-200, 1914.

Geographic Names Division, *Antarctica,* 3rd ed., U.S. Army Topographic Command, Washington, D. C., 1969.

Gibson, J. D., Fork-tailed swift at Macquarie Island, *Emu, 59,* 64, 1959.

Gibson, J. D., Third report of the New South Wales albatross study group (1962) summarizing activities to date, *Emu, 63,* 215-223, 1963.

Gibson, J. D., The wandering albatross (*Diomedea exulans*): Results of banding and observations in New South Wales coastal waters and the Tasman Sea, *Notornis, 14,* 47-57, 1967.

Gibson-Hill, C. A., Notes on the Cape hen *Procellaria aequinoctialis, Ibis, 91,* 422-426, 1949.

Gill, F. B., Observations on the pelagic distribution of seabirds in the western Indian Ocean, *Proc. U.S. Nat. Mus., 123,* 1-33, 1967.

Gillham, M. E., *Sub-Antarctic Sanctuary, Summertime on Macquarie Island,* 223 pp., Victor Gollancz, London, 1967.

Guillard, R., and J. Prévost, Observations écologiques à la colonie de manchots empereurs de Pointe Géologie (Terre Adélie) en 1963, *Oiseau, 34,* no. spec., 33-51, 1964.

Gwynn, A. M., The egg-laying and incubation periods of rockhopper, macaroni, and gentoo penguins, *A.N.A.R.E. Rep., Ser. B, 1,* 1-29, 1953*a.*

Gwynn, A. M., Some additions to the Macquarie Island list of birds, *Emu, 53,* 150-152, 1953*b.*

Haga, R., Birds and seals around Japanese Syowa base on Prince Harald Coast, Antarctica (Preliminary), *Nankyoku Shiryo (Antarctic Rec.), 11,* 146-148, 1961.

Hagen, Y., Birds of Tristan da Cunha, *Results Norw. Sci. Exped. Tristan da Cunha 1937-1938, 20,* 1-248, 1952.

Hall, R., Field notes on the birds of Kerguelen Island, *Ibis, 42,* 1-34, 1900.

Hanson, N., Extracts from the private diary of the late Nicolai Hanson, in *Report on the Collections of Natural History Made in the Antarctic Regions During the Voyage of the 'Southern Cross,'* edited by R. B. Sharpe, pp. 79-105, British Museum (Natural History), London, 1902.

Harper, P. C., The field identification and distribution of the thin-billed prion (*Pachyptila belcheri*) and the Antarctic prion (*Pachyptila desolata*), *Notornis, 19,* 140-175, 1972.

Harper, P. C., The field identification and supplementary notes on the soft-plumaged petrel (*Pterodroma mollis* Gould, 1844), *Notornis, 20,* 193-201, 1973.

Harper, P. C., G. E. Watson, and J. P. Angle, New records of the Kerguelen petrel (*Pterodroma brevirostris*) in the South Atlantic and Pacific oceans, *Notornis, 19,* 56-60, 1972.

Harrison, P. P. O., *Sea Birds of the South Pacific, A Handbook for Passengers and Seafarers,* 144 pp., Royal Naval Bird Watching Society, Narberth, England, 1962.

Hatherton, T., E. W. Dawson, and F. C. Kinsky, Balleny Islands Reconnaissance Expedition, 1964, *N. Z. J. Geol. Geophys., 8,* 164-179, 1965.

Holdgate, M., *Mountains in the Sea; The Story of the Gough Island Expedition,* St. Martin's Press, New York, 1958.

Holdgate, M. W., Observations of birds and seals at Anvers Island, Palmer Archipelago, in 1955-57, *Brit. Antarctic Surv. Bull., 2,* 45-51, 1963*a*.

Holdgate, M. W., Observations in the South Sandwich Islands, 1962, *Polar Rec., 11,* 394-405, 1963*b*.

Holdgate, M. W., Occurrence of stray land birds in Drake Passage and the South Orkney Islands, *Brit. Antarctic Surv. Bull., 6,* 77, 1965*a*.

Holdgate, M. W., The fauna of the Tristan da Cunha islands, 3, The Biological Report of the Royal Society Expedition to Tristan da Cunha 1962, *Phil. Trans. Roy. Soc. London, Ser. B, 249,* 361-402, 1965*b*.

Holdgate, M. W., P. J. Tilbrook, and R. W. Vaughan, Biology of Bouvetøya, *Brit. Antarctic Surv. Bull., 15,* 1-7, 1968.

Holgersen, H., Antarctic and sub-antarctic birds, *Sci. Result. Norw. Antarctic Exped. 1927-28, 2*(23), 1-100, 1945.

Holgersen, H., On the birds of Peter I Island, *Proc. 10th Int. Ornithol. Congr. Uppsala 1950,* 614-616, 1951.

Holgersen, H., Ornithology of the 'Brategg' Expedition, *Sci. Result. 'Brategg' Exped. 1947-48, 4,* 1-80, 1957.

Holgersen, H., Fugl i norske biland, 1, Peter I's Oy, *Sterna, 3,* 215-223, 1959.

Holgersen, H., Fugl i norske biland, 2, Bouvet-øya, *Sterna, 4,* 1-9, 1960.

Huber, L. N., Notes on the migration of the Wilson's storm petrel *Oceanites oceanicus* near Eniwetok Atoll western Pacific Ocean, *Notornis, 18,* 38-42, 1971.

Hudson, R., Adult survival estimates for two antarctic petrels, *Brit. Antarctic Surv. Bull., 8,* 63-73, 1966.

Hudson, R., The great skua in the Caribbean, *Bird Study, 15,* 33-34, 1968*a.*

Hudson, R., The white-phase giant petrels of the South Orkney Islands, *Ardea, 56,* 178-183, 1968*b.*

Humphrey, P. S., D. Bridge, P. W. Reynolds, and R. T. Peterson, *Preliminary Smithsonian Manual; Birds of Isla Grande (Tierra del Fuego)*, 411 pp., Smithsonian Institution, Washington, D. C., 1970.

Imber, M. J., The food of grey faced petrels (*Pterodroma macroptera gouldi* (Hutton)), with special reference to diurnal vertical migration of their prey, *J. Anim. Ecol., 42,* 645-662, 1973.

Isenmann, P., Contribution à la biologie de reproduction du pétrel des neiges (*Pagodroma nivea* Forster); Le problème de la petite et de la grande forme, *Oiseau, 40,* no. spec., 99-134, 1970*a.*

Isenmann, P., Note sur la biologie de reproduction comparée de damiers du Cap *Daption capensis* aux Orcades du Sud et en Terre Adélie, *Oiseau, 40,* no. spec., 135-141, 1970*b.*

Isenmann, P., Contribution à l'éthologie et à l'écologie du manchot empereur (*Aptenodytes forsteri* Gray) à la colonie de Pointe Géologie (Terre Adélie), *Oiseau, 41,* no. spec., 9-64, 1971.

Isenmann, P., and E. P. Jouventin, Eco-éthologie du manchot empereur (*Aptenodytes forsteri*) et comparaison avec le manchot Adélie (*Pygoscelis adeliae*) et le manchot royal (*Aptenodytes patagonica*) conséquences du problème du territoire sur l'organisation sociale à la colonie, *Oiseau, 40,* 136-159, 1970.

Isenmann, P., E. P. Jouventin, J. Prévost, and M. van Beveren, Note sur le contrôle de quelques espèces d'oiseaux bagnés en Terre Adélie de 1968 à 1970, *Oiseau, 41,* no. spec., 1-8, 1971.

Jameson, W., *The Wandering Albatross,* 99 pp., Hart-Davis, London, 1958.

Jarman, M., Experiments on the emperor penguin, *Aptenodytes forsteri,* in various thermal environments, *Brit. Antarctic Surv. Bull., 33-34,* 57-63, 1973.

Jeffries, C. J. S., A vagrant American egret at South Georgia, *Brit. Antarctic Surv. Bull., 6,* 77-78, 1965.

Jehl, J. R., Jr., The distribution of marine birds in Chilean waters in winter, *Auk, 90,* 114-135, 1973.

Jehl, J. R., Jr., The distribution and ecology of marine birds over the continental shelf of Argentina in winter, *Trans. San Diego Soc. Natur. Hist., 17,* 217-234, 1974.

Jenkins, J., Does the great shearwater reach the southwest Pacific?, *Notornis, 15,* 214-215, 1968.

Johnsgard, P. A., *Handbook of Waterfowl Behavior,* 378 pp., Cornell University Press, Ithaca, N. Y., 1965.

Johnson, A. W., *The Birds of Chile and Adjacent Regions of Argentina, Bolivia and Peru,* vol. 1, 398 pp., Platt, Buenos Aires, 1965.

Johnson, A. W., *The Birds of Chile and Adjacent Regions of Argentina, Bolivia and Peru,* vol. 2, 447 pp., Platt, Buenos Aires, 1967.

Johnston, G. C., Predation by southern skua on rabbits on Macquarie Island, *Emu, 73,* 25-26, 1973.

Johnstone, G. W., Field characters and behaviour at sea of giant petrels in relation to their oceanic distribution, *Emu, 74,* 209-218, 1974.

Johnstone, G. W., and M. D. Murray, Dominican gulls in Australian antarctic territory, *Aust. Bird Bander, 10,* 59-60, 1972.

Jones, N. V., The sheathbill, *Chionis alba* (Gmelin), at Signy Island, South Orkney Islands, *Brit. Antarctic Surv. Bull., 2,* 53-71, 1963.

Jouanin, C., Le matériel ornithologue de la mission 'Passage de Vénus sur le Soleil' (1874), Station de l'Ile Saint-Paul, *Bull. Mus. Nat. Hist. Natur., Ser. 2, 25,* 529-540, 1953.

Jouanin, C., and P. Paulian, Migrateurs continentaux dans les îles Nouvelle-Amsterdam et Kerguelen, *Oiseau, 24,* 136-142, 1954.

Jouanin, C., and P. Paulian, Recherches sur des ossements d'oiseaux provenant de l'Ile Nouvelle-Amsterdam (Océan Indian), *Proc. 12th Int. Ornithol. Congr. Helsinki 1958,* 368-372, 1960.

Jouanin, C., and J. Prévost, Captures de manchots innattendus en Terre Adélie et considerations systématiques sur *E. chrysolophus schlegeli* Finsch, *Oiseau, 23,* 281-287, 1953.

Jouventin, P., Comportement et structure sociale chez le manchot empereur, *Terre Vie, 25,* 510-586, 1971.

Keith, K., and M. P. Hines, New and rare species of birds at Macquarie Island during 1956 and 1957, *CSIRO Wildlife Res., 3,* 50-53, 1958.

Kemp, S., and A. L. Nelson, The South Sandwich Islands, *Discovery Rep., 3,* 133-198, 1931.

Kidder, J. H., and E. Coues, Contributions to the natural history of Kerguelen Island, 1, Ornithology, *U.S. Nat. Mus. Bull., 2,* 1-51, 1875.

Kidder, J. H., and E. Coues, A study of *Chionis minor* with reference to its structure and systematic position, *U.S. Nat. Mus. Bull., 3,* 85-116, 1876.

King, W. B., G. E. Watson, and P. J. Gould, An application of automatic data processing to the study of seabirds, *Proc. U.S. Nat. Mus., 123,* 1-29, 1967.

Kinsky, F. C., The southern black-backed gull (*Larus dominicanus*) Lichtenstein; Measurements, plumage colour, and moult cycle, *Rec. Dominion Mus., 4,* 149-219, 1963.

Konovalov, G. V., Observations of birds in Queen Maud Land, *Sov. Antarctic Exped. Inform. Bull., 4,* 156-158, 1964.

Konovalov, G. V., and O. G. Shulyatin, Unique bird colony in Antarctica (in Russian), *Piroda, 10,* 100-101, 1964.

Kooyman, G. L., C. M. Drabek, R. Elsner, and W. B. Campbell, Diving behavior in the emperor penguin *Aptenodytes forsteri, Auk, 88,* 775-795, 1971.

Korotkevich, E. S., Birds of eastern Antarctica (in Russian), *Probl. Arktiki Antarktiki, 1,* 95-108, 1959.

Korotkevich, E. S., Observations on the birds during the first wintering of the Soviet Antarctic Expedition in 1956-1957, *Sov. Antarctic Exped. Inform. Bull., 1,* 149-152, 1964.

Korotkevich, Y. S., and V. G. Ledenev, Investigations in Enderby Land, *Sov. Antarctic Exped. Inform. Bull., 4*(33), 65-68, 1970.

Kullenberg, B., Uber Verbreitung and Wanderungen von vier *Sterna*-Arten, *Ark. Zool., 38A*(17), 1-80, 1946.

Kuroda, N., On the melanic phase of the McCormick great skua, *Yamashina Chorui Kenkyusho Kenkyu Hokoku (Misc. Rep. Yamashina Inst. Ornithol. Zool.), 3,* 212-217, 1962.

Lacan, F., Observations écologiques sur le pétrel de Wilson (*Oceanites oceanicus*) en Terre Adélie, *Oiseau, 41,* no. spec., 65-89, 1971.

La Grange, J. J., Notes on the birds and mammals on Marion Island and Antarctica (S.A.N.A.E.), *J. S. Afr. Biol. Soc., 3,* 27-84, 1962.

Lathbury, G., H.M.S. Endurance—Passage to the Antarctic—October 1969-May 1970 observation of seabirds, *Sea Swallow, 22,* 10-15, 1973.

Law, P. G., and T. Burstall, Macquarie Island, *A.N.A.R.E. Interim Rep., 14,* 1-40, 1953.

Le Morvan, P., J.-L. Mougin, and J. Prévost, Ecologie du skua antarctique (*Stercorarius skua maccormicki*) dans l'Archipel de Pointe Géologie (Terre Adélie), *Oiseau, 37,* 193-220, 1967.

LeResche, R. E., and W. J. L. Sladen, Establishment of pair and breeding site bonds by known age Adélie penguins (*Pygoscelis adeliae*), *Anim. Behav., 18,* 517-526, 1970.

Levick, G. M., *Antarctic Penguins: A Study of Their Social Habits,* 140 pp., William Heinemann, London, 1914.

Levick, G. M., Natural history of the Adélie penguin, *Natur. Hist. Rep. Brit. Antarctic Terra Nova Exped. 1910, 1*(2), 55-84, 1915.

Lindholm, E., Bar-tailed godwit at Macquarie Island, *Emu, 52,* 213, 1952.

Lönnberg, E., Contributions to the fauna of South Georgia, *Kgl. Svenska Vetenskapsakad. Handl., 40,* 1-104, 1906.

Loranchet, J., Observations biologiques sur les oiseaux des Iles Kerguelen, *Rev. Fr. Ornithol. Sci. Pract., 4,* 113-116, 153-157, 190-192, 207-210, 240-242, 256-259, 305-307, 326-331, 1915-1916.

Loveridge, A., A sheathbill, *Chionis alba* (Gmelin) on St. Helena, *Bull. Brit. Ornithol. Club, 89,* 48-49, 1969.

Løvenskiold, H. L., The snow petrel *Pagodroma nivea* in Dronning Maud Land, *Ibis, 102,* 132-134, 1960.

Lowe, P. R., and N. B. Kinnear, Birds, *Natur. Hist. Rep. Brit. Antarctic Terra Nova Exped. 1910, 4,* 103-193, 1930.

Loy, W., Ornithological profile from Iceland to Antarctica, *Gerfaut, 52,* 626-640, 1962.

Luna Perez, J. C., Visita a la roquería de pingüines emperador de Bahía Austral (Mar de Weddell), *Inst. Antartico Argent. Contrib., 70,* 1-19, 1963.

MacKenzie, D., The birds and seals of Bishop and Clerk islets, Macquarie Island, *Emu, 67,* 241-245, 1968.

Maher, W. J., Breeding biology of the snow petrel near Cape Hallett, Antarctica, *Condor, 64,* 488-499, 1962.

Margni, R. A., and O. D. Castrelos, Bacterias aisladas del arbol respiratorio superior de aves antarticas capturadas en Bahía Esperanza, *Inst. Antartico Argent. Contrib., 75,* 1-13, 1963.

Matsuda, T., Ecological observations on the breeding behaviour of Adélie penguin (*Pygoscelis adeliae*) at Ongulkaven Island near Syowa base Antarctic continent, *Nankyoku Shiryo (Antarctic Rec.), 20,* 1-7, 1964.

Matthews, G., *Pterodroma externa tristani* subsp. nov., *Bull. Brit. Ornithol. Club, 52,* 63, 1931.

Matthews, L. H., The birds of South Georgia, *Discovery Rep., 1,* 561-592, 1929.

Matthews, L. H., The origin of stomach oil in the petrels, with comparative observations on the avian proventriculus, *Ibis, 91,* 373-392. 1949.

Matthews, L. A., *Wandering Albatross: Adventures Among the Albatrosses and Petrels in the Southern Ocean,* McGibbon and Kee, London, 1951.

Merilees, W. J., Bird observations—Macquarie Island 1967, *Notornis, 18,* 55-57, 1971*a*.

Merilees, W. J., Three song thrushes at Macquarie Island, *Notornis, 18,* 87-90, 1971*b*.

Milon, P., Aspects de l'Ile de la Possession, 3, Oiseaux, mammifères, *Terres Australes Antarctiques Fr., 19-20,* 30-32, 1962.

Milon, P., Comment, in *Biologie Antarctique, Proceedings of the First SCAR Symposium on Antarctic Biology,* edited by R. Carrick, M. W. Holdgate, and J. Prévost, p. 378, Hermann, Paris, 1964.

Milon, P., and C. Jouanin, Contribution à l'ornithologie de l'Ile Kerguelen, *Oiseau, 23,* 4-53, 1953.

Morris, R. O., Bird life in N. W. South Georgia, *Sea Swallow, 15,* 43-49, 1962.

Mörzer Bruyns, W. F. J., and K. H. Voous, Notes on sea birds, 7, White-faced storm petrels (*Pelagodroma marina*) in the Indian Ocean, *Ardea, 52,* 223-224, 1964.

Mougin, J.-L., Etude écologique des deux espèces de fulmars le fulmar atlantique (*Fulmarus glacialis*) et le fulmar antarctique (*Fulmarus glacialoides*), *Oiseau, 37,* 57-103, 1967.

Mougin, J.-L., Etude écologique de quatre espèces de pétrels antarctiques, *Oiseau, 38,* no. spec., 2-52, 1968.

Mougin, J.-L., Les pétrels de l'Ile de la Possession (Archipel Crozet), *Sci. Nature, 95,* 25-35, 1969*a*.

Mougin, J.-L., Reconnaissance effectuée dans la zone sud-ouest de

l'Ile de la Possession (Iles Crozet) (Octobre 1968), *Terres Australes Antarctiques Fr., 47,* 42-53, 1969*b*.
Mougin, J.-L., Notes écologiques sur le pétrel de Kerguelen *Pterodroma brevirostris* de l'Ile de la Possession (Archipel Crozet), *Oiseau, 39,* no. spec., 58-81, 1969*c*.
Mougin, J.-L., Observations écologiques sur les grands albatros (*Diomedea exulans*) de l'Ile de la Possession (Archipel Crozet) en 1968, *Oiseau, 40,* no. spec., 16-36, 1970*a*.
Mougin, J.-L., Les albatros fuligineux *Phoebetria palpebrata* et *P. fusca* de l'Ile de la Possession (Archipel Crozet), *Oiseau, 40,* no. spec., 37-61, 1970*b*.
Mougin, J.-L., Le pétrel à menton blanc *Procellaria aequinoctialis* de l'Ile de la Possession (Archipel Crozet), *Oiseau, 40,* no. spec., 62-96, 1970*c*.
Mougin, J.-L., Note complémentaire sur le pétrel à menton blanc *Procellaria aequinoctialis* de l'Ile de la Possession (Archipel Crozet), *Oiseau, 41,* 82-83, 1971.
Mougin, J.-L., Enregistrements continus de températures internes chez quelques Spheniscidae, 1, Le manchot papou *Pygoscelis papua* de l'Ile de la Possession (Archipel Crozet), *Oiseau, 42,* no. spec., 84-110, 1972.
Mougin, J.-L., Enregistrements continus de températures internes chez quelques Spheniscidae, 2, Le manchot royal *Aptenodytes patagonica* de l'Ile de la Possession (Archipel Crozet), *Com. Nat. Fr. Rech. Antarctiques, 33,* 29-56, 1974.
Müller-Schwarze, D., Behavior of antarctic penguins and seals, in *Research in the Antarctic,* edited by L. O. Quam, pp. 259-276, American Association for the Advancement of Science, Washington, D. C., 1971.
Murphy, R. C., Observations on birds of the South Atlantic, *Auk, 31,* 439-457, 1914.
Murphy, R. C., Notes on American subantarctic cormorants, *Bull. Amer. Mus. Natur. Hist., 35,* 31-84, 1916*a*.
Murphy, R. C., Anatidae of South Georgia, *Auk, 33,* 270-277, 1916*b*.
Murphy, R. C., A study of Atlantic *Oceanites, Bull. Amer. Mus. Natur. Hist., 38,* 117-146, 1918.
Murphy, R. C., Notes sur *Anthus antarcticus, Hornero, 3,* 56-59, 1923.
Murphy, R. C., *Oceanic Birds of South America,* vols. 1, 2, 1245 pp., American Museum of Natural History, New York, 1936.
Murphy, R. C., Birds collected during the Whitney South Sea Expedition, 37, On pan-antarctic terns, *Amer. Mus. Nov., 977,* 1-17, 1938.
Murphy, R. C., Systematics and distribution of Antarctic petrels, in *Biologie Antarctique, Proceedings of the First SCAR Symposium on Antarctic Biology,* edited by R. Carrick, M. W. Holdgate, and J. Prévost, pp. 349-358, Hermann, Paris, 1964.
Murphy, R. C., and F. Harper, A review of the diving petrels, *Bull.*

Amer. Mus. Natur. Hist., 44, 495-554, 1921.
Murphy, R. C., and J. M. Pennoyer, Larger petrels of the genus *Pterodroma, Amer. Mus. Nov., 1580,* 1-43, 1952.
Napier, R. B., Erect-crested and rockhopper penguins interbreeding in the Falkland Islands, *Brit. Antarctic Surv. Bull., 16,* 71-72, 1968.
Norris, A. Y., Observations of seabirds in the Tasman Sea and in New Zealand waters in October and November, 1962, *Notornis, 12,* 80-105, 1965.
Novatti, R., Notas sobre una roquería de pingüin emperador en el Mar de Weddell, *Inst. Antartico Argent. Contrib., 34,* 1-11, 1959.
Novatti, R., Distribucion pelagica de aves en el Mar de Weddell, *Inst. Antartico Argent. Contrib., 67,* 1-22, 1962.
Oliver, W. R. B., Occurrence of the Mediterranean shearwater in New Zealand, *Emu, 34,* 23-25, 1934.
Oliver, W. R. B., *New Zealand Birds,* 2nd ed., 661 pp., A. H. and A. W. Reed, Wellington, 1955.
Olrog, C. C., Observaciones sobre avifauna antártica y de alta mar desde el Rio de la Plata hasta los 60° de latitud sur, *Acta Zool. Lilloana, 15,* 19-33, 1958.
Olrog, C. C., Lista y distribucion de las aves argentinas, *Opera Lilloana, 9,* 1-377, 1963.
Ornithological Society of New Zealand, *Annotated Checklist of the Birds of New Zealand Including the Birds of the Ross Dependency,* 96 pp., A. H. and A. W. Reed, Wellington, 1970.
Orton, M. N., A brief survey of the fauna of the Windmill Islands, Wilkes Land, Antarctica, *Emu, 63,* 14-22, 1963.
Orton, M. N., Notes on Antarctic petrels, *Thalassoica antarctica, Emu, 67,* 225-229, 1968.
Ozawa, K., The life of the Antarctic Ocean, 2, Observations of oceanic birds in the southern ocean (in Japanese with English summary), *J. Tokyo Univ. Fish., 1,* 325-328, 1956.
Ozawa, K., Distribution of sea birds in austral summer season in the southern ocean, *Nankyoku Shiryo (Antarctic Rec.), 29,* 1-36, 1967*a*.
Ozawa, K., Summer distribution of chinstrap penguin in the Antarctic, *Mer, 5,* 95-99, 1967*b*.
Paulian, P., Pinnipèdes, cétacés, oiseaux des îles Kerguelen et Amsterdam, *Mem. Inst. Sci. Madagascar, Ser. A, 8,* 111-234, 1953.
Paulian, P., Addition à l'avifaune de l'Ile Amsterdam, *Oiseau, 26,* 65-66, 1956.
Peale, T. R., *United States Exploring Expedition During the Years 1838, 1839, 1840, 1841, 1842 Under the Command of Charles Wilkes, U.S.N.,* vol. 8, *Mammalia and Ornithology,* C. Sherman, Philadelphia, Pa., 1848.
Pefaur, J. E., and R. Murua, Estudios ecologicos en Isla Robert (Shetland del Sur), 7, Aves de la Peninsula de Isla Robert, *Inst. Antartico Chileno Ser. Cient., 2,* 11-23, 1972.

Penney, R. L., Territorial and social behavior in the Adélie penguin, in *Antarctic Bird Studies, Antarctic Res. Ser.*, vol. 12, edited by O. L. Austin, Jr., pp. 83-131, AGU, Washington, D. C., 1968.

Penney, R. L., and W. J. L. Sladen, The use of Teflon for banding penguins, *J. Wildl. Manage., 30,* 847-850, 1966.

Pereyra, J. A., Descripcion de un nuevo ejemplar de ralido de la Isla Georgia del Sud, *Hornero, 8,* 484-489, 1944.

Perkins, J. E., Biology at Little America III, The West Base of the United States Antarctic Service Expedition 1939-1941, *Proc. Amer. Phil. Soc., 89,* 270-284, 1945.

Peterson, R. T., and G. E. Watson, Franklin's gull and bridled tern in southern Chile, *Auk, 88,* 670-671, 1971.

Pettingill, O. S., Jr., Crèche behavior and individual recognition in a colony of rockhopper penguins, *Wilson Bull., 72,* 213-221, 1960.

Pettingill, O. S., Jr., Penguins ashore at the Falkland Islands, *Living Bird, 3,* 45-64, 1964.

Phillips, J. C., *A Natural History of the Ducks,* vol. 2, pp. 103-113, 281-284, 339-343, Houghton Mifflin, Boston, 1923.

Pinder, R., The Cape pigeon, *Daption capensis* Linnaeus, at Signy Island, South Orkney Islands, *Brit. Antarctic Surv. Bull., 8,* 19-47, 1966.

Pisano V., E., Observaciones fito-ecologicas en las Islas Diego Ramírez, *An. Inst. Patagonia, 3,* 161-169, 1972.

Prévost, J., Notes sur l'écologie des pétrels de Terre Adélie, *Alauda, 21,* 205-222, 1953.

Prévost, J., *Ecologie du Manchot Empereur,* 204 pp., Hermann, Paris, 1961.

Prévost, J., Densités de peuplement et biomasses des vertébrés terrestres de l'Archipel de Pointe Géologie, Terre Adélie [Aves, Pinnipedia], *Terre Vie, 17,* 35-49, 1963.

Prévost, J., Remarques écologiques sur quelques procellariens antarctiques, *Oiseau, 34,* no. spec., 91-112, 1964.

Prévost, J., A propos des pétrels des neiges de la Terre Adélie, *Oiseau, 39,* no. spec., 33-49, 1969.

Prévost, J., Relation d'une visite à l'Ile de l'Est, Archipel Crozet, en 1969, *Oiseau, 40,* no. spec., 1-15, 1970.

Prévost, J., and J.-L. Mougin, *Guide des Oiseaux et Mammifères des Terres Australes et Antarctiques Françaises, Guides Natur.,* 230 pp., Delachaux et Niestlé Editeurs, Neuchâtel, Switzerland, 1970.

Prévost, J., and J. Sapin-Jaloustre, A propos des premiers mesures de topographie thermique chez les spheniscides de la Terre Adélie, *Oiseau, 34,* no. spec., 52-90, 1964.

Prévost, J., and J. Sapin-Jaloustre, Ecologie des manchots antarctiques, in *Biogeography and Ecology in Antarctica, Monogr. Biol.,* vol. 15, edited by J. Van Mieghem and P. Van Oye, pp. 551-648, Junk, The Hague, 1965.

Prévost, J., and V. Vilter, Histologie de la sécrétion oesophagienne du manchot empereur, *Proc. 13th Int. Ornithol. Congr. Ithaca 1962,* 1085-1094, 1963.

Pryor, M. A., Silver-gray fulmar and Antarctic petrel of Haswell Islet, *Sov. Antarctic Exped. Inform. Bull., 5,* 281-282, 1965.

Pryor, M. E., The avifauna of Haswell Island, Antarctica, in *Antarctic Bird Studies, Antarctic Res. Ser.,* vol. 12, edited by O. L. Austin, Jr., pp. 57-82, AGU, Washington, D. C., 1968.

Racovitza, E. G., La vie des animaux et des plantes dans l'Antarctique, *Bull. Soc. Belge Geogr., 24,* 177-230, 1900.

Rand, A. L., The origin of the land birds of Tristan da Cunha, *Fieldiana Zool., 37,* 139-166, 1955.

Rand, R. W., Notes on the birds of Marion Island, *Ibis, 96,* 173-206, 1954.

Rand, R. W., The penguins of Marion Island, *Ostrich, 26,* 57-69, 1955.

Rand, R. W., Cormorants on Marion Island, *Ostrich, 27,* 127-133, 1956.

Rand, R. W., Seabirds south of Madagascar, *Ostrich, 33,* 48-51, 1962.

Rand, R. W., Seabirds in the southern Indian Ocean, *Ostrich, 34,* 121-128, 1963.

Rankin, N., *Antarctic Isle, Wildlife in South Georgia,* 383 pp., Collins, London, 1951.

Reichenow, A., Vögel des Weltmeeres, die Meeresvögel de östlichen Erdhälfte, *Deut. Südpolar-Exped. 1901-1903, 9, Zool., 1,* 435-535, 1908.

Reid, B. E., The Cape Hallett Adélie penguin rookery, its size, composition and structure, *Rec. Dominion Mus., 5,* 11-37, 1964.

Reid, B. E., The growth and development of the south polar skua (*Catharacta maccormicki*), *Notornis, 13,* 71-89, 1966.

Reynolds, P. W., Notes on the birds of Cape Horn, *Ibis, 77,* 65-101, 1935.

Richdale, L. E., The titi wainui or fairy prion *Pachyptila turtur* (Kuhl), *Trans. Roy. Soc. N. Z., 74*(1), 32-48; *74*(2), 165-181; 1944*a*.

Richdale, L. E., The parara or broad-billed prion (*Pachyptila vittata*), *Emu, 43,* 191-217, 1944*b*.

Richdale, L. E., The kuaka or diving petrel *Pelecanoides urinatrix* (Gmelin), *Emu, 43*(1), 24-48; *43*(2), 97-107; 1945.

Richdale, L. E., Duration of parental attentiveness in the sooty shearwater, *Ibis, 96,* 586-600, 1954.

Richdale, L. E., Biology of the sooty shearwater *Puffinus griseus, Proc. Zool. Soc. London, 141,* 1-117, 1963.

Richdale, L. E., Notes on the mottled petrel *Pterodroma inexpectata* and other petrels, *Ibis, 106,* 110-114, 1964.

Richdale, L. E., Breeding behaviour of the narrow-billed prion and the broad-billed prion on Whero Island, New Zealand, *Trans. Zool. Soc. London, 31,* 87-155, 1965.

Risebrough, R. W., and G. M. Carmignani, Chlorinated hydrocarbons in antarctic birds, in *Proceedings of the Colloquium on Conservation Problems in Antarctica,* edited by B. C. Parker, pp. 63-80, Virginia Polytechnic Institute and State University, Blacksburg, 1972.

Roberts, B., The life cycle of Wilson's petrel *Oceanites oceanicus* (Kuhl), *Brit. Graham Land Exped. 1934-37 Sci. Rep., 1,* 141-194, 1940*a.*

Roberts, B., The breeding behaviour of penguins with special reference to *Pygoscelis papua* (Forster), *Brit. Graham Land Exped. 1934-37 Sci. Rep., 1,* 195-254, 1940*b.*

Roberts, B. B., A bibliography of antarctic ornithology, *Brit. Graham Land Exped. 1934-1937 Sci. Rep., 9,* 337-367, 1941.

Robertson, C. J. R., R. S. Abel, and F. C. Kinsky, First New Zealand record of Magellanic penguin (*Spheniscus magellanicus*), *Notornis, 19,* 111-113, 1972.

Routh, M., Ornithological observations in the antarctic seas, *Ibis, 91,* 577-606, 1949.

Rowan, A. N., H. F. I. Elliott, and M. K. Rowan, The 'spectacled' form of the Shoemaker *Procellaria aequinoctialis* in the Tristan da Cunha group, *Ibis, 93,* 169-179, 1951.

Rowan, M. K., The yellow-nosed albatross, *Diomedea chlororhynchos,* at its breeding grounds in the Tristan da Cunha group, *Ostrich, 22,* 139-155, 1951.

Salomonsen, F., Migratory movements of the Arctic tern (*Sterna paradisaea* Pontoppidan) in the southern ocean, *Biol. Medd., 24,* 1-42, 1967.

Sapin-Jaloustre, J., *Ecologie du Manchot Adélie,* 211 pp., Hermann, Paris, 1960.

Saunders, H., Reports on the collections of birds made during the voyage of H.M.S. 'Challenger,' 5, On the Laridae collected during the expedition, *Proc. Zool. Soc. London,* 794-800, 1877.

Sclater, W. L., *Bull. Brit. Ornithol. Club, 27,* 94, 1911.

Segonzac, M., La nidification du puffin à pieds pales (*Puffinus carneipes*) à l'Ile Saint Paul, *Oiseau, 40,* 131-135, 1970.

Segonzac, M., Données récentes sur la faune des îles Saint-Paul et Nouvelle Amsterdam, *Oiseau, 42,* 3-68, 1972.

Serventy, D. L., White egret at Macquarie Island, *Emu, 52,* 66-67, 1952*a.*

Serventy, D. L., Movements of the Wilson storm-petrel in Australian seas, *Emu, 52,* 105-116, 1952*b.*

Serventy, D. L., V. Serventy, and J. Warham, *The Handbook of Australian Sea-Birds,* 254 pp., A. H. and A. W. Reed, Wellington, 1971.

Sharpe, R. B., Birds, An account of the petrological, botanical and zoological collections made in Kerguelen's land and Rodriguez during the Transit of Venus expeditions 1874-1875, *Phil. Trans. Roy. Soc., 168,* extra vol., 101-162, 1879.

Shaughnessy, P. D., The genetics of plumage phase dimorphism of the southern giant petrel *Macronectes giganteus, Heredity, 25,* 501-506, 1970.

Shaughnessy, P. D., Frequency of the white phase of the southern giant petrel, *Macronectes giganteus* (Gmelin), *Aust. J. Zool., 19,* 77-83, 1971.

Simpson, K. G., First record of a grey plover at Macquarie Island, *Emu,* [illegible].

Siple, P. A., and A. Lindsey, Ornithology of the Second Byrd Antarctic Expedition, *Auk, 54,* 147-159, 1937.

Sladen, W. J. L., Arctic skua in the Antarctic, *Ibis, 94,* 543, 1952.

Sladen, W. J. L., Pomarine skua in the Antarctic, *Ibis, 96,* 315-316, 1954.

Sladen, W. J. L., The pygoscelid penguins, 1, Methods of study; 2, The Adélie penguin *Pygoscelis adeliae* (Hombron and Jacquinot), *Falkland Isl. Depend. Surv. Sci. Rep., 17,* 1-97, 1958.

Sladen, W. J. L., The distribution of the Adélie and chinstrap penguins, in *Biologie Antarctique, Proceedings of the First SCAR Symposium on Antarctic Biology,* edited by R. Carrick, M. W. Holdgate, and J. Prévost, pp. 359-365, Hermann, Paris, 1964.

Sladen, W. J. L., and R. E. LeResche, New and developing techniques in antarctic ornithology, in *Antarctic Ecology,* vol. 2, edited by M. W. Holdgate, pp. 585-596, Academic, New York, 1970.

Sladen, W. J. L., R. C. Wood, and E. P. Monaghan, The Usarp bird banding program, 1958-1965, in *Antarctic Bird Studies, Antarctic Res. Ser.,* vol. 12, edited by O. L. Austin, Jr., pp. 213-262, AGU, Washington, D. C., 1968.

Slater, P., *A Field Guide to Australian Birds, Non-Passerines,* 428 pp., Livingston, Wynnewood, Pa., 1970.

Solyanik, G. A., Some bird observations on Bouvet Island, *Sov. Antarctic Exped. Inform. Bull., 2,* 97-100, 1964.

Sorensen, J. H., The light-mantled sooty albatross at Campbell Island, *Cape Exped. Ser. Bull., 8,* 1-30, 1950.

Southern, H. N., Addendum: The status of *Procellaria conspicillata, Ibis, 93,* 174-179, 1951.

Spellerberg, I. F., Incubation temperatures and thermoregulation in the McCormick skua, *Condor, 71,* 59-67, 1969.

Spellerberg, I. F., Body measurements and colour phases of the McCormick skua *Catharacta maccormicki, Notornis, 17,* 280-285, 1970.

Spellerberg, I. F., Aspects of McCormick skua breeding biology, *Ibis, 113,* 357-363, 1971*a*.

Spellerberg, I. F., Arrival and departure of birds at McMurdo Sound, Antarctica, *Emu, 71,* 167-171, 1971*b*.

Spenceley, G. B., The South Georgia teal, *Anas georgica, Wildfowl Trust Ninth Annu. Rep. 1956-1957,* 196-198, 1958.

Stirling, I., and D. J. Greenwood, The emperor penguin colony at Cape Washington in the western Ross Sea, Antarctica, *Notornis. 17,* 277-279, 1970.

Stonehouse, B., The emperor penguin *Aptenodytes forsteri* Gray, 1,

Breeding behavior and development, *Falkland Isl. Depend. Surv. Sci. Rep., 6,* 1-33, 1953.
Stonehouse, B., The brown skua *Catharacta skua lönnbergi* (Mathews) of South Georgia, *Falkland Isl. Depend. Surv. Sci. Rep., 14,* 1-25, 1956.
Stonehouse, B., Notes on the ringing and the breeding distribution of the giant petrel *Macronectes giganteus, Ibis, 100,* 204-208, 1958.
Stonehouse, B., The king penguin *Aptenodytes patagonica* of South Georgia, 1, Breeding behavior and development, *Falkland Isl. Depend. Surv. Sci. Rep., 23,* 1-81, 1960.
Stonehouse, B., Bird life, in *Antarctic Research,* edited by R. Priestley, R. J. Adie, and G. de Q. Robin, pp. 219-239, Butterworths, London, 1964.
Stonehouse, B., Birds and mammals, in *Antarctica,* edited by T. Hatherton, pp. 153-186, Methuen, London, 1965.
Stonehouse, B., The general biology and thermal balances of penguins, in *Advances in Ecological Research,* vol. 4, edited by J. B. Cragg, pp. 131-196, Academic, New York, 1967.
Stonehouse, B., Geographic variation in gentoo penguins, *Pygoscelis papua, Ibis, 112,* 52-57, 1970*a.*
Stonehouse, B., Adaptation in polar and subpolar penguins (Spheniscidae), in *Antarctic Ecology,* vol. 1, edited by M. W. Holdgate, pp. 526-541, Academic, New York, 1970*b.*
Stonehouse, B. (Ed.), *The Biology of Penguins,* Macmillan, London, 1975.
Strange, I. J., A breeding colony of *Pachyptila turtur* in the Falkland Islands, *Ibis, 110,* 358-359, 1968.
Stroud, E. D., Some notes on the birds of Deception Island, South Shetlands, *Sea Swallow, 6,* 13-15, 1953.
Studer, T., *Die Forschungsreise S.M.S., 'Gazelle' in der Jahren 1874 bis 1876,* vol. 3, *Zoologie und Geologie,* 322 pp., Ernst Siegfried Mittler und Sohn, Berlin, 1889.
Swales, M. K., The sea-birds of Gough Island, *Ibis, 107,* 17-42, 1965.
Swales, M. K., and R. C. Murphy, A specimen of *Larus pipixcan* from Tristan da Cunha, *Ibis, 107,* 394, 1965.
Szijj, L. J., Notes on the winter distribution of birds in the western Antarctic and adjacent Pacific waters, *Auk, 84,* 366-378, 1967.
Taylor, R. H., The Adélie penguin *Pygoscelis adeliae* at Cape Royds, *Ibis, 104,* 176-204, 1962.
Third Antarctic Treaty Consultative Meeting, Agreed Measures for the Conservation of the Antarctic Fauna and Flora, Brussels, June 1964.
Thomas, D. G., The Dominican gull in Tasmania, *Emu, 66,* 296, 1967.
Thoresen, A. C., Observations on the breeding behaviour of the diving petrel *Pelecanoides u. urinatrix* (Gmelin), *Notornis, 16,* 241-260, 1969.
Tickell, W. L. N., Notes from the South Orkneys and South

Georgia, *Ibis, 102,* 612-614, 1960.

Tickell, W. L. N., The dove prion, *Pachyptila desolata* Gmelin, *Falkland Isl. Depend. Surv. Sci. Rep., 33,* 1-55, 1962.

Tickell, W. L. N., Feeding preferences of the albatrosses *Diomedea melanophris* and *D. chrysostoma* at South Georgia, in *Biologie Antarctique, Proceedings of the First SCAR Symposium on Antarctic Biology,* edited by R. Carrick, M. W. Holdgate, and J. Prévost, pp. 383-387, Hermann, Paris, 1964.

Tickell, W. L. N., New records for South Georgia, *Ibis, 107,* 388-389, 1965.

Tickell, W. L. N., Movements of black-browed and grey-headed albatrosses in the South Atlantic, *Emu, 66,* 357-367, 1967.

Tickell, W. L. N., The biology of the great albatrosses, *Diomedea exulans* and *Diomedea epomophora,* in *Antarctic Bird Studies, Antarctic Res. Ser.,* vol. 12, edited by O. L. Austin, Jr., pp. 1-55, AGU, Washington, D. C., 1968.

Tickell, W. L. N., and J. D. Gibson, Movements of wandering albatrosses *Diomedea exulans, Emu, 68,* 7-20, 1968.

Tickell, W. L. N., and R. Pinder, Breeding frequency in the albatrosses *Diomedea melanophris* and *D. chrysostoma, Nature, 213,* 315-316, 1967.

Tickell, W. L. N., and C. D. Scotland, Recoveries of ringed giant petrels *Macronectes giganteus, Ibis, 103a,* 260-266, 1961.

Tickell, W. L. N., and R. W. Woods, Ornithological observations at sea in the South Atlantic Ocean, 1954-1964, *Brit. Antarctic Surv. Bull., 31,* 63-84, 1972.

Trawa, G., Note préliminaire sur la vascularisation des membres des spheniscides de Terre Adélie, *Oiseau, 40,* no. spec., 142-156, 1970.

U.S. Naval Oceanographic Office, Ice observations, Oceanographic Office Observers Manual, *Publ. 606d,* 42 pp., Washington, D. C., 1968.

Vanhöffen, E., Tiere und Pflanzen von St. Paul und Neu Amsterdam, *Deut. Südpolar-Exped. 1901-1903, 2,* 399-410, 1912.

van Oordt, G. J., Over de noordelijkste verspreidingsgrens van enkele antarctische zeevögelsoorten in den Atlantischen Ocean, *Ardea, 28,* 1-4, 1939.

van Oordt, G. J., and J. P. Kruijt, On the pelagic distribution of some Procellariiformes in the Atlantic and southern oceans, *Ibis, 95,* 615-637, 1953.

van Oordt, G. J., and J. P. Kruijt, Birds, observed on a voyage in the South Atlantic and southern oceans in 1951/1952, *Ardea, 42,* 245-280, 1954.

van Zinderen Bakker, E. M., Jr., Comparative avian ecology, in *Marion and Prince Edward Islands Report on the South African Biological and Geological Expedition/1965-1966,* edited by E. M. van Zinderen Bakker, Sr., J. M. Winterbottom, and R. A. Dyer, pp. 161-172, A. A. Balkema, Capetown, 1971*a*.

van Zinderen Bakker, E. M., Jr., The genus *Diomedea,* in *Marion and Prince Edward Islands Report on the South African Biological and Geological Expedition/1965-1966,* edited by E. M. van Zinderen Bakker, Sr., J. M. Winterbottom, and R. A. Dyer, pp. 273-282, A. A. Balkema, Capetown, 1971*b*.

van Zinderen Bakker, E. M., Jr., Birds observed at sea, in *Marion and Prince Edward Islands Report on the South African Biological and Geological Expedition/1965-1966,* edited by E. M. van Zinderen Bakker, Sr., J. M. Winterbottom, and R. A. Dyer, pp. 249-250, A. A. Balkema, Capetown, 1971*c*.

van Zinderen Bakker, E. M., Jr., A behaviour analysis of the gentoo penguin, in *Marion and Prince Edward Islands Report on the South African Biological and Geological Expedition/1965-1966,* edited by E. M. van Zinderen Bakker, Sr., J. M. Winterbottom, and R. A. Dyer, pp. 251-272, A. A. Balkema, Capetown, 1971*d*.

Velain, C., Passage de Vénus sur le soleil (9 Décembre 1874); Expédition française aux îles Saint Paul et Amsterdam; Zoologie; Observations générales sur la faune des deux îles suivies d'une description des mollusques, *Arch. Zool. Exp. Gen., 6,* 1-144, 1877.

Verrill, G. E., On some birds and eggs collected by Mr. Geo. Comer at Gough Island, Kerguelen Island and the island of South Georgia with extracts from his notes, including a meteorological record for about six months at Gough Island, *Trans. Conn. Acad. Arts Sci., 9,* 430-478, 1895.

Voisin, J.-F., Les pétrels géants (*Macronectes halli* et *Macronectes giganteus*) de l'Ile de la Possession, *Oiseau, 38,* no. spec., 95-122, 1968.

Voisin, J.-F., On the specific status of the Kerguelen shag and its affinities, *Notornis, 17,* 286-290, 1970.

Voisin, J.-F., Note sur les manchots royaux (*Aptenodytes patagonica*) de l'Ile de la Possession, *Oiseau, 41,* 176-180, 1971.

Voous, K. H., The morphological, anatomical and distributional relationship of the arctic and antarctic fulmars (Aves, Procellariidae), *Ardea, 37,* 113-122, 1949.

Voous, K. H., Royal penguin (*Eudyptes schlegeli*) on Marion Island, *Ardea, 51,* 251, 1963.

Voous, K. H., Antarctic birds, in *Biogeography and Ecology in Antarctica, Monogr. Biol.,* vol. 15, edited by J. Van Mieghem and P. Van Oye, pp. 649-689, Junk, The Hague, 1965.

Warham, J., The breeding of the great-winged petrel *Pterodroma macroptera, Ibis, 98,* 171-185, 1956.

Warham, J., A spine-tailed swift at Macquarie Island, *Emu, 61,* 189-190, 1961.

Warham, J., The biology of the giant petrel *Macronectes giganteus, Auk, 79,* 139-160, 1962.

Warham, J., The rockhopper penguin, *Eudyptes chrysocome,* at Macquarie Island, *Auk, 80,* 229-256, 1963.

Warham, J., The white-headed petrel *Pterodroma lessoni* at Macquarie Island, *Emu, 67,* 1-22, 1967.

Warham, J., Notes on some Macquarie Island birds, *Notornis, 16,* 190-197, 1969.

Warham, J., Aspects of breeding behaviour in the royal penguin (*Eudyptes chrysolophus schlegeli*), *Notornis, 18,* 91-115, 1971*a.*

Warham, J., Body temperature of petrels, *Condor, 73,* 214-219, 1971*b.*

Warham, J., Breeding seasons and sexual dimorphism in rockhopper penguins, *Auk, 89,* 86-105, 1972.

Warham, J., and W. R. P. Bourne, Additional notes on albatross identification, *Amer. Birds, 28,* 598-603, 1974.

Warham, J., and B. R. Keeley, New and rare birds at Snares Island during 1968-1969, *Notornis, 16,* 221-224, 1969.

Warham, J., W. R. P. Bourne, and H. F. I. Elliott, Albatross identification in the North Atlantic, *Brit. Birds, 59,* 376-384, 1966.

Watson, G. E., The status of the black noddy in the Tristan da Cunha group, *Bull. Brit. Ornithol. Club, 89,* 105-107, 1969.

Watson, G. E., Molting greater shearwaters (*Puffinus gravis*) off Tierra del Fuego, *Auk, 88,* 440-442, 1971.

Watson, G. E., and A. B. Amerson, Jr., Instructions for collecting bird parasites, *Smithson. Inst. Inform. Leafl., 477,* 1-10, 1967.

Watson, G. E., et al., Birds of the Antarctic and Subantarctic, *Antarctic Map Folio Ser., 14,* 1-18, 1971.

Weller, M. W., Ecological studies of the South Georgia pintail (*Anas g. georgica*), *Antarctic J. U.S., 7,* 77-78, 1972.

Weller, M. W., Ecology and behaviour of the South Georgian pintail *Anas g. georgica, Ibis, 117,* 217-231, 1975.

Weller, M. W., and R. L. Howard, Breeding of speckled teal *Anas flavirostris* on South Georgia, *Brit. Antarctic Surv. Bull., 30,* 65-68, 1972.

Wilkinson, J., South Sandwich Islands—Bird life, *Sea Swallow, 9,* 18-20, 1956.

Wilkinson, J., A second visit to the South Sandwich Islands, *Sea Swallow, 10,* 22, 1957.

Wilson, A. E., and M. K. Swales, Flightless moorhens (*Porphyriornis comeri*) from Gough Island breed in captivity, *Avicultural Mag., 64,* 43-45, 1958.

Wilson, E. A., Aves, *Nat. Antarctic Exped. 1901-1904, 2, Zool., 2,* 1-121, 1907.

Winterbottom, J. M., Tristan da Cunha birds, *Ibis, 100,* 285, 1958.

Winterbottom, J. M., The position of Marion Island in the sub-antarctic avifauna, in *Marion and Prince Edward Islands Report on the South African Biological and Geological Expedition/1965-1966,* edited by E. M. van Zinderen Bakker, Sr., J. M. Winterbottom, and R. A. Dyer, pp. 241-248, A. A. Balkema, Capetown, 1971.

Woods, R. W., Great shearwater *Puffinus gravis* breeding in the Falkland Islands, *Ibis, 112,* 259-260, 1970.

Young, E. C., The breeding behaviour of the south polar skua *Catharacta maccormicki, Ibis, 105,* 203-233, 1963*a*.

Young, E. C., Feeding habits of the south polar skua *Catharacta maccormicki, Ibis, 105,* 301-318, 1963*b*.

VARIANT NAMES

A common goal of scientific and vernacular nomenclature is the use of standard names to facilitate communication. Regrettably, scientists do not yet agree completely on the Latin names of all antarctic and subantarctic birds. On the other hand, it is surprising that there is currently so much agreement on English names of widespread species that occur in American, African, Australian, and New Zealand sectors of the southern oceans. Argentine and Chilean Spanish names for the same species still differ in many cases.

In this list the species are arranged in the same standard taxonomic sequence as they are in the text. The first entry is the English name adopted in the handbook. There follow, in order, variant English names (including those for distinctive subspecies), the scientific name used in the handbook, Latin synonyms and subspecies names (the latter cited in full in the text but treated here only as binomials), and those French and Spanish vernacular names that are readily available in such basic references as *Prévost and Mougin* [1970], *Olrog* [1963], and *Johnson* [1965, 1967]. The list is not exhaustive but is at least intended to cover all names adopted in major English works that have dealt with antarctic and subantarctic birds since publication of the first edition of *Alexander*'s [1954] *Birds of the Ocean* in 1928.

Emperor penguin, *Aptenodytes forsteri,* manchot empereur, pingüino emperador.

King penguin, *Aptenodytes patagonicus, A. patagonica,* manchot royal, pingüino rey.

Adélie penguin, *Pygoscelis adeliae,* manchot adélie, pingüino de Adélia.

Chinstrap penguin, bearded penguin, ringed penguin, *Pygoscelis antarctica,* manchot à jugulaire, pingüino de barbijo.

Gentoo penguin, *Pygoscelis papua, P. ellsworthi,* manchot papou, pingüino de pico rojo.

Rockhopper penguin, *Eudyptes crestatus, E. chrysocome, E. moseleyi,* gorfou sauteur, pingüino de penacho amarillo.

Macaroni penguin, royal penguin, *Eudyptes chrysolophus, E. schlegeli,* gorfou doré, gorfou macaroni, gorfou de Schlegel, pingüino de penacho anaranjado.

Crested penguin, thick-billed penguin, Snares-crested penguin, erect-crested penguin, Victoria penguin, big-crested penguin, *Eudyptes pachyrhynchus, E. sclateri, E. robustus, E. atratus.*

Magellanic penguin, *Spheniscus magellanicus,* pingüino de Magallanes, pingüino del Sur.

Wandering albatross, snowy albatross, *Diomedea exulans, D. dabbenena, D. chionoptera,* grand albatros, albatros errante.

Royal albatross, *Diomedea epomophora,* albatros royal, albatros real.

Black-browed albatross, black-browed mollymauk, *Diomedea melanophris, D. impavida,* albatros à sourcils noirs, albatros ojeroso, albatros chico, albatros de ceja negra.

Gray-headed albatross, gray-headed mollymauk, *Diomedea chrysostoma,* albatros à tête grise, albatros de cabeza gris.

Yellow-nosed albatross, yellow-nosed mollymauk, yellow-billed albatross, *Diomedea chlororhynchos,* albatros à bec jaune, albatros de pico amarillo, albatros chlororrinco.

White-capped albatross, white-capped mollymauk, shy albatross, Salvin's albatross, Chatham Island albatross, gray-backed albatross, Layard's albatross, *Diomedea cauta, D. salvini, D. eremita,* albatros piliblanco, albatros de frente blanca.

Sooty albatross, *Phoebetria fusca,* albatros fuligineux à dos sombre, albatros oscuro.

Light-mantled sooty albatross, *Phoebetria palpebrata,* albatros fuligineux à dos clair, albatros oscuro de manto claro.

Southern giant fulmar, southern giant petrel, nelly, stinker, *Macronectes giganteus,* pétrel géant antarctique, petrel gigante.

Northern giant fulmar, northern giant petrel, nelly, *Macronectes halli,* pétrel géant subantarctique.

Southern fulmar, antarctic fulmar, silver gray petrel, silver gray fulmar, *Fulmarus glacialoides, Priocella antarctica,* fulmar antarctique, petrel plateado.

Antarctic petrel, *Thalassoica antarctica,* pétrel antarctique, petrel antártico.

Cape pigeon, Cape petrel, pintado petrel, *Daption capense, D. capensis, D. australe,* damier du Cap, petrel damero, petrel común, damero del Cabo, petrel moteado.

Snow petrel, *Pagodroma nivea, P. confusa, P. major,* pétrel des neiges, petrel blanco, petrel de las nieves.

Narrow-billed prion, slender-billed prion, thin-billed prion, *Pachyptila belcheri,* prion de Belcher, petrel-ballena picofino, petrel-ballena de pico delgado, petrel-paloma de pico delgado.

Antarctic prion, dove prion, *Pachyptila desolata, P. banksi, P. altera,* prion de la Désolation, petrel-ballena picoancho, petrel-ballena de pico ancho, petrel-paloma antártico.

Broad-billed prion, lesser broad-billed prion, medium-billed prion, Salvin's prion, *Pachyptila vittata, P. forsteri, P. salvini, P. macgillivrayi,* prion de Forster, prion de Salvin, petrel-ballena picogrueso.

Fulmar prion, thick-billed prion, *Pachyptila crassirostris, P. eatoni.*

Fairy prion, *Pachyptila turtur,* petit prion, petrel-ballena picocorto, petrel-ballena chico.

Blue petrel, *Halobaena caerulea,* pétrel bleu, petrel azulado.

Great-winged petrel, gray-faced petrel, long-winged petrel, *Pterodroma macroptera, P. gouldi,* pétrel noir.

White-headed petrel, *Pterodroma lessoni,* pétrel à tête blanche, fardela gargantiblanca, fardela de frente blanca, petrel gris y blanco.

Atlantic petrel, Schlegel's petrel, hooded petrel, *Pterodroma incerta,* pétrel de Schlegel, fardela alinegra, petrel pardo y blanco.

Kerguelen petrel, *Pterodroma brevirostris, P. lugens, P. kidderi,* pétrel de Kerguelen, fardela apizarrada, petrel apizarrada.

Soft-plumaged petrel, *Pterodroma mollis, P. deceptornis,* pétrel soyeux, fardela coronigris, petrel de corona gris.

Mottled petrel, scaled petrel, Peale's petrel, *Pterodroma inexpectata.*

Gould's petrel, white-winged petrel, *Pterodroma leucoptera, P. brevipes,* pétrel de Gould, fardela blanca (de Masafuera), fardela gris.

Cook's petrel, blue-footed petrel, *Pterodroma cookii,* pétrel de Cook, fardela blanca (de Cook or de Masatierra), fardela chica.

Juan Fernandez petrel, black-capped petrel, white-necked petrel, *Pterodroma externa, P. tristani,* fardela blanca de Juan Fernandez, petrel de corona negra.

White-chinned petrel, shoemaker, Cape hen, *Procellaria aequinoctialis, P. conspicillata,* pétrel à menton blanc, petrel negro, petrel de barba blanco, petrel de pico amarillo, fardela negra grande.

Gray petrel, brown petrel, great gray shearwater, pediunker, black-tailed shearwater, *Procellaria cinerea, Adamastor cinereus,* pétrel gris, petrel gris, fardela gris.

Cory's shearwater, North Atlantic shearwater, Mediterranean shearwater, Kuhl's shearwater, *Calonectris diomedea, C. disputans, Puffinus diomedea, P. kuhli,* puffin cendré, pardela cenicienta.

Sooty shearwater, *Puffinus griseus,* pardela oscura, fardela negra común.

Short-tailed shearwater, slender-billed shearwater, Tasmanian muttonbird, *Puffinus tenuirostris,* pardela picofino, fardela australiana.

Flesh-footed shearwater, fleshy-footed shearwater, pale-footed shearwater, *Puffinus carneipes,* puffin à pieds pâles, pardela negruzca, fardela negra de patas pálidas.

Pink-footed shearwater, *Puffinus creatopus,* fardela blanca común, pardela parde.

Wedge-tailed shearwater, *Puffinus pacificus,* fouquet, pardela pacífica.

Greater shearwater, *Puffinus gravis,* puffin majeur, pardela capirotada.

Little shearwater, dusky shearwater, allied shearwater, *Puffinus assimilis, P. elegans, P. kempi, P. myrtae,* petit puffin, pardela chica, petrel plomizo.

Wilson's storm petrel, *Oceanites oceanicus, O. exasperatus,* pétrel de Wilson, paiño común, petrel de tormenta, petrel chico de las tormentas, golondrina de mar común.

Leach's storm petrel, Leach's fork-tailed petrel, *Oceanodroma leucorhoa,* pétrel cul-blanc, paiño de Leach.

Black-bellied storm petrel, black-streaked storm petrel, *Fregetta tropica, F. melanoleuco,* pétrel tempête à ventre noir, paiño ventrinegro, petrel de las tormentas de vientre negro, golondrina de mar de vientre negro.

White-bellied storm petrel, *Fregetta grallaria,* paiño ventriblanco, golondrina de mar de vientre blanco.

Gray-backed storm petrel, *Garrodia nereis,* pétrel tempête à croupion gris, paiño gris, petrel gris de las tormentas, golondrina de mar subantártico.

White-faced storm petrel, frigate petrel, *Pelagodroma marina,* pétrel frégate, paiño cariblanco, petrel blanco de las tormentas.

South Georgia diving petrel, *Pelecanoides georgicus,* pétrel plongeur de Géorgie du Sud, potoyunco georgico, petrel zambullidor georgico.

Kerguelen diving petrel, subantarctic diving petrel, common diving petrel, *Pelecanoides (urinatrix) exsul, P. urinatrix, P. dacunhae,* pétrel plongeur commun, potoyunco malvinero, patoyunco de los canales, petrel zambullidor malvinero.

Blue-eyed shag, blue-eyed cormorant, *Phalacrocorax atriceps, P. nivalis, P. georgianus, P. bransfieldensis, Leucocarbo atriceps,* cormorán imperial.

King shag, king cormorant, Kerguelen shag, *Phalacrocorax albiventer, P. verrucosus, P. melanogenis, P. purpurascens, Leucocarbo albiventer,* cormoran à ventre blanc, cormorán de Kerguelen, cormorán blanco, cormorán imperial de las Malvinas, biguà blanco, viguà de vientre blanco.

Great cormorant, black shag, *Phalacrocorax carbo,* gran cormoran, cormorán grande.

Gannet, Australian gannet, Cape gannet, *Morus bassanus, Sula bassana, M. serrator, M. capensis,* fou de Bassan.

Frigatebird, *Fregeta* sp., frégate, rabihorcado.

White-faced heron, *Ardea novaehollandiae, Notophoyx novaehollandiae.*

Great egret, common egret, American egret, white heron, great white egret, *Egretta alba, Casmerodius albus,* grande aigrette, garceta grande, garza blanca.

Yellow-billed egret, plumed egret, *Egretta intermedia, Mesophoyx intermedius.*

Cattle egret, tickbird, *Bubulcus ibis, Ardeola ibis,* héron garde-boeuf, garcilla bueyera.

Snowy egret, *Egretta thula, Leucophoyx thula,* garceta blanca, garza chica.

Little egret, *Egretta garzetta,* aigrette garzette, garceta europea, garceta común.

Black-crowned night heron, *Nycticorax nycticorax,* héron bihoreau, cuaco, martineta, huairavo común.

Black-necked swan, *Cygnus melanocoryphus,* cisne cuellinegro, cisne de cuello negro.

Upland goose, Magellan goose, *Chloephaga picta,* cauquén común, caiquén.

Gray duck, Australian duck, black duck, *Anas superciliosa, A. poecilorhyncha.*

Black duck, *Anas rubripes.*

Mallard duck, *Anas platyrhynchos,* canard colvert, anade real.

Yellow-billed pintail, yellow-billed teal, South Georgia teal, *Anas georgica, A. spinicauda,* pato maicero, pato jergón grande.

Kerguelen pintail, *Anas (acuta) eatoni,* canard d'Eaton.

Speckled teal, yellow-billed teal, Chilean green-winged teal, South American green-winged teal, *Anas flavirostris,* cerceta barcina, pato barrero, pato jergón chico.

Green-winged teal, European teal, *Anas crecca,* sarcelle d'hiver, cerceta aliverde.

Gray teal, *Anas giberrifrons.*

Chiloe wigeon, *Anas sibilatrix,* pato overo, pato real.

Garganey teal, *Anas querquedula,* sarcelle d'été, cerceta carretona.

Argentine ruddy duck, lake duck, *Oxyura vittata,* pato-malvasia mediano, pato rano de pico delgado.

Marsh harrier, swamp harrier, *Circus aeruginosus, C. approximans.*

Banded rail, *Rallus philippensis, R. macquariensis,*

Inaccessible Island flightless rail, island cock, *Atlantisia rogersi.*

Weka, woodhen, *Gallirallus australis.*

Gough moorhen, *Gallinula nesiotes, G. comeri, Porphyriornis nesiotes.*

Purple gallinule, *Porphyrula martinica, P. georgica,* polla de agua azul, tagüita purpúrea.

Eurasian coot, *Fulica atra,* foulque macroule, focha común.

Red-gartered coot, *Fulica armillata,* focha piquirroja, tagua común, gallareta de pico rojo.

Corn crake, *Crex crex,* râle de genêt, guión de codornices.

Black-bellied plover, gray plover, *Pluvialis squatarola, Squatarola squatarola, Charadrius squatarolus,* pluvier argenté, chorlito gris.

Chilean dotterel, winter plover, rufous-chested dotterel, *Zonibyx modestus,* chorlito pechicolorado, chorlo negro.

White-tailed plover, *Chettusia leucura,* pluvier à queue blanche.

Willet, *Catoptrophorus semipalmatus,* archibebe aliblanco, playero grande de alas blancas.

Upland sandpiper, upland plover, Bartram's sandpiper, *Bartramia longicauda,* bartramie à longue queue, batitú.

Common curlew, *Numenius arquata,* courlis cendré, zarapito real.

Eastern long-billed curlew, *Numenius madagascariensis.*

Whimbrel, *Numenius phaeopus,* courlis corlieu, zarapito trinador.

Bar-tailed godwit, *Limosa lapponica,* barge rousse, aguja colipinta.

Greenshank, *Tringa nebularia,* chevalier aboyeur, archibebe claro.

Common sandpiper, *Actitis hypoleucos, Tringa hypoleuca,* chevalier guignette, andarríos chico.

Spotted sandpiper, *Actitis macularia,* chevalier grivelé, andarríos manchado, playero manchado.

Ruddy turnstone, *Arenaria interpres,* tournepierre à collier, vuelvepiedras, chorlo vuelvepiedras.

Japanese snipe, *Gallinago hardwicki, Capella hardwicki.*

Knot, red knot, eastern knot, *Calidris canutus,* bécasseau maubèche, correlimos gordo, playero ártico.

White-rumped sandpiper, *Calidris fuscicollis, Erolia fuscicollis,* bécasseau de Bonaparte, correlimos lomiblanco, playero de lomo blanco.

Sharp-tailed sandpiper, Siberian pectoral sandpiper, *Calidris*

acuminata, Erolia acuminata, bécasseau à queue pointue.
Curlew sandpiper, *Calidris ferruginea,* bécasseau cocorli, correlimos zarapitín.
Sanderling, *Calidris alba, Crocethia alba,* bécasseau sanderling, correlimos blanco, playero común.
Red phalarope, gray phalarope, *Phalaropus fulicarius,* phalarope à bec large, faláropo picogrueso, pollito de mar rojizo.
Wilson's phalarope, *Steganopus tricolor,* falároро tricolor, pollito de mar de Wilson.
American sheathbill, pale-faced sheathbill, snowy sheathbill, *Chionis alba,* paloma antártica.
Lesser sheathbill, black-faced sheathbill, *Chionis minor, C. nasicornis, C. marionensis, C. crozettensis,* petit bec-en-fourreau.
South polar skua, McCormick's skua, Antarctic skua, *Catharacta maccormicki, C. skua,* skua antarctique, salteador polar.
Brown skua, southern skua, skua, *Catharacta lonnbergi, C. skua, C. hamiltoni, C. chilensis, C. antarcticus,* skua subantarctique, salteador pardo, salteador oscuro, págalo grande.
Pomarine jaeger, pomatarhine skua, *Stercorarius pomarinus,* labbe pomarin, págalo pomarino.
Parasitic jaeger, Arctic skua, Richardson skua, *Stercorarius parasiticus,* labbe parasite, págalo parásito, salteador chico.
Southern black-backed gull, Dominican gull, kelp gull, *Larus dominicanus,* goéland dominicain, gaviota dominicana, gaviota cocinera, gaviota común.
Band-tailed gull, Belcher's gull, *Larus belcheri, L. atlanticus,* gaviota colanegra, gaviota peruana.
Franklin's gull, *Larus pipixcan,* gaviota menor, gaviota de Franklin.
Silver gull, red-billed gull, Hartlaub's gull, *Larus novaehollandiae, L. scopulinus, L. hartlaubi.*
Antarctic tern, wreathed tern, swallow-tailed tern, *Sterna vittata, S. georgiae, S. gaini, S. tristanensis, S. bethunei, S. macquariensis,* sterne subantarctique, charrán antártico, gaviotín antártico.
Kerguelen tern, *Sterna virgata, S. mercuri,* sterne de Kerguelen.
Arctic tern, *Sterna paradisaea, S. macrura, S. antistropha,* sterne arctique, charrán ártico, gaviotín ártico.
Bridled tern, brown-winged tern, *Sterna anaethetus,* charrán frentiblanco.
Brown noddy, common noddy, *Anous stolidus,* tiñosa común, gaviotín pardo.
Black noddy, white-capped noddy, lesser noddy, *Anous tenuirostris, A. minutus, Megalopterus minutus,* tiñosa negra.
Red-crowned parakeet, *Cyanorhamphus novaezelandiae, C. erythrotis.*
Pigeon, rock dove, *Columba livia,* pigeon (biset), paloma bravía.
Turtledove, *Streptopelia* sp., tourterelle, tórtola.
Madagascar turtledove, *Streptopelia turata,* tourterelle malgache.
Eurasian turtledove, *Streptopelia turtur,* tourterelle des bois, tórtola común.
Fork-tailed swift, *Apus pacificus.*

Spine-tailed swift, *Chaetura caudacuta, Hirunaapus caudacutus,* martinet épineux, rabitojo.
Swift, martinet, vincejo.
Broad-billed roller, dollar bird, *Eurystomus glaucurus, E. pacificus,* rolle africain.
Barn swallow, *Hirundo rustica,* hirondelle de cheminée, golondrina común, golondrina bermeja.
Bank swallow, sand martin, *Riparia riparia,* hirondelle de rivage, avión zapador.
Tristan thrush, starchy, *Nesocichla eremita, N. gordoni, N. procax.*
Blackbird, *Turdus merula,* merle noir, mirlo común.
Song thrush, *Turdus philomelos,* grive musicienne, zorzal común.
South Georgia pipit, *Anthus antarcticus,* cachirla antártica.
Song sparrow, *Melospiza melodia,* pinson, chingolo (French and Spanish generic names only).
Wagtail, *Motacilla* sp., bergeronnette, lavandera.
Correndera pipit, *Anthus correndera,* cachirla común, bailarín chico común.
Pipit, *Anthus* sp., pipit, bisbita.
New Zealand pipit, *Anthus novaeseelandiae.*
Starling, *Sturnus vulgaris,* étourneau sansonnet, estornino pinto.
White eye, silvereye, *Zosterops lateralis.*
European goldfinch, *Carduelis carduelis,* chardonneret élégant, jilguero.
Redpoll, *Acanthis flammea, Carduelis flammea,* sizerín flammé, pardillo sizerín.
Tristan bunting, canary, *Nesospiza acunhae, N. questi.*
Wilkins' bunting, grosbeak bunting, big-billed bunting, *Nesospiza wilkinsi, Crithagroides wilkinsi, N. dunnei.*
Gough bunting, *Rowettia goughensis.*
House sparrow, English sparrow, *Passer domesticus,* moineau domestique, gorrión común.

INDEX

The English and Latin names of species and subspecies in this index are those that are actually used in the species accounts. Only the first page of the account and the plate or plates where the bird appears are cited. Current well-known English and scientific synonyms that are not used in the text as well as French and Spanish vernacular names are cited in the variant names list.

ENGLISH NAMES

Albatross, black-browed, Pl. 4, p. 88
- gray-headed, Pl. 4, p. 91
- light-mantled sooty, Pl. 3, p. 97
- royal, Pl. 3, p. 88
- sooty, Pl. 3, p. 95
- wandering, Pl. 3, p. 85
- white-capped, Pl. 4, p. 95
- yellow-nosed, Pl. 4, p. 93

Blackbird, p. 237

Bunting, Gough, Pl. 11, p. 245
- Tristan, Pl. 11, p. 242
- Wilkins', Pl. 11, p. 244

Coot, Eurasian, p. 192
- red-gartered, p. 192

Cormorant, see Shag
- great, p. 172

Crake, corn, p. 193

Curlew, common, p. 195
- eastern long-billed, p. 195

Diving petrel, Kerguelen, Pl. 7, p. 164
- South Georgia, Pl. 7, p. 162
- subantarctic, p. 164

Dotterel, Chilean, p. 193

Duck, Argentine ruddy, p. 186
- black, Pl. 8, p. 178
- gray, Pl. 8, p. 178
- mallard, p. 179

Egret, cattle, p. 174
- great, p. 174
- little, p. 175
- snowy, p. 175
- yellow-billed, p. 174

Frigatebird, p. 173

Fulmar, northern giant, Pl. 3, p. 100
- southern, Pl. 5, 6; p. 105
- southern giant, Pl. 3, p. 100

Gallinule, purple, p. 192

Gannet, p. 172

Godwit, bar-tailed, p. 195

Goldfinch, European, p. 241

Goose, upland, p. 177

Greenshank, p. 196

Gull, band-tailed, p. 218
- Franklin's, p. 219
- silver, p. 220
- southern black-backed, Pl. 9, 10; p. 215

Harrier, marsh, p. 187

Heron, black-crowned night, p. 175
- white-faced, p. 174

Jaeger, pomarine, Pl. 9, p. 213
- parasitic, Pl. 10, p. 214

Knot, p. 198

Mollymauk, see Albatross

Mollymawk, see Albatross

Moorhen, Gough, Pl. 11, p. 191

Noddy, black, p. 232
- brown, Pl. 10, p. 230

Parakeet, red-crowned, p. 232

Penguin, Adélie, Pl. 1, p. 70
- chinstrap, Pl. 1, p. 72
- crested, Pl. 2, p. 83
- emperor, Pl. 1, p. 64
- erect-crested, Pl. 2, p. 83
- gentoo, Pl. 1, p. 74
- king, Pl. 1, p. 66
- macaroni, Pl. 2, p. 80
- Magellanic, Pl. 1, p. 84

rockhopper, Pl. 2, p. 76
royal, Pl. 2, p. 80
Snares-crested, Pl. 2, p. 83
Petrel, Antarctic, Pl. 5, 6; p. 107
Atlantic, Pl. 5, p. 131
blue, Pl. 7, p. 125
Cook's, p. 138
Gould's, p. 137
gray, Pl. 5, p. 141
great-winged, Pl. 5, p. 127
Juan Fernandez, p. 139
Kerguelen, Pl. 6, p. 132
mottled, Pl. 6, p. 136
snow, Pl. 6, p. 112
soft-plumaged, Pl. 6, p. 134
white-chinned, Pl. 6, p. 139
white-headed, Pl. 5, p. 129
Phalarope, red, p. 200
Wilson's, p. 201
Pigeon, p. 232
Cape, Pl. 5, p. 109
Pintail, Kerguelen, Pl. 8, p. 181
yellow-billed, Pl. 8, p. 180
Pipit, p. 239
Correndera, p. 237
New Zealand, p. 239
South Georgia, Pl. 8, p. 237
Plover, black-bellied, p. 193
white-tailed, p. 194
Prion, Antarctic, Pl. 7, p. 117
broad-billed, Pl. 7, p. 117
fairy, Pl. 7, p. 123
fulmar, Pl. 7, p. 121
lesser broad-billed, Pl. 7, p. 119
narrow-billed, Pl. 7, p. 115
Salvin's, Pl. 7, p. 119
Rail, banded, p. 188
Inaccessible Island flightless, Pl. 11, p. 188
Redpoll, Pl. 8, p. 241
Roller, broad-billed, p. 234
Sanderling, p. 200
Sandpiper, common, p. 196
curlew, p. 199
sharp-tailed, p. 199
spotted, p. 197
upland, p. 194
white-rumped, p. 198
Shag, blue-eyed, Pl. 2, 8; p. 167
Kerguelen, Pl. 2, p. 169
king, Pl. 2, p. 169
Shearwater, Cory's, p. 143
flesh-footed, p. 145
greater, Pl. 6, p. 148
little, Pl. 6, p. 150
pink-footed, p. 145
short-tailed, Pl. 5, p. 145
sooty, Pl. 5, 6; p. 143
wedge-tailed, p. 145
Sheathbill, American, Pl. 8, p. 202
lesser, Pl. 8, p. 204
Skua, brown, Pl. 9, p. 210
south polar, Pl. 9, p. 207
Snipe, Japanese, p. 198
Sparrow, house, p. 246
song, p. 237
Starling, Pl. 8, p. 239
Storm petrel, black-bellied, Pl. 7, p. 155
gray-backed, Pl. 7, p. 158
Leach's, p. 155
white-bellied, p. 155
white-faced, Pl. 7, p. 160
Wilson's, Pl. 7, p. 152
Swallow, bank, p. 235
barn, p. 234
Swan, black-necked, p. 177
Swift, p. 234
fork-tailed, p. 233
spine-tailed, p. 233
Teal, Garganey, p. 186
gray, p. 185
green-winged, pp. 183, 184
speckled, p. 183
Tern, Antarctic, Pl. 9, 10; p. 221
Arctic, Pl. 10, p. 227
black-capped, p. 221
bridled, p. 230
Kerguelen, Pl. 10, p. 225
Thrush, song, p. 237
Tristan, Pl. 11, p. 235
Turnstone, ruddy, p. 197
Turtledove, p. 232
Eurasian, p. 232
Madagascar, p. 232
Wagtail, p. 237

Weka, Pl. 8, p. 189
Whale bird, see Prion
Whimbrel, p. 195
White eye, p. 241
Wigeon, Chiloe, p. 185
Willet, p. 194

LATIN NAMES

Acanthis flammea, Pl. 8, p. 241
Actitis hypoleucos, p. 196
macularia, p. 197
Anas acuta, p. 181
(a.) eatoni, Pl. 8, p. 181
crecca, pp. 183, 184
flavirostris, p. 183
georgica, Pl. 8, p. 180
g. spinicauda, p. 180
gibberifrons, p. 185
platyrhynchos, p. 179
querquedula, p. 186
rubripes, p. 178
sibilatrix, p. 185
superciliosa, Pl. 8, p. 178
Anous stolidus, Pl. 10, p. 230
tenuirostris, p. 232
Anthus sp., p. 239
antarcticus, Pl. 8, p. 237
correndera, p. 237
Aptenodytes forsteri, Pl. 1, p. 64
patagonicus, Pl. 1, p. 66
Apus pacificus, p. 233
Ardea novaehollandiae, p. 174
Arenaria interpres, p. 197
Atlantisia rogersi, Pl. 11, p. 188
Bartramia longicauda, p. 194
Bubulcus ibis, p. 174
Bulweria, see *Pterodroma*
Calidris acuminata, p. 199
alba, p. 200
canutus, p. 198
ferruginea, p. 199
fuscicollis, p. 198
Calonectris diomedea, p. 143
Carduelis carduelis, p. 241
Catharacta antarctica, p. 210
chilensis, p. 210
hamiltoni, p. 210
lonnbergi, Pl. 9, p. 210
maccormicki, Pl. 9, p. 207
skua, p. 210
Cataptrophorus semipalmatus, p. 194
Chaetura caudacuta, p. 233
Chettusia leucura, p. 194
Chionis alba, Pl. 8, p. 202
minor, Pl. 8, p. 204
m. crozettensis, p. 204
m. marionensis, p. 204
m. minor, p. 204
m. nasicornis, p. 204
Chloephaga picta, p. 177
Circus aeruginosus, p. 187
a. approximans, p. 187
Columba livia, p. 232
Crex crex, p. 193
Cyanorhamphus novaezelandiae erythrotis, p. 232
Cygnus melanocoryphus, p. 177
Daption capense, Pl. 5, p. 109
c. australe, p. 109
Diomedea cauta, Pl. 4, p. 95
c. eremita, p. 95
c. salvini, p. 95
chlororhynchos, Pl. 4, p. 93
chrysostoma, Pl. 4, p. 91
epomophora, Pl. 3, p. 88
exulans, Pl. 3, p. 85
e. dabbenena, p. 85
melanophris, Pl. 4, p. 88
m. impavida, p. 88
Egretta alba, p. 174
garzetta, p. 175
intermedia, p. 174
thula, p. 175
Eudyptes chrysolophus, Pl. 2, p. 80
c. schlegeli, Pl. 2, p. 80
crestatus, Pl. 2, p. 75
c. moseleyi, Pl. 2, p. 76
pachyrhynchus, Pl. 2, p. 83
p. robustus, Pl. 2, p. 83
p. sclateri, Pl. 2, p. 83

Eurystomus glaucurus, p. 234
Fregata sp., p. 173
Fregetta grallaria, p. 155
tropica, Pl. 7, p. 155
Fulica sp., p. 192
armillata, p. 192
atra, p. 192
Fulmarus glacialoides, Pl. 5, 6; p. 105
Gallinago hardwicki, p. 198
Gallinula nesiotes, Pl. 11, p. 191
n. comeri, Pl. 11, p. 191
Gallirallus australis, Pl. 8, p. 189
Garrodia nereis, Pl. 7, p. 158
Halobaena caerulea, Pl. 7, p. 125
Hirundo rustica, p. 234
Larus belcheri, p. 218
b. atlanticus, p. 218
dominicanus, Pl. 9, 10; p. 215
novaehollandiae, p. 220
pipixcan, p. 219
Limosa lapponica, p. 195
Macronectes giganteus, Pl. 3, p. 100
halli, Pl. 3, p. 100
Melanodera, p. 245
Melospiza melodia, p. 237
Morus bassanus, p. 172
b. capensis, p. 172
b. serrator, p. 172
Motacilla, p. 237
Nesocichla eremita, Pl. 11, p. 235
e. gordoni, Pl. 11, p. 235
e. procax, Pl. 11, p. 235
Nesospiza acunhae, Pl. 11, p. 242
a. questi, p. 242
wilkinsi, Pl. 11, p. 244
w. dunnei, p. 244
Numenius arquata, p. 195
madagascariensis, p. 195
phaeopus, p. 195
Nycticorax nycticorax, p. 175
Oceanites oceanicus, Pl. 7, p. 152
o. exasperatus, p. 152
Oceanodroma leucorhoa, p. 155
Oxyura vittata, p. 186
Pachyptila belcheri, Pl. 7, p. 115
crassirostris, Pl. 7, p. 121
c. eatoni, p. 121
desolata, Pl. 7, p. 117
turtur, Pl. 7, p. 123
vittata, Pl. 7, p. 119
v. salvini, Pl. 7, p. 119
Pagodroma nivea, Pl. 6, p. 112
n. confusa, p. 112
Passer domesticus, p. 246
Pelagodroma marina, Pl. 7, p. 160
Pelecanoides georgicus, Pl. 7, p. 162
urinatrix, p. 164
u. dacunhae, p. 164
(u.) exsul, Pl. 7, p. 164
Phalacrocorax albiventer, Pl. 2, p. 169
a. melanogenis, p. 169
a. verrucosus, Pl. 2, p. 169
atriceps, Pl. 2, 8; p. 167
a. nivalis, p. 167
carbo, p. 172
Phalaropus fulicarius, p. 200
Phoebetria fusca, Pl. 3, p. 95
palpebrata, Pl. 3, p. 97
Pluvialis squatarola, p. 193
Porphyrula martinica, p. 192
Procellaria, also see *Puffinus*
Procellaria aequinoctialis, Pl. 6, p. 139
a. conspicillata, Pl. 6, p. 139
cinerea, Pl. 5, p. 141
Pterodroma brevirostris, Pl. 6, p. 132
cookii, p. 138
externa, p. 139
incerta, Pl. 5, p. 131
inexpectata, Pl. 6, p. 136
lessoni, Pl. 5, p. 129
leucoptera, p. 137
macroptera, Pl. 5, p. 127
m. gouldi, p. 127
mollis, Pl. 6, p. 134
Puffinus assimilis, Pl. 6, p. 150
a. elegans, Pl. 6, p. 150
a. kempi, p. 150
a. myrtae, p. 150
carneipes, p. 145
creatopus, p. 145
gravis, Pl. 6, p. 148

griseus, Pl. 5, 6; p. 143
pacificus, p. 145
tenuirostris, Pl. 5, p. 145
Pygoscelis adeliae, Pl. 1, p. 70
antarctica, Pl. 1, p. 72
papua, Pl. 1, p. 74
p. ellsworthi, p. 74
Riparia riparia, p. 235
Rowettia goughensis, Pl. 11, p. 245
Spheniscus magellanicus, Pl. 1, p. 84
Steganopus tricolor, p. 201
Stercorarius parasiticus, Pl. 10, p. 214
pomarinus, Pl. 9, p. 213
Sterna sp., p. 221
anaethetus, p. 230
paradisaea, Pl. 10, p. 227
virgata, Pl. 10, p. 225
v. mercuri, p. 225
vittata, Pl. 9, p. 221
Streptopelia sp., p. 232
picturata, p. 232
turtur, p. 232
Sturnus vulgaris, Pl. 8, p. 239
Thalassoica antarctica, Pl. 5, 6; p. 107
Tringa nebularia, p. 196
Turdus merula, p. 237
philomelos, p. 237
Zonibyx modestus, p. 193
Zosterops lateralis, p. 241